CAMOUPEDIA

A Compendium of Research
on Art, Architecture and
Camouflage

CAMOU

Roy R. Behrens

PEDIA

a compendium of research on art, architecture and camouflage

bobolink books \ 2009

Dedicated to the memory of
Everett Longley Warner (1877-1963),
American Impressionist painter
and US Navy camoufleur

CAMOUPEDIA:
A Compendium of Research on Art, Architecture and Camouflage

published by
Bobolink Books
2022 X Avenue
Dysart, Iowa 52224-9767
<www.bobolinkbooks.com>

First edition
ISBN 0-9713244-6-8

Printed in the United States of America

"Shape may, from this point of view, be regarded as a kind of stylistic 'mean' lying between the extremes of chaotic overdifferentiation [*dazzle*] and primordial homogeneity [*blending*]. Thus while a balanced relationship, as established within the style, produces an impression of well-organized patterns, exaggerated differentiation gives rise to disorganized ones, and intensified uninformity results in unorganized ones."

—Leonard B. Meyer
Emotion and Meaning in Music

contents

CAMOUFLAGE

ART, SCIENCE, AND POPULAR CULTURE

SATURDAY, APRIL 22, 2006
KAMERICK ART BUILDING

THIS IS an impromptu and largely unfunded event, consisting of a one-day interdisciplinary mini-conference on any and all dimensions of camouflage. It includes but is not limited to artistic, scientific, natural, military, psychological, magical, and fashion-related aspects of "fooling the eye." The event will be held in the Kamerick Art Building on the campus of the University of Northern Iowa (Cedar Falls), beginning at 8:30 am on Saturday, April 22, 2006, and ending at 8:30 pm that same evening. There is a $20 registration fee for conference participants (payable on site at registration), with the exception of all students, UNI faculty, and all session presenters, who will be admitted free. Breaks will be scheduled at mealtimes, and all participants will be responsible for their own meals, as well as their lodging. This is a pre-announcement of the event, and a call for proposals from faculty and graduate students (send title and brief description) for thirty minute slide talks, panel discussions or other events. Please e-mail all proposals to Prof Roy R. Behrens (UNI graphic design professor)the conference organizer, at <ballast@netins.net>. The deadline for responses is December 1, 2005.

UNI DEPARTMENT OF ART

Poster design by Megan Maurer, UNI graphic design student

introduction

In a prehistoric cave painting, there is a drawing of a man disguised as an animal. From all appearances, deception has always been critical to daily survival—for human and non-human creatures alike—and, judging from its current ubiquity, there is no end in immediate sight.

Deception is a furtive snake, a pretender of dead leaves, a flatfish on an ocean floor. It is always, only, just itself. Yet it takes on new colors and patterns with ease, to keep pace with the sinister, ongoing march of malevolent surveillance. To recall William Shakespeare: How easy is a bush a bear.

A pioneering champion of modern camouflage was Abbott Handerson Thayer, a well-known American painter from New Hampshire. In 1896, in a scientific journal, he unveiled his discovery of "The Law Which Underlies Protective Coloration."

As a classically-trained painter, Thayer was adept at shading, by which a completely flat drawing can look three-dimensional. He discovered that animals have light-colored bellies because, in his words, they are "countershaded." They are darkest on their upper surfaces, which are most commonly sunlit, and lightest on their undersides. An effectively camouflaged prey may appear flat and insubstantial.

An alternate term for countershading is "Thayer's Law." But Thayer's scientific research, according to Theodore Roosevelt, was compromised by his "artistic temperament." His pronouncements

This is the text, revised somewhat, of an address by the author at *Camouflage: Art, Science and Popular Culture*, an international conference, held April 22-23, 2006, at the University of Northern Iowa, Cedar Falls.

Above Conference logo. *Facing page* The conference call for proposals.

were provocative in part because he boasted that Nature behaves like an artist, not a scientist. The concealing coloration of animals, said Thayer, "belongs to the realm of pictorial art and can only be interpreted by painters" (Thayer 1909). His writings were also annoying because, at the end of the 19th century, camouflage was thought to be unmanly or effeminate. A genuine soldier stands his ground. He would rather fight than switch. *Bring 'em on.*

H.G. Wells (1917): "My impression—it may be quite an unjust one—was that some of our British colonels misunderstand and dislike camouflage... Let me, for the purposes of illustration, flash a caricature upon the screen of a certain Colonel X. of the old school, who is still, for want of proper combing out, in a position of responsibility at the front. Let me explain clearly that I have never met him, that I have no one in my mind, but that here and there I fancied I saw his influence. He is, you know, a quintessential army man, a good sportsman, owns four thousand acres, hunts. Never reads much or any of that rot. Doesn't believe much in any of this modern stuff. No. And particularly he doesn't like camouflage. He thinks a gun should look like a gun. He thinks a horse should look like a horse and that a soldier or a camp should be 'smart' before anything else. This camouflage fills him with a bashful hostility. He thinks it almost as objectionable as double entendre. It is antipathetic to his simple straightforward sporting tastes. It is as if one handed over that nice white grandstand at Epsom for decorative improvement to Mr. Roger Fry. It is like meeting a pretty woman who turns out to be so clever and educated and all that, and says things a man can't make head or tail of... At any rate very many of the British tents look as though they had been dabbed over by a protesting man muttering 'foolery' as he did it. With a good telescope the chief points of interest in the present British front in France would be visible from Mars."

To paraphrase American actor and politician Arnold Schwarzenegger, it is "girlie" to dress in a ghillie.

It was Thayer who blazed a conceptual path between military camouflage and protective coloration in nature. As early as the Spanish-American War, he and a loyal American friend, the painter George de Forest Brush, proposed that a fighting ship should also be countershaded. A few years later, Brush and his wife would develop a plan for a transparent airplane.

Modern camouflage became feasible in reply to the German use of the airplane as a way to pinpoint targets for long-range bombardment. Although its paternity is debatable, the originators of World War I camouflage were undoubtedly French artists, who were serving at the time in the infantry. Dressed in paint-streaked, hooded cloaks (called *cagoules*), French gun crews were "discovered by hostile airmen [only] with the greatest difficulty and with considerable risk because of the necessity of swooping low in order to unmask the camouflage" (Blechman 2004:126).

By far the most famous eyewitness account of modern camouflage is reported in Gertrude Stein's autobiography, which she impishly mistitled *The Autobiography of Alice B. Toklas*. During the first winter of the Great War, as she and Pablo Picasso were walking at night on the Boulevard Raspail, "All of a sudden down the street came some big cannon, the first any of us had seen painted, that is camouflaged. Pablo stopped, he was spellbound. '*C'est nous qui avons fait ca*,' he said, 'it is we who have created that'" (Stein 1933).

In spite of contributions made by artists to wartime camouflage, there has always been ample hostility in the exchanges between artists and the military. Late in World War I, an American Army officer exclaimed, "Oh, God, as if we didn't have enough trouble! They send us artists!" Even the French public was unconvinced of the practicality of abstract paint splattering, as shown by the fact that the moniker for French camoufleurs was *les barbouilleurs*, the childish scribblers or smearers (Kahn 1984).

Among the American camoufleurs was Henry Berry, who later detailed his experience in an autobiography titled *Make the Kaiser Dance* (Berry 1978). During World War I, while Berry was serving in France, John J. Pershing, the American Commanding General, drove up suddenly unannounced, and Berry (holding a sketchbook of camouflage plans) and his unit were required to stand for inspection. When Pershing asked Berry what he had in his hand, he replied, "Oh, this is just my sketchbook." "Sketchbook, sketchbook," Pershing thundered, "what the hell do you think this is, an art school? You're in the United States Army, soldier. Give me that sketchbook." Berry handed it over, and of course he never saw it again.

A year earlier, Pershing had been present in the US at another inspection, at which he was the tour guide for US President Woodrow Wilson and John Singer Sargent, the acclaimed society painter. Wilson, his wife, and the painter had been driven to a training camp on the outskirts of Washington, DC, to inspect the amusing accomplishments of a new military unit, consisting mostly of artists, called the American Camouflage Corps. In a still existing photograph of that inspection, Pershing is shown at the moment he says to President Wilson that a camouflaged soldier is within ten feet. Soon after, a large rock in the foreground moves, and a sniper pops out to salute them. An imitation rock made of *papier mâché*, it had served as the lid of a foxhole.

For many people, perhaps the most curious episode in camouflage history is the optical disruption of

Greville Rickard (1942:197): "Camouflage involves a great amount of dovetailing between science and art, between knowledge and actual application... The designer must expand to be part scientist, part engineer; the scientist must unbend to be part artist; he must visualize and he must be practical."

Above A disruptive naval camouflage scheme (known as dazzle painting or dazzle camouflage), as applied to the *USS Santa Olivia* in 1918. Courtesy US Naval Historical Center (NH 100996).

ships by means of a method called *dazzle painting* that began in World War I. This tactic was developed by a British naval officer and seascape painter named Norman Wilkinson. His later American counterpart was Everett Longley Warner, an artist who supervised other artists in distorting the appearance of American ships during both World Wars.

Above The box and a typical exercise in a Milton Bradley family game from the 1960s, called *Camouflage*. The players were asked to find drawings of objects, such as this clover, in a background of otherwise meaningless lines. Author's collection.

In thinking about camouflage, most people assume that it only occurs when a figure blends in with a background. On the contrary, it is most likely to result from a high difference strategy that consists of a blending of blending with dazzle, in what is called *coincident disruption*. The originator of that term was a British zoologist, military camoufleur and scientific illustrator named Hugh B. Cott, who said of art and camouflage (respectively) that "the one makes something unreal recognizable [while] the other makes something real unrecognizable" (Cott 1940). In cognitive psychology, shapes that are encrypted by coincident disruption are better known as *embedded figures*.

In the US in the 1960s, there was a television game called "Camouflage" in which contestants were challenged to search for hidden figures. These began as simple line drawings of common objects such as shoes, hats, keys or mittens, but were then made hard to see by superimposing a maze of meandering, meaningless lines. Later, this same television show was turned into a parlor game called "Camouflage: A Game of Fun, Skill and Perception." It enabled the Cleavers and other American families to play along with the game show by attaching a sheet of plastic to their television screen, or by recreating the game in their own living rooms, without a television. On the cover of the box are cartoon drawings of nine contestants or family members, with a letter from the word CAMOUFLAGE embedded in each.

In 1925, Hungarian-born Bauhaus architect Marcel Breuer bought his first bicycle, as therapeutic relief from the rigors of inventing furniture. Preoccupied with the design process, he presumably looked at the strong tubular steel frame of his new bicycle, and saw instead the ingredients of a new lightweight chair. *Voila!* He invented his now-famous Wassily chair using materials and production techniques that until then had been associated with bicycles.

Later, in the mid-30s, as Breuer prepared to immigrate from Germany to England, he addressed a package to his friends in London. Opening it, they were surprised and disgusted to find a copy of Adolf Hitler's *Mein Kampf*. Regarding this a tasteless joke, they threw it in the trash. Soon after however, when Breuer arrived, he asked for the book, explaining that, in order to leave Germany with more money than was legally allowed, he had interleaved the Führer's book with banknotes, thinking that no one would ever look there.

When I told this story to my wife, she shared a similar incident from her high school days at a Catholic girls school. One day during lunch, she and three other students were smoking cigarettes outside, behind the school, when a nun suddenly appeared. All the other girls were caught red-handed when they discarded their still-lit cigarettes on the ground. But my resourceful wife escaped, for lack of evidence—she had hidden her lit cigarette in her egg salad sandwich.

In daily experience, camouflage could be described as disguising one thing as two, two things as one, three as two, and so on. Most often, it causes confusion between an object and its background (called *blending*), between one kind of object and another (*mimicry*), or it makes a single thing break up into meaningless unrelated components (*dazzle*).

In category formation, we use these identical tactics in producing what artists and scientists call innovative, inventive or creative patterns. This is true throughout the widest range of human experience, since categories such as "art" and "science" have come from comparable measures—as has of course the notion of "human experience." The affinity of camouflage with innovation may be its most durable asset.

George Steiner: "At every level, from brute camouflage to poetic vision, the linguistic capacity to conceal, misinform, leave ambiguous, hypothesize, invent, is indispensable to the equilibrium of human consciousness."

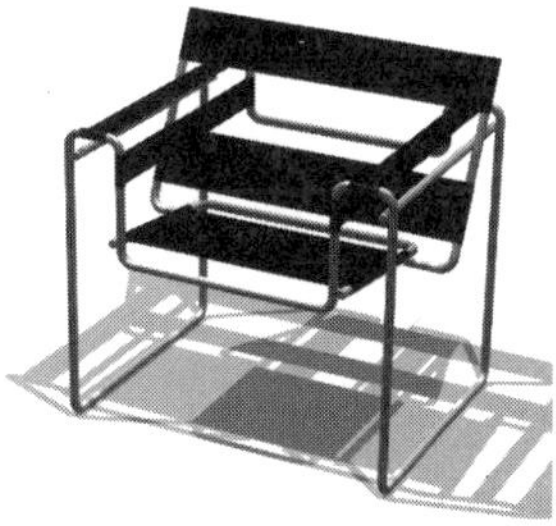

Above The famous Wassily chair, devised in 1925 by Marcel Breuer while he was a student at the German Bauhaus.

Robert Scott Root-Bernstein (1991): "Until we understand how people form and recognize patterns, use metaphors, make analogies, abstract, develop aesthetic sensibility, play act, play, manipulate, model, think kinesthetically—and how to correlate those tools of thought through transformational thinking—we will never succeed in understanding how inventive people invent, or how to train others to emulate them."

The two-part poem on this and the following page was written by American poet Marvin Bell especially for the opening session of *Camouflage: Art, Science and Popular Culture*, an international conference at the University of Northern Iowa.

preamble

The Book of the Dead Man (Camouflage)

Live as if you were already dead.
—Zen admonition

1. About the Dead Man and Camouflage

When the dead man wears his camouflage suit, he hides in plain sight.
The dead man, in plain sight, disrupts the scene but cannot be seen.
His chocolate-chip-cookie shirt mimics the leaves in a breeze.
His frog-skin dress, his bumpy earth nature, leave us lost and alone,
his mottled apparel sends us in circles.
His displacements distract and disabuse us, he is a slick beguiler.
Everything the dead man does is a slight disruption of normality.
He is the optical trickster, the optimum space-saver, the one to watch for.
He is of a stripe that flusters convention, he is the one to watch out for.
That we thought him gone only proves his wily knowledge.
The dead man has lain unseen among the relics of embalmed time.
He was always here, always there, right in front of us, timely.
For it was not in the dead man's future to be preserved.
It was his fate to blend in, to appear in the form of, to become...
Now he lives unseen among the lilies, the pines, the sweet corn.
It was the dead man's native desire to appear not to be.

Facing page Drawing submitted for US Patent No. 301,289, titled "Camouflage Fabric," invented by M.A. McIlhinney (1989).

2. More About the Dead Man and Camouflage

The dead man knows that camouflage is all in the mind.
He has seen in the human need for shape the undoing of shape.
He has witnessed the displacement of up-and-down, across and slantwise.
He has curled the straight lines and unbent the curves, he has split the wishbone
and painted outside the lines.
The dead man has undone the map by which to get there.
It is not what the dead man looks like, but what he no longer resembles.
For he hath reappeared in no disguise but as himself.
Call him disheveled, call him disposed, call him shiftless, he is.
For he hath been made and remade in the form of his surroundings.
He hath become all things that he looketh like.
Hence, he has been stepped on by those who could not see him.
He has been knelt upon by those who looked in vain.
The dead man bestirs in a background that looks inert.
The dead man is the ultimate camouflage.
He is everywhere, but where is he?

—Marvin Bell (2006)

Marvin Bell taught forty years for the Writers' Workshop at The University of Iowa, holding the position of Flannery O'Connor Professor of Letters. The author of numerous books of poetry, he was Iowa's first Poet Laureate. His literary honors include awards from the Academy of American Poets, the American Academy of Arts and Letters, *Poetry* and the *American Poetry Review*, as well as Guggenheim and NEA fellowships and Senior Fulbright appointments to Yugoslavia and Australia.

I think it likely that there are no finer galleries of abstract art than the cabinet drawers of the tropical butterfly collector. Each "work" is a symbol, if I must not say of emotion, then of vivid life...It is often, I believe, the fascination of this abstract color and design, as much as an interest in biology or a love of nature, that allures the ardent lepidopterist, although all these may be combined; he has his favorite genera and dotes upon his different species of *Vanessa* and *Parnassius*, as the modernist does upon his examples of Matisse or Ben Nicholson. The one-time schoolboy collector will in later life be transfixed with emotion for a moment at the sight of a Camberwell Beauty or a swallowtail—I speak from experience.

—Alister Hardy (1965)

Sir Alister Hardy (1896-1985) was a distinguished British biologist. "Equally drawn to science and art" (his own words in Hardy (1976a)), he was chosen for camouflage training during World War I by British painter Solomon J. Solomon—an error, as it turned out, because his name had been confused with that of another artist named Hardy. Nevertheless, Alister Hardy went on to become a prominent marine biologist, with a particular interest in protective coloration.

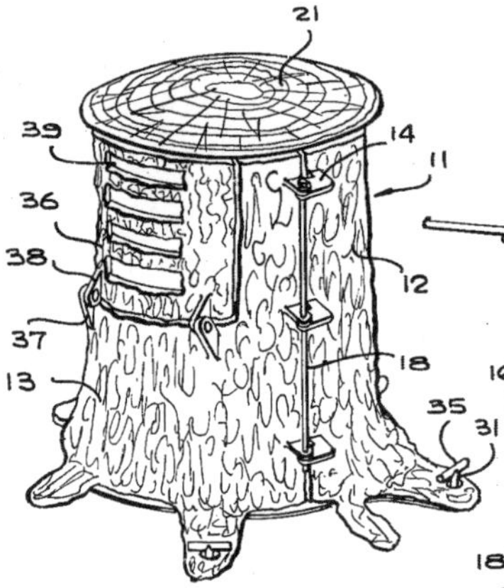
21
39
14
11
36
12
38
37
18
13
35
31

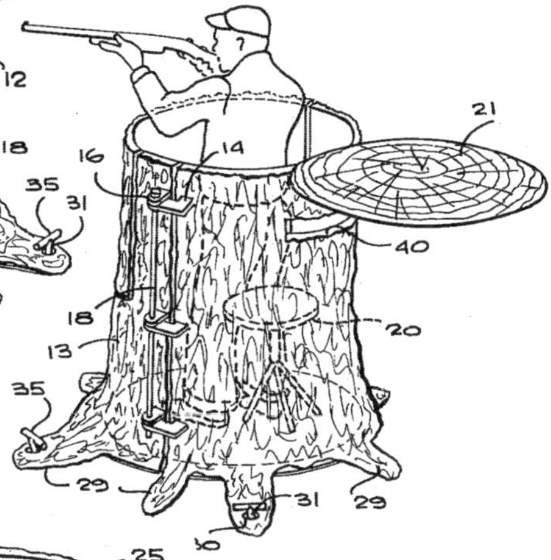
21
16
14
40
18
20
13
35
29
31
29
30

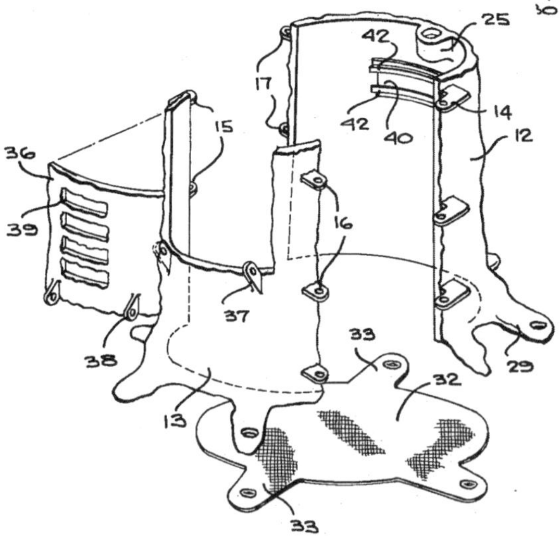
25
42
17
14
15
42
40
12
36
16
39
37
38
33
32
29
13
33

A

Myer Abel
(January 5, 1904-1948)
American painter, printmaker

Originally from Cincinnati OH, Abel attended that city's Art Academy, then studied in Paris at the Académe Moderne and (with Fernand Léger) at the Académie Julian. Returning to Ohio, he taught at his alma mater (1930-42). According to Shaw (1985): "Once he was even instrumental in getting a young man [his former student, graphic designer Noel Martin*] who had just been drafted into the army, a spot in a camouflage school where his talent had some outlet or at the least some company.

While serving his country in World War II, Captain Abel became attached to the China-Burma-India Theater of Operations in an unusual military unit made up of artists studying camouflage."

*Confirmed in conversation with Noel Martin at AAC, c1987 [RB].

•

Louis Abel-Truchet
(December 29, 1857-September 9, 1918)
French painter

Abel-Truchet was born in Versailles. At the outbreak of World War I, he volunteered for active duty at age 57. He is listed in Coutin (1988) as a member of the French camouflage forces, and described by Kahn (1984:18) as "second-in-command under [Lucien-Victor Guirand de] Scévola." He died on the battlefield in 1918. Posthumously he was awarded the Legion of Honor, and the Croix de Guerre.

Facing page Drawings for US Patent 2,992,503 titled "Game Blind," invented by H.L. Webb (1961).

Joseph B. Abrahams

(July 4, 1884-July 1, 1969)
Lithuanian-born American calligrapher, designer

Martha Banta (1987:221-222): "By the close of the nineteenth century new forms of protective, ceremonial disguise had come into play—methods for altering appearance that concealed identities and revealed presences, and that acted to repress particulars of individuality in order to emphasize associations with type. Some of these forms became means for self-protection. One such was the technique of camouflage that found favor in the art world prior to its adoption by the military during World War I."

Born in Grodno, Lithuania, of Polish ancestry, he came to the US at age six, and was naturalized in 1905. While living in New York, he attended the National Academy of Design (1898-1905) and the Art Students League (1900-1905), then worked as a reporter for the *New York Tribune*. He was also an art editor for magazines, and a research assistant at both the New York Public Library and the Smithsonian Institution. He designed the bronze doors of Temple Emanu-El, as well as the interiors of the Century Theatre, the Ziegfeld Follies and various New York restaurants, including Sherry's and Delmonico's. In subsequent years, he made illuminated award citations for prominent civic leaders, among them Bernard Baruch and, beginning with Woodrow Wilson, nearly all US Presidents until his death.

New York Times ("Joseph B. Abrahams" 1969): "In World War I, he served in the camouflage service." Specifically, he was a field supervisor in the civilian camouflage service in New York. He is quoted in the same article as having said: "I'd like to have done the Declaration of Independence. It should have been an illuminated manuscript. Can you imagine, it's just written in ordinary handwriting."

•

Above (top) Photograph of Herbert Adams (n.d.) by Peter A. Juley & Son. Courtesy Photograph Archives, Smithsonian American Art Museum (J0001165). (bottom) The medal insignia of Army Air Force aviators, the first version of which was designed by Adams.

Herbert (Samuel) Adams

(January 28, 1858-May 21, 1945)
American sculptor

Adams was a prominent sculptor at the turn of the century. Although too old for active duty during World War I, he publicly supported the efforts of Barry Faulkner, Sherry Fry, and others who formed the New York Camouflage Society. Among his other accomplishments, Adams designed the aviator badge for the Army Air Forces in 1918. Initially a cloth shoulder patch with wings and a shield, since World War II it has been issued as a medal.

According to Faulkner (1957), E.H. Blashfield was President of the New York Camouflage Society, but Adams and Ernest Peixotto were among the other artists who "devoted their time and energies in furthering its interests."

•

Hal Adelquist
(1914-1981)
American artist, designer

Adelquist was a long-time employee at the Walt Disney Studio, where he was partly responsible for the design and development of the Mickey Mouse Club. During World War II, he also led the Disney animators and set designers who contributed to the camouflage of the Lockhead Aircraft factory in Burbank CA.

This is confirmed by a letter to the author from David R. Smith, Disney Studio archivist (Smith 1978), in which he wrote: "[During WWII, a] Walt Disney Camouflage Unit was formed at the Disney Studio in Burbank. It consisted of 10-12 men, led by Hal Adelquist. The men worked on their own time, during the evenings, on their primary project, which was to design the camouflage for the Lockhead Aircraft plant."

•

Aerial reconnaissance
Observation from the air

Camouflage has been employed throughout human history. While it did not originate during World War I, it did become an urgent concern, prompting the formation of the first units devoted to deception. The explanation most often given for the adoption of camouflage is that it was a consequence of the coincident emergence of three wartime technologies: aerial observation, photography, and long range bombardment. Here are some of the ways that people have talked about this:

Fry (1917:68): "Camouflage had its birth during the war [WWI]. It is one of the consequences of the growing importance of airplane observation in warfare. It is humbug, disguise and concealment of military things, the art of making them look like something harmless and uninteresting to an observer."

Right World War I airplane equipped with an aerial camera, with the photographer seated behind the pilot. US Signal Corps photo.

H.G. Wells (1917:111): "One consequence of the growing importance of the airplane in warfare is the development of a new military art, the art of camouflage. Camouflage is humbugging disguise, it is making things—and especially in this connection, military things—seem not what they are, but something peaceful and rural, something harmless and quite uninteresting to airplane observers.

It is the art of making big guns look like haystacks and tents like level patches of field. Also it includes the art of making attractive models of guns, camps, trenches and the like that are not bona fide guns, camps, or trenches at all, so that the airplane bomb-dropper and the airplane observer may waste his time and energies and the enemy gunfire be misdirected."

Peixotto (1917:2): "Before every important battle nowadays flights of airplanes soar aloft, fight off assailants, and photograph as accurately as possible the enemy positions. These photographs, when developed and examined under a magnifying glass, reveal every secret that cannot be carefully hidden."

According to Rickard (1942:192), World War I reconnaissance planes began photographing as early as 4:30 every morning when the weather was sufficiently clear. The photographs were closely compared from day to day, in search of inconsistencies. He notes that "At least 95 percent of our work had to do with perpendicular camouflage, the combating of photography on the part of the aerial camera."

Gale (1968:33): "As our observation over the enemy lines increased, aerial photography enabled our map makers to produce more accurate maps of the German trench system, his strong points and communications, as well as his gun positions. We learned to interpret air photographs, and by examination of these could tell used from unused trenches. It taught us, also, the necessity for and the potentialities of camouflage and concealment. Furthermore, we learned to make disused trenches and gun positions appear to be occupied. Dummy artillery positions were created, and into these roving guns were moved and fired at night. The

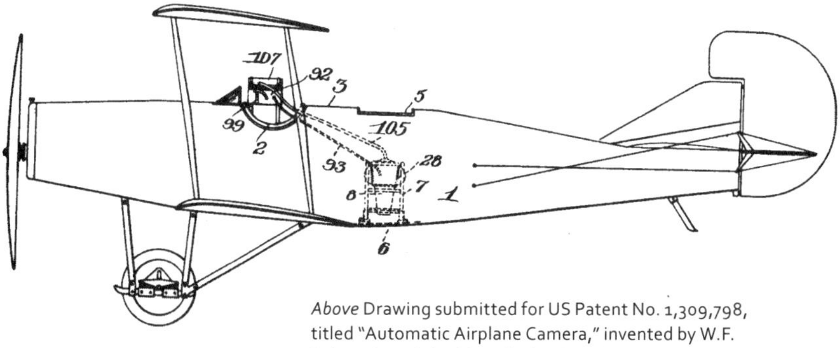

Above Drawing submitted for US Patent No. 1,309,798, titled "Automatic Airplane Camera," invented by W.F. Folmer (1919). In this and other WWI-era patents, one person was able to function both as pilot and photographer.

scars from muzzle-blast, and other signs of usage, helped us to deceive hostile reconnaissances."

World War I artist Harry Townsend describing his first flight in a captured German Fokker, quoted in Cornebise (1991:237): "What a picture around, under and above me. I have often tried to imagine it all from photos I have seen and from what I have always known the color and the look of the landscape to be. But here was something so different, so unreal, and altogether beautiful, that it was another world, and granted an aviator had a real appreciation for the beautiful (and he must soon get it in such surroundings), I can see how he would become perfectly enamored of it all there in the air. There are no words to describe it."

•

Peter Agostini
(1913-1993)
American sculptor

A figurative sculptor associated with Abstract Expressionism and Pop Art, Agostini was involved with the Works Projects Administration (WPA) and the Office of War Information in the late 1930s and early 1940s. In conjunction with the latter, he taught camouflage to American GIs. See Archives of American Art (1995:2).

•

Emile Aillaud

(January 18, 1902-1988)
French architect

Aillaud designed a cluster of 18 tower-like apartment buildings, known as Tours Aillaud (also called the Cloud Towers) in Nanterre, a suburb of Paris. The tallest has thirty floors. Completed in 1977, its exterior surfaces were painted with camouflage patterns designed by Fabio Rietti. For photos and more information, see Blechman (2004:386-387).

•

Above Prior to World War I, the French public despised Cubism, which they regarded as influenced by the Germans, their perennial enemies. The American public found it equally absurd, as shown by this cartoon from the *New York Evening Sun* (1913), in which Cubism is compared to a "crazy quilt" (a quilt made of leftover fragments from other projects).

New York Times ("Plenty of Camouflage..." 1918:63): "The wonders of camouflage, as it is practiced by the men of the fighting armies abroad, are much pictured and described for American readers, but any New Yorker who wants to inspect its experiments for himself has only to take a ferry boat trip across the harbor on almost any bright day. There he will see at anchor, or coming in, or going out, numerous ships whose painted sides reveal such wild extravagances of form and color as make the landsman open his eyes with amazement and mystification."

Ángeles Agrela

(1966-)
Spanish artist

Born in Úbeda, Jaén, Spain, she has been exhibiting since the mid-1990s. Around 2000, as described and illustrated by Méndez Baiges (2007), she created various staged photographs in which figures were partly embedded in domestic settings. Often the subject is wearing a patterned outfit that echoes the colors and patterns of common aspects of the home, such as a parquet wood floor, a wallpaper pattern, and so on. In her writings about this, Méndez Baiges characterizes Agrela's work as commentaries on the invisibility of housewives, who blend in with their household surroundings, in the course of which they are taken for granted and "merit scant social or family attention."

•

Robert (Ingersoll) Aitken

(May 8, 1878-1949)
American sculptor

Aitken studied at the Mark Hopkins Institute of Sculpture in San Francisco, where he also later taught. He is described in Nibbe (n.d.) as having "enlisted in [the] camouflage corps" during World War I. A prominent public sculptor for government

Above When Allied ships were first dazzle painted in 1917, the public (who saw the ships in the harbors, despite wartime precautions) thought that the patterns reminded them of Cubist paintings. This was especially true of the *RMS Mauretania*, a majestic ocean liner and sister ship of the *Lusitania*. According to British poet Alfred Noyes, dazzle painting was "crazy quilt" camouflage. Photo from *London Illustrated News*, January 4, 1919.

buildings and other official settings, he created the nine figures above the entrance to the US Supreme Court Building, inscribed with "Equal Justice Under Law." In advance of his camouflage service, he was commissioned to design a fountain and several statues at the Missouri State Capitol in Jefferson City. The building itself (completed in 1917) was co-designed by Evarts Tracy, while the sculpture on its dome is by Sherry Edmundson Fry, both of whom would later serve, along with Aiken, as American camoufleurs in France.

•

Eric Aldwinckle
(January 22, 1909-January 13, 1980)
British-born Canadian war artist

At the outset of World War II, Aldwinckle was a conscientious objector, and was assigned to designing war posters, some of which became well known. He later worked as a camouflage designer in Halifax, Nova Scotia. In 1942, when he heard that Ottawa was seeking volunteers for its new War Artist program, he applied for and received a commission in the Royal Canadian Air Force. One of his other accomplishments was the design of the Great Seal of Canada for the Canadian Mint.

Harold Altman

(1924-2003)
American printmaker

Altman was a student at Black Mountain College, where he worked with Bauhaus painter Josef Albers. During World War II, he served in the 937th Engineer Camouflage unit (1942-45). Later, he was a professor of art at Pennsylvania State University.

•

Below When New Hampshire artist Barry Faulkner published his autobiography, *Sketches from an Artist's Life* (Faulkner 1973), he included a vivid account of his days in the American Camouflage Corps. He also published this photograph of himself (second from the right) and other members of the unit.

Burges Johnson (1920:292-293): "Who [were] our camoufleurs [in World War I]?... Thirty-six percent of the men in our American camouflage unit were architects in civil life, and this is the largest contribution of any single profession. Only seven percent were artists. Ranging along between come engineers, landscape gardeners, professional men, chauffers, and what not."

Karle (Otto) Amend

(1889-January 2, 1944)
American artist, theatre set designer

Originally from Ohio, Amend was trained as a set designer. He designed stage sets for various Broadway plays, including *Smile At Me* and *I Must Love Someone*, and was founder and proprietor of Amend Scenic Studios.

New York Times ("Karle Amend" 1944): "In the first World War he served as a camouflage instructor at the Great Lakes Naval Training Station."

•

American Camouflage Corps

World War I camouflage unit

The American Camouflage Corps was a US Army unit formed in 1918 as Company F of the 25th Engineers, and was shortly thereafter reorganized as Company A of the 40th Engineers.

According to Faulkner (1957), the first enlistments for this unit took place on September 4, 1918, and the first four persons to enlist were Sherry Fry, Evert Herter, William Twigg-Smith and Barry Faulkner. They trained initially at Camp American University, on the outskirts of Washington DC, housed in a single tent between two marble buildings. Within a few days, they occupied three tents, one of which was used by Lieutenant Homer Saint-Gaudens (son of Augustus Saint-Gaudens, and Faulkner's former roommate at Harvard) and Sergeant Major Dwight Bridge. Another tent was for orderlies, while the third housed enlisted men, who by then also included Laurence Grant, Henry Sutter, Harry Thrasher and "Casey" Jones.

In his autobiography, Faulkner (1973:85-87) recalls that it was his friend Fry who, in 1917, when the US entered the war, showed him a photograph of a train and railway station that French camoufleurs had painted with disruptive shapes. Fry also told him about the French camouflage unit, largely made up of artists. As a cousin of Abbott H. Thayer, whose writings on camouflage were well known, Faulkner decided to use that connection as (in his words) "bogus prestige" in his and Fry's efforts to organize the New York Camouflage Society, before the US had entered the war.

•

American Society of Marine Camoufleurs
World War I naval camouflage group

Covert (2007:53): "Included in the [Rhode Island School of Design's Fleet Library] collection [of World War I American naval camouflage plans] is a copy of the constitution of the American Society of Marine Camoufleurs. This group was located in District 2, New York, but it included non-resident members as well. Annual dues were one dollar and members met monthly. Members were 'Any man who shall be appointed to the position of camoufleur in the US Shipping Board'."

•

Adelbert Ames II
(1880-1955)
American artist, optical physiologist, and psychologist

Ames, a Harvard law graduate, served in the US Army Signal Corps during World War I, first as an aerial observer and photographer, then as supervisor of a machine shop where prototypes of optical devices were invented, constructed and tested. It was not until the mid-1930s that he began to design the laboratory set-ups (called the Ames Demonstrations in Perception) for which he is now widely known. Nearly all his demonstrations pertain to aspects of vision that are equally fundamental to the design of camouflage. This is especially true of dazzle camouflage, which uses anamorphosis and related perspective distortion techniques. See Ittelson (1968) and Behrens (1994, 1999).

•

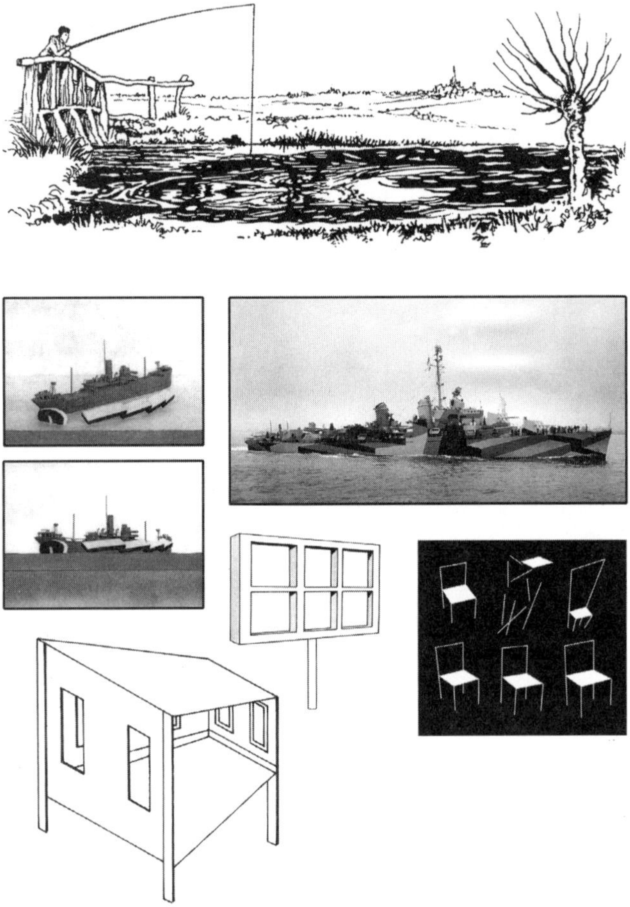

Above One of the most fascinating aspects of ship camouflage, as practiced by Everett L. Warner, Frederick Waugh and others, was the use of anamorphosis or forced perspective to make a ship's surface appear warped. Such methods were nothing new, as is evidenced by the 19th century puzzle picture at the top of this page: Viewed straight on, the drawing merely shows a man fishing in a pond, but when viewed from either side, the water in the pond becomes an elongated portrait of Ludwig von Beethoven. Public domain. Below that on the left are two views of a ship model, designed by Warner and his team, and, on the right, a photo of the camouflage scheme they designed during World War II for the *USS Charles S. Sperry*. Photo courtesy the US National Archives (19-N-66384). In 1919, Warner discussed his deliberate use of anamorphosis in an essay in which he described how the perceived shape of any room can be dramatically altered by applying forced perspective shapes on its walls (Warner 1919:218). Exactly these kinds of experiments were made in the 1930s and 40s, including (as shown in the bottom three diagrams) distorted rooms that appear normal, trapezoidal windows, and chairs that only look like chairs from a designated peephole. These were built not by Warner, but by Adelbert Ames II, an artist, optical physiologist and psychologist at Dartmouth College, hence they are widely referred to as the Ames Demonstrations in perception.

Linwood P(ayson) Ames
(dates undetermined)
American artist

Ames was a US Army camouflage instructor, at a time when training in the subject was required of New York State soldiers, beginning in May 1918 at Spartanburg SC. He was under the command of Major General John F. O'Ryan.

New York Times ("Camouflage Course for NY Soldiers" 1918): "[Ames]...then a Sergeant in the Divisional Headquarters Troop, had attracted the favorable notice of headquarters because of his feats in camouflage. Under his direction all sorts of 'stunts' were put through. A 'dead horse' was rigged up, which afforded concealment for three expert riflemen; a badly battered 'tree stump' sprung into being, to do its bit as a listening post; and, finally, a huge canvas tent was erected, under brush and earth, till it looked exactly like a hill, but a hill that gave effective concealment to 400 sharpshooters.

These examples of camouflage aroused great interest among the men. General O'Ryan was quick to note the effect of the work and decided to launch a course in camouflage—something that, in the case of other divisions, has generally been left for overseas training."

Above As shown by this WWI photograph of the *USS Robinson* (c1918), high contrast, disruptive dazzle designs were applied to ships to increase the difficulty of aiming at them accurately through a U-boat periscope. These schemes made frequent use of anamorphic distortion, in much the same way as in the Ames Demonstrations. Courtesy US Naval Historical Center (NH 94140).

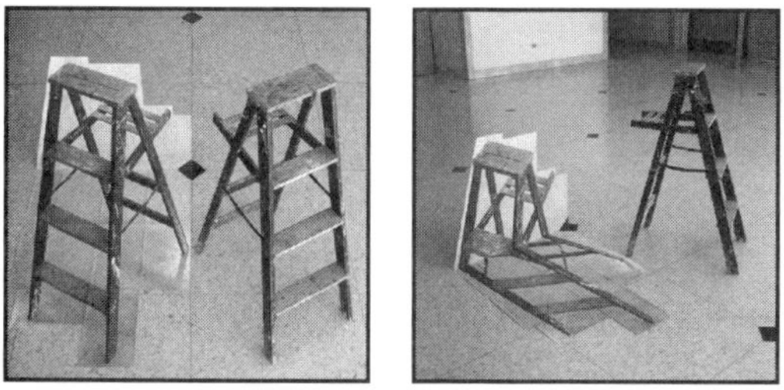

© Richard Koenig

Anamorphosis is often used by contemporary artists, one of whom is American photographer Richard Koenig, who made the two photographs of a ladder shown at left (c2006). The photo on the far left is the intended result, an effect that—like the Ames Demonstrations—only works from a single point of view. In the nearer photo, everything is in the same position as before; only the camera angle has changed.

Anamorphosis
Perspective-based distortion techniques

In anamorphic artworks, the image looks distorted when seen from a customary viewing position but correctly proportioned when seen from the side

Norman Wilkinson (1919): "...if a vessel can be seen at all it does not matter how visible she is, providing her course remains a matter of question to the attacker."

Above Ship camouflage model using anamorphosis, designed by American artist Frederick Waugh for *Proteus*, a US coal ship. *Left* The same ship in 1918, with its camouflage applied. Crowell (1921:508): "By reversing the perspective, as was done in the design painted on the *Proteus*, the ship was made to seem to head northeast (assuming the observer is looking north as he views the picture), although as a matter of fact the model is steering southeast."

Of this project, Everett L. Warner (n.d.) later remarked: "The eye is so accustomed to the normal operation of the laws of perspective that if you were to see a group of telegraph poles, and they had been so graded in height that the nearest one was much the shortest, you would be likely to think that the tall one in the distance was the nearest. I remember this design well [as] it cost me a box of cigars. When the model came in, it looked like such a difficult problem that I offered *a box of cigars* if any one of the designers could fool me with a design. Mr. Waugh won the box of cigars, but the joke was on him as he does not smoke. An effective design, but one belonging to [an] early period before we had entered the realm of *solid geometry*."

Abraham Rattner, as quoted by Culkin (1980): "The execution of camouflage varies in quality and efficiency because all camoufleurs (as well as all artists in civil life) do not have the same degree of talent. If camouflage is badly executed the result can be disastrous."

(often by means of a peephole). This kind of perspective distortion was purposely used as early as 1485 by Leonardo da Vinci, who included in his notebooks three drawings of his experiments with it, in an offshoot of linear perspective called *anamorphosis* or "anamorphic projection." Throughout history, it has also been used in theatrical set design, where it is commonly known as *forced perspective*. As early as the sixteenth century, a group of Dutch artists made anamorphic "peepshows" called *perspectyfkas* or perspective cabinets, of which a handful still survive and are now on display in museums.

In perceptual psychology, many of the Ames Demonstrations relied on anamorphic distortion, but the tradition is also implicitly linked to certain varieties of camouflage, especially dazzle painting.

•

William Anastasi
(1933-)
American artist

Born in Philadelphia, Anastasi is a New York-based minimalist and conceptual artist. In 1966, he proposed an exhibition called *Blind*, which would have been accomplished by painting the entire interior surfaces (walls, floor, ceiling) of a New York art gallery with a Vietnam-era military camouflage pattern. The exhibition did not happen then, but was later carried out in March 2003, coincident with the Iraq War, at the White Box Annex Gallery in New York. See photos and further details in Blechman (2004), and Méndez Baiges (2007).

•

Douglas (Shenton) Annand
(1903-1976)
Australian graphic designer

Born in Toowoomba, Queensland, Australia, he grew up in Brisbane. In 1930, he moved to Sydney to work as a freelance designer. He is especially remembered for his corporate work, magazine covers, travel posters, murals, coins—including the "kangaroo penny"—and for having art directed the Australian Pavillion for the 1939 New York World's Fair.

MacAulay (1993): "In 1941-44 Annand was a camouflage artist with the Royal Australian Air Force." He is also listed by Elias (2003) as belonging to the Sydney Camouflage Group.

•

Architects
Their suitability for camouflage

Throughout World Wars I and II, there was no end to the public disputes (usually by letters to the editor) about which professions are compatible with the practice of camouflage and which are not. The usual participants in this were artists, architects, theatre designers, biologists, chemists and physicists. Being experienced in management, and residing somewhat between art and science, architects were often recommended. Here are a two confirmations of that:

Above The use of the term "dazzle painting" for ship camouflage was initiated by British artist Norman Wilkinson, who originated the method in 1917. In the US, it acquired various synonyms, including "razzle dazzle," a term that was already commonly used in reference to certain rides at amusement parks, as shown in these two photos from the 1890s. Photos courtesy the US Library of Congress, Prints and Photographs.

Saint-Gaudens (1919:26): "Our best officers were architects. They not only understood the principles of form and color, but they had been faced with clients who would have the linen closet, the stairs, and the chimney all in the same place. Less pestered with the practical side of life, they tempered their art to a line of brass tacks.

For our non-commissioned officers and privates, the moving picture and stage property men and carpenters were, by all odds, the most successful. An ability to handle those superior in rank and a resourcefulness at all hours was theirs."

Tracy (1919:176): "It was found, not only by our higher officers, but by our allies, that skilled architects as a rule made the best field camoufleurs, although there are brilliant exceptions, as appeared in our own organization."

•

Giuseppe Arcimboldo
(1537-1593)
Italian artist

Renaissance artist Arcimboldo is remembered for his visual puns, for example, portraits made entirely by combining fruits, vegetables, fish or other elements. Since the early 20th Century, there has been increased attention to his work, in part because the Surrealists believed that he had anticipated their own preoccupation with grotesque dream imagery. His work is especially relevant to the concept of mimicry. See Wickler (1968), Gregory and Gombrich (1973), Hulton (1987), and Schwartz (1996).

•

Hella Czech (in Portmann 1967): "The subject of this book [*Animal Forms and Patterns*] is animal *Gestalt*."

Walter W(endell) Arnett
(May 12, 1912-July 10, 1998)
American artist

Born in Salyersville KY, he attended the Art Academy of Cincinnati (1932-33), then worked as an advertising artist. In 1942, he enlisted in the US Army and was assigned to the 603rd Engineer Camouflage Battalion, where he completed his training at Fort Meade, and was then assigned to the 23rd Headquarters Special Troops, also known as the Ghost Army, a clandestine unit in Europe.

•

Rudolf Arnheim
(July 15, 1904-June 9, 2007)
German-born American psychologist, art theorist

Born in Berlin, he was a student of the Gestalt psychologists at the Psychological Institute at the University of Berlin, where his research focused on perceptual psychology, art and music. He left Germany for Italy when the Nazis came to power, moved to England, then to the US, arriving in 1940. He taught at the New School for Social Research, at Sarah Lawrence College, and, from 1969 to 1972, at Harvard University, where he held the position of Professor of the Psychology of Art. In 1954, he published a major, influential book on vision in relation to art, titled *Art and Visual Perception: A Psychology of the Creative Eye* (Arnheim 1954). In it, he discussed at length the various inherent tendencies that are fundamental to both art and camouflage. In his lifetime, he published more than fifteen books on related subjects, among them *Visual Thinking*, *The Dynamics of Architectural Form*, *The Power of the Center*, *Parables of Sun Light*, and *The Split and the Structure*.

•

Boris Aronson
(1900-1979)
American set designer

Aronson was born in Russia, in Kiev, where his father was the chief rabbi. He attended art school and became interested in set design. He was particularly influenced by Vsevolod Meyerhold and other Russian Constructivists. After living briefly in Europe, he came to the US in 1923, where he worked as a set designer in New York. Between 1951 and 1976, he received six Tony Awards, and is credited with the set designs for *The Rose Tattoo*, *The Diary of Anne Frank*, *Orpheus Descending*, *J.B.*, *Fiddler on the Roof*, *Cabaret* and *Zorba*. He was a member of the Camouflage Society of Professional Stage Designers. See Naversen (1989).

Artists

Their suitability for camouflage

Artists, in contrast to architects and theatre designers, are less likely to work in groups. So, while Abbott H. Thayer believed that camouflage "belongs to the realm of pictorial art and can only be interpreted by painters," there was no shortage of people who thought otherwise. Here are various opinions on that:

Alister Hardy (1965:129): "We should note that the successful camouflage officers of the two world wars were artists rather than scientists, or sometimes scientists with artistic inclinations."

Peter Rodyenko (1936): "Contrary to popular belief, it is the opinion of this writer that professional artists do not make good camouflage officers...The average artist, because of his training, is liable to attend, personally, to details which should be left to subordinates, and thus lose track of other projects also under his jurisdiction. In addition to this temperamental handicap, camouflage has developed to a stage where color amounts to but little. Among artists there is seldom found a thorough knowledge of construction, optics, atmospheric phenomena, mathematics and geometry, the psychology of the human mind, in addition to familiarity with map reading and the tactics of the various arms. This cannot be taught quickly. Architects, civil engineers and other technically trained men, specially interested in camouflage, offer better possibilities."

Chesney (1941:42): "Why a successful painter or scenic artist, say, who has perhaps never been in an airplane should be thought to be the most suitable person to undertake three-dimensional work of this nature, is rather astonishing."

•

Above It is puzzling that in Feuer (1999), British camoufleur Norman Wilkinson is quoted as saying that "The [WWI] U-boats did not have range finders to use with their periscopes. Therefore, before a torpedo was fired, the target's speed, direction and distance had to be carefully estimated." In contrast, in his lecture notes, US camoufleur Everett L. Warner mentions range finders, and includes this photograph of a ship as seen through a coincident range finder. Also, there is considerable mention of submarine range finders in Skerrett (1919:102), who states that "The means of range-finding and speed-measuring are certain cross hairs or graduated lines on the field of the periscope, as a rule, and these lines must stand out in sufficient contrast against parts of the object-vessel so that the eye can check them up as the observed craft passes across the field." The confusion may be in the difference between a coincident range finder and one that uses only cross hairs and graduated lines. US Navy photograph.

Above A sequence of images showing (from top down) a military vehicle with no camouflage, one with disruptive camouflage, and the confusion of object and background that comes from coincident disruption. Adapted from a series of lithographs in a French publication titled *La Guerre Documentée* (Paris c1920).

G(eorge) Bradford Ashworth
(1892-?)
American set designer

According to Naversen (1989), Ashworth was a member of the Camouflage Society of Professional Stage Designers.

•

Michel Aubrey
(1959-)
French artist and filmmaker

Aubrey is a French artist, filmmaker and camouflage scholar. There is a short statement by him about "counterfeit camouflage" (the fabrication of counterfeit camouflage artifacts, of which there are apparently many) in Blechman (2004:192-195). His own astonishing works of art, many of which pertain to camouflage, have been reproduced in an exquisite catalog (Aubrey 2001).

Henry Moore (quoted in Blechman 2004:274): "[During World War II, having already served in the previous war, if called up again, he hoped it would be] something less destructive than taking part in the actual fighting and killing. There ought to be ways of being used even as a sculptor—in making of splints, etc., or jobs connected with plastic surgery—though the most likely thing I suppose is camouflage work."

...

Pierre-Louis Gautier (WWI French naval camoufleur, quoted in Blechman 2004:274): "We were receiving ship profiles from all over the navy. Using this, workers in the camouflage section of the navy boat yards constructed little wooden models of all sorts of ships that needed camouflaging."

•

Émile Aubry
(April 18, 1880-1964)
French painter

Aubry was born in Sétif, Algeria. He studied in Paris with Jean-Léon Gérome and Carolus Duran. Listed in Coutin (1988) as a World War I French camoufleur, he was awarded the Legion of Honor.

•

Joseph-Marius Avy
(September 21, 1871-December 29, 1939)
French painter, illustrator, set designer

He is listed in Coutin (1988) as a World War I French camoufleur, for which he was awarded the Legion of Honor, and the Croix de Guerre.

•

Above During World War I, more or less the same methods were used for designing and testing camouflage schemes by artists in France, England and the US. Shown here painting models (c1918) are artists at the design subsection of the US Navy camouflage division. The artists are: (left to right) John Gregory, unidentified, Frederick Waugh, and M. Nash. Courtesy US National Archives (165-WW-70D-1). For photographs of comparable WWI French and British naval camouflage studios, see Blechman (2004:166-167) and Wilkinson (1969).

Lemuel Ayres
(1915-1955)
American theatre designer, producer

Ayes studied at Princeton University and the University of Iowa, and was the set designer for such well-known Broadway productions as *Oklahoma*, *Kismet* and *Camino Real*. According to Naverson (1989), he was a member of the Camouflage Society of Professional Stage Designers.

•

Anthony M. Ayrton
(?-1940)
British painter

Tony Ayrton was the son of British architect Maxwell Ayrton and the older brother of the sculptor and illustrator Michael Ayrton. He is mentioned by Geoffrey Barkas (1952), with whom he served in the North African desert during World War II. He is also cited in Hartcup (1980) and Goodden (2007). He died in Tunisa of an illness early in the war.

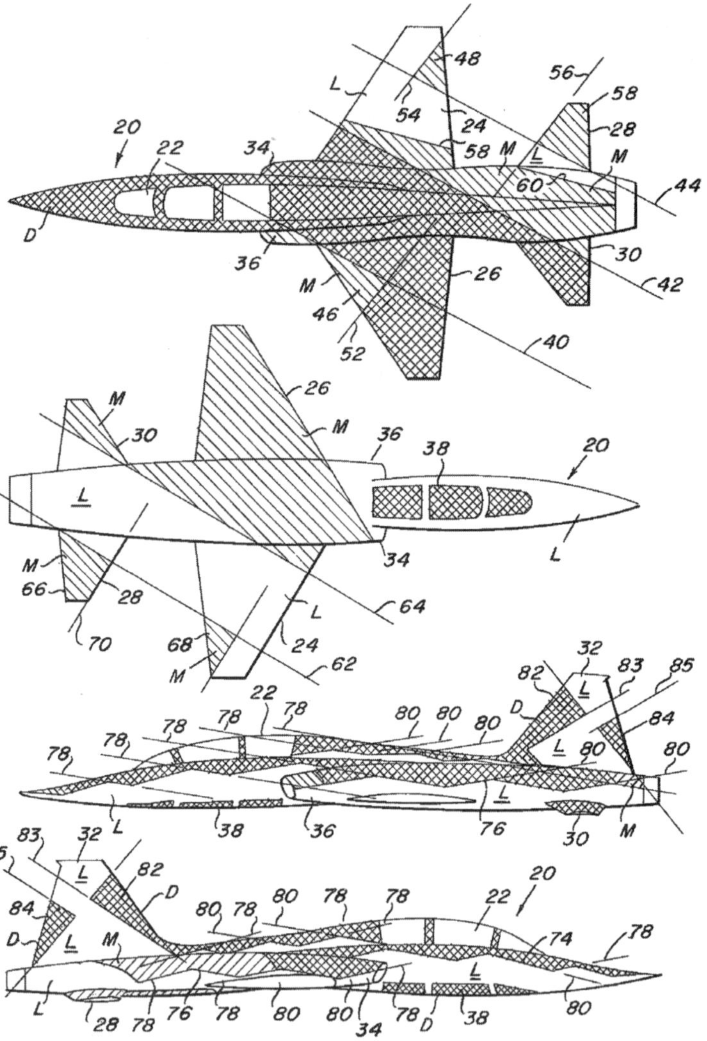

20
22
34
36
D
L
M
24
26
28
30
40
42
44
46
48
52
54
56
58
60
38
62
64
66
68
70
32
74
76
78
80
82
83
84
85

B

Background Picturing
Figure-ground pattern resemblance

The idea of background picturing was originated c1896 by American artist Abbott H. Thayer, who believed that an animal's surface pattern is "a picture of such background as one might see if the animal were transparent" (Thayer 1896a). He did not mean a literal picture, but rather, as his son Gerald later clarified, an abstracted "picture-pattern" (Thayer 1908). Based on this idea, Thayer recommended that anyone could create appropriate, functional camouflage, for any specific surrounding, by the following method: "[A person] has only to cut out a stencil of the soldier, ship, cannon or whatever figure he wishes to conceal, and look through this stencil from the viewpoint under consideration, to learn just what costume from that viewpoint would most tend to conceal this figure" (Thayer 1918). Perhaps the best explanation of how background picturing works in combination with countershading is an essay by Gerald Thayer, published after his father's death (Thayer 1923). In Bogle (1998), there is a photograph of a cut-out silhouette of an Australian soldier, as devised by William Dakin.

•

Baden (Fletcher Smyth) Baden-Powell
(May 22, 1860-January 8, 1941)
British military aviator

During World War I, Baden Baden-Powell served in France and Belgium in 1917-18 as a Camouflage Officer in XIV Corps. His brother was Robert Baden-Powell, founder of the Boy Scouts.

Facing page Drawings submitted for US Patent No. 4,089,491, titled "Camouflaged Aircraft, Surface Vessel or Vehicle or the Like," invented by American artist and camoufleur Keith Ferris (1978).

Robert (Stephenson Smyth) Baden-Powell

(February 22, 1857-October 3, 1937)
British army officer, author

Founder of the Boy Scouts, Baden-Powell also was at times a spy for the British Secret Intelligence Service. In one instance, he pretended to be a lepidopterist in what was then Dalmatia (now Croatia). Walking about with a net and sketchbook, he positioned himself on hills overlooking strategic fortresses, and pretended to make drawings of butterflies. In truth, within his drawings, he was concealing tiny diagrams of the structures and their armaments. He stated in his book, *My Adventures as a Spy* (Baden-Powell 1915): "...for anyone who is tired of life, the thrilling life of a spy should be the very finest recuperator!"

Above Abbott H. Thayer's demonstration of the *background picturing* inherent in even the "war paint" of indigenous peoples.

Below Demonstrating the same principle is a sheet of canvas with cut-out silhouettes of skunks, filled in with photographs of the animal's typical setting. As reproduced in Thayer (1918).

•

Baffle Painting

World War I-era name for dazzle camouflage

The application of conspicuous, abstract geometric shapes on a ship's surface, as a deterrent against German submarine attacks, it is now more commonly known as dazzle painting. When comparable strategies were widely adopted, other names came into use, such as *razzle dazzle, jazz painting, parti-coloring, zébrage, eccentric painting* and *crazy quilt camouflage.*

Porter (1921:252): "A slight protection was also afforded by camouflage, and ships were painted so as to render themselves less visible. To be sure, no ship can ever disguise itself, but the art of 'baffle painting' consists in giving to a ship the impression of a different size from what it actually is, or in giving the impression that it is turning to one side or the other, so that an enemy [in a U-boat] would wrongly calculate the point at which he should dive."

Baltimore Sun ("Baffle Painting" 1918): "'Baffle painting' has been developed as a substitute [for figure-ground blending] to deceive a submarine commander as to the size and form of a ship and her course and speed.

Lieut. Com. Norman Wilkinson, Royal Naval Volunteer Reserve, the inventor of 'baffle painting,' came to the conclusion after long experiment that the moment a submarine comes to the surface within striking distance no method of painting would render a ship sufficiently invisible to escape being seen."

•

Marcel-Adolphe Bain
(November 21, 1878-1937)
French painter

A landscape painter, Bain is listed in Coutin (1988) as a World War I French camoufleur. He was awarded the Legion of Honor.

•

William (James) Bain, Sr.
(March 27, 1896-1985)
Canadian-American architect

Bain was born in New Westminster, British Columbia. In 1921, having earned an architectural degree from the University of Pennsylvania, he returned to Seattle, where he established his own practice.

MacIntosh (1998): "In 1942, the State of Washington named [William] Bain camouflage director, in charge of hiding critical operations from potential air attacks...

To hide Boeing's Seattle assembly plant, Bain fabricated a fake complete rooftop town that included small versions of houses, streets, cars and trees. In other parts of the country, critical buildings often received church spires or similar disguises. Sometimes a simple tarp with painted buildings and streets masked large operations. Bain's solution, however, was as extensive and detailed as a movie set."

•

Edward Bainbridge-Copnall
(August 29, 1903-1973)
British sculptor

He is mentioned in Barkas (1952:26) as a World War II camouflage officer who "was to join us later in the desert [in North Africa]." He was president of the Royal Society of British Sculptors.

•

(William) Billy Baldwin
(dates undetermined)
American interior designer

Baldwin served in the World War II American unit called the Ghost Army, along with Bill Blass, Ellsworth Kelly, Art Kane and various others. He is mentioned several times in Blass (2002).

•

Above World War II prototypes of disruptive airplane camouflage, designed by American artist McClelland Barclay, but never adopted. Photos courtesy US Naval Historical Center (NH 96158; NH 96146; and NH 96158). On the *facing page* is one of three drawings from a Barclay US Patent (No. 2190691) dated 1940.

Thomas R(aymond) Ball
(February 12, 1896-June 16, 1943)
American artist, US Congressman

Born in New York, he studied at the Art Students League. During World War I, he served in the Depot Battalion, Seventh New York Infantry (1917), and in France with the Camouflage Section of the 40th Engineers (1918-1919). After the war, he lived in Old Lyme CT, and was a member of the Connecticut House of Representatives (1927-37), and the US House of Representatives (1939-41).

•

André-Victor Ballet
(June 30, 1885-?)
French painter, designer

Born in Paris. He is listed in Coutin (1988) as a World War I French camoufleur.

•

Charles-George Barbéris
(January 1, 1888-?)
French sculptor

Born in Paris, Barbéris was mostly an animal painter. He is listed in Coutin (1988) and Ducasse (1962) as a World War I French camoufleur.

•

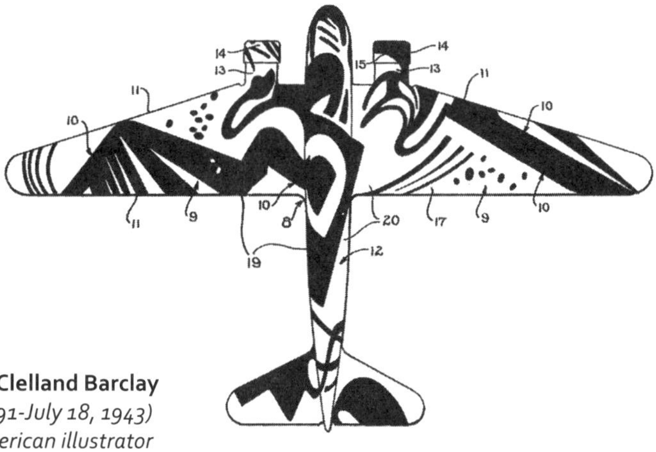

McClelland Barclay
(1891-July 18, 1943)
American illustrator

Originally from St. Louis, Barclay studied at the Art Institute of Chicago, then with George Bridgman and Thomas Fogarty at the Art Students League in New York. As early as 1912, his work was often featured in major US magazines, including *Cosmopolitan*, *Saturday Evening Post* and *Ladies' Home Journal*. During World War I, he produced recruiting posters for the US government, and worked with William Andrew MacKay in designing ship camouflage for the US Shipping Board.

In the 1920s and 30s, Barclay was widely known as an advertising illustrator, designing posters for the film industry, painting "pin-up girl" illustrations, and developing marketing images for leading manufacturers, the most popular of which were his paintings of sexy models for General Motors' "Body by Fisher" sales campaign.

In 1938, he was appointed an Assistant Naval Constructor with the US Naval Reserve. He worked on camouflage, and obtained a US patent for dazzle-like aircraft camouflage titled "Camouflaging" (Barclay 1940). In that year, an article in the *New York Times* ("Nazis Say" 1940) reported that his patent was "for the camouflaging of airplanes by painting the wings with designs that were said to conceal the shape and make it difficult to judge the position of an airplane." Some of his designs were applied to prototypes and tested, but were never actually implemented. During World War II, Barclay continued to experiment with camouflage and to design recruiting posters. Appointed a Lieutenant Commander, he was a passenger on a US Navy vessel that was struck by a torpedo near the Solomon Islands on July 18, 1943 (Barclay 1943, 1944).

See also Ledyard Towle. David Jones. Harley Earl.

Geoffrey Barkas

(1896-1979)
British filmmaker

Born in the UK, Barkas spent five years in the infantry during World War I, then lived briefly in Canada. After moving to Hollywood, he became a cinematographer, producer and director, mostly of documentary films. During World War II, he was named Director of Camouflage for the Middle East Forces of the Royal Engineers. While serving in that capacity in 1941-42, he oversaw an extraordinary wartime deception at El Alamein in the North African desert. In a now-famous large-scale diversion, he and his unit created the illusion of a massive build-up of forces, the details of which he documented in *The Camouflage Story* (Barkas 1952). Accounts of this are also found in Reit (1978), Hartcup (1980), and Goodden (2007).

Alister Mackenzie (1920:130): "In golf architecture and camouflage a knowledge of psychology is of enormous value."

...

Rasmussen (1964:94): "But the houses he [Le Corbusier] designed [in the late 1920s] were quite different—an attempt to create a Cubist framework for everyday life. They were color compositions without weight, just as intangible as the [WWI] camouflaged ships."

•

Above WWI-era magazine illustration showing a dazzle painted ship. Artist and source unknown.

Bion [or Bjorn] (Hall) Barnett, Jr.

(September 18, 1887-?)
American painter

Although an American, he is listed by Coutin (1988) as a French camoufleur during World War I. Born to a prominent Jacksonville FL family, he was the son of Florida banker Bion Hall Barnett, and the twin brother of William L'Engle Barnett. He attended Trinity College in CT, and is listed in Falk (1999). He studied in New York with George Bridgman and Birge Harrison at the Art Students League, in Provincetown with Charles Hawthorne, and in Paris at the Académie Julian.

New York Times ("Art Notes" 1917a): "Bion Barnett is now a camoufleur..." He was the uncle of American writer Madeleine L'Engle.

Ginzl (2001:97): "[Barnett] volunteered during the First World War to serve in France with the American Expeditionary Force. After the war, he remained in France and studied at several Paris art academies.

He married a French pianist and became a noted landscape artist, winning a number of awards for his paintings."

•

Robert Barnhardt
(dates undetermined)
American lighting designer

He is listed by Naversen (1989) as a member of the Camouflage Society of Professional Stage Designers.

•

Wilhelmina Barns-Graham
(June 8, 1912-January 24, 2004)
British painter

Born in St. Andrews, Scotland, she was educated at the Edinburgh College of Art. She was married to writer and critic David Lewis, and taught briefly at Leeds Art School. According to the Times Online Obituary: "[During World War II] As a species of war work she took a job making camouflage nets in a nearby ICI factory, where [painter Ben] Nicholson, [sculptor Barbara] Hepworth, Stokes and Mellis had already worked, but she had to give this up after a year because of dermatitis."

•

Charles W. Barr
(October 19, 1906-June 25, 2000)
American architect and educator

Born in Pittsburgh PA, he studied landscape architecture at the University of Massachusetts at Amherst (BS 1929), Recreational Forestry at Michigan State University (1940), and City Planning at MIT (MA 1949). In 1930, he joined the faculty at Michigan State University, and remained on the faculty until his retirement. According to an obituary on the MSU web site: "During World War II, as a civil employee, he was given the rank of captain. He was proud to have taught courses on Combat Camouflage and Mathematics to Air Force personnel."

•

Jean-Louis Barrault
(dates undetermined)
French comedian

In a biography, it is said that Marcel Damboise sculpted a bust of Barrault, with whom he became friends during World War II, while both were serving in the French Army in a camouflage unit.

•

Above A converted patrol yacht, the *USS Gem* (c1917), in an experimental low visibility camouflage scheme, developed and recommended by the Submarine Defense Association. Photo courtesy US Naval Historical Center (NH 79363).

Lindon W(allace) Bates
(1858-April 22, 1924)
American civil engineer

Bates was a prominent engineer who made important contributions to the design of railways and waterways, including the Panama Canal. During World War I, he headed the Engineering Committee of the Submarine Defense Association. His willingness to serve in that capacity may have been motivated in part by the death of one of his sons, Lindon W. Bates, Jr., also an accomplished engineer, who died in 1915 when the *Lusitania* was sunk by a German U-boat. Subsequently, another son, Lindell T. Bates, made important contributions to WWI ship camouflage. According to Skerrett (1919:175), the younger Bates would have been one of the heads of the US Naval Camouflage Section in 1918 "had he not been engaged upon work of the utmost significance in perfecting an automatic course indicator, designed to revolutionize the practice of zigzagging and thus, in another department of deception, to help to outwit the submarine." Later, under his direction, a patrol yacht, loaned by William Ziegler, Jr., and named the *USS Gem*, was used by the Submarine Defense Association for an experimental camouflage scheme.

New York Times (May 11, 1915, p. 2): "Lindell T. Bates, son of Lindon W. Bates of New York, Vice Chairman of the American Commission for the Relief of Belgium, was arrested at Kinsale [County Cork, Ireland] yesterday on a charge of espionage while searching for the body of his brother, Lindon W. Bates, Jr., who is believed to have perished on the *Lusitania*. Newton B. Knox, an American mining engineer, who was with Mr. Bates, was taken into custody at the same time. The Sergeant who made the arrests accused them of being officers of a German submarine. After being taken before a Captain they were detained at the barracks half an hour, until United States Consul Frost, at Queenstown, vouched for their innocence. Their search of the coast revealed no trace of the body of L. W. Bates, Jr."

•

James Bateman
(March 22, 1893-August 2, 1959)
British painter and printmaker

Bateman was born in Kendal, Westmoreland, England. He studied at Leeds, the Royal College of Art, and Slade School of Art, where his primary interest was in sculpture. As a result of battlefield injuries in World War I, he gave up sculpture for painting and wood engraving. He is particularly known for rural landscapes. According to the Tate Archives Online, he designed camouflage at Warwickshire in World War II.

•

René-Albert Baucour
(August 10, 1878-1948)
French sculptor

Born in Paris, he is listed in Coutin (1988) as a World War I French camoufleur. One of his sculptures, titled *First Thrill* (1921), is in the famous Parc Montsouris public park in Paris.

•

Jacques Baugnies
(1874-1925)
French painter

Born in Paris, he studied with Jean-Léon Gérome. Much of his work consisted of historical scenes. He is listed in Coutin (1988) as a World War I French camoufleur. According to Kurth (2001:72), he was romantically connected with American dancer Isadora Duncan in Paris, where Duncan was first presented to French society at the salon of Baugnies' mother. Along with two other men (at the same time), he was "the momentary focus of her sexual longings and her ticket to Paris society."

•

Frederick Beddington
(1896-1979)
British artist

Beddington was born in England, and educated at Wellington College. At age 18, during World War I, he volunteered for the British Army, where he became a celebrated sniper on the Western Front in France. Following the war, he studied at the Slade School of Art, eventually making his living as a stockbroker and art dealer at Wildenstein's.

During World War II, he was in charge of the first British Camouflage Development and Training Center at Farnham Castle (see Goodden, 2007). He was the brother of (John Louis) Jack Beddington, an advertising expert who commissioned many of the well-known Shell petroleum posters in the years between the wars. See Long (1979) and Goodden.

•

Alfred Felix Landon Beeston
(February 20, 1911-September 29, 1995)
British orientalist

Beeston served in the British Intelligence Corps during World War II, in the course of which he taught Arabic to Steven Sykes and other camoufleurs.

Sykes (1990:24-27) recalls that Freddie Beeston was "a very likeable Intelligence Corps Officer, bespectacled and owl-like, who knew classical Arabic from university days, but he had never been to an Arabic speaking country." He also describes the comic awkwardness of Beeston's first encounter with "live Arabs" and his attempt to speak to them. He writes, "The fellaheen were halted in their tracks. We all thought his words were proving incomprehensible—but not a bit of it. They were astonished into silence by the pure classic style of speech coming from this rotund Englishman. It was, I suppose, as if the stevedores of a British port were addressed by a foreigner in pure Chaucerian, or perhaps Shakespearean, English."

•

Alan Beeton
(February 8, 1880-December 20, 1942)
British painter

Born in England, he studied history at Trinity College, Cambridge, then served in the British camouflage section during World War I. He once made a series of four paintings, titled *Posing*, *Composing*, *Reposing*, and *Decomposing*.

Innes (2005:147): "During the war [WWII] the [US] army borrowed part of the Futurama model [designed by Norman Bel Geddes] for a camouflage school, where it was used to demonstrate techniques of deceptive coloration and the effectiveness of visual illusion...[In addition] Bel Geddes himself designed camouflage patterns for military vehicles and aircraft."

Above Cover of *False Colors: Art, Design and Modern Camouflage* by Roy R. Behrens (Behrens 2002), an account of the extensive role played by artists in the development of camouflage during both World Wars.

Kurt Vonnegut [in the voice of a fictional character named Rabo Karabekian] (1987:1-2): "I was not born a cyclops. I was deprived of my left eye while commanding a platoon of Army Engineers, curiously enough artists of one sort or another in civilian life...We were specialists in camouflage, but at that time we were fighting for our lives as ordinary infantry. The unit was composed of artists, since it was the theory of someone in the Army that we would be especially good at camouflage."

Hartcup (1980:20): "[Alan Beeton] was one of the first to appreciate that the camera could be deceived by texture, but not by color." Chesney (1943:99): "At Aire was Beeton, who made a speciality of aerial observation of our work..."

Rutter (1999): "It was an artist who discovered the defects in this system and brought camouflage to its second stage of development, based on tone rather than color, on how things look to the camera rather than how they look to the human eye. The [British] Army's debt to Solomon J. Solomon is only equaled, if not exceeded, by its debt to the young portrait painter Alan Beeton, who was the pioneer of the subtle and highly scientific camouflage practiced in 1918." See also "Alan Beeton" (n.d.).

•

Norman (Melancton) Bel Geddes
(April 27, 1893-May 8, 1958)
American theatrical and industrial designer

Bel Geddes was born and raised in Adrian MI. He moved to New York, where he worked as a set designer at the Metropolitan Opera, then settled in Hollywood, where he designed film sets for Cecil B. De Mille and D.W. Griffith. He eventually became one of the country's leading industrial designers (a new profession at the time). In his autobiography (Bel Geddes, 1960:350), there is a chronological list of his commissions including those of designing, in 1942, a "new self-camouflage technique for the United States Army Engineers" and a "camouflaged factory for Nash Sikorsky at New Orleans."

•

M.C.F. Bell
(dates undetermined)
British painter

He is mentioned in Barkas (1952:26) as a World War II camouflage officer.

•

Camille Beltrand
(1877-?)
French printmaker

Beltrand was a wood engraver who illustrated Paul Valéry's *Eupalinos or The Architect*. He is listed in Coutin (1988) as a World War I French camoufleur.

•

Alon Bement

(August 15, 1876-1954)
American artist

Born in Ashfield MA, he studied in Paris with Léon Bonnat and Benjamin Constant, then taught design and painting at Columbia University, and the Maryland Institute of the Arts. He is especially remembered for having influenced American painter Georgia O'Keeffe, beginning with a summer workshop in 1912 at the University of Virginia. It was he who introduced her to the writings of Arthur Wesley Dow on composition. She continued to work with Bement for several years, presumably at Columbia Teachers College. During World War I, he was a camoufleur for the US Shipping Board. His article titled "The Report of the U-16" (Bement 1919a) is a persuasive account of the value of dazzle painting.

Above Just as it rarely made sense to use figure-ground blending in camouflaging a ship, it was probably equally fruitless to use blatantly disruptive strategies in the camouflage of industrial sites.

Norman Wilkinson (in a letter on April 4, 1918): "The person in charge of this [British naval camouflage] service must be an artist, there is no doubt about that. In general, camouflage, on land or at sea is not the job of a scientist. It was developed entirely by artists and is essentially their field of work. Scientists have, up to now, contributed almost nothing at all to this important area of war conduct."

Bement (1917:458): "There seems to be much confusion concerning the origin of the word 'camouflage'. There are rumors afloat that it was the product of the underworld of Paris, and came to its present position of secure respectability through ways both devious and obscure. There is also current an opinion that it was used to denote a mixture of styles: a young architect thinking to execute a design in pure Gothic and finding Romanesque creeping in was said to have 'camouflaged' his drawing. Still other authorities say that it meant to draw a veil; and for a long time in the French theatre camouflage has meant 'make-up'. But whatever it did mean, or however obscure its origin, there is not the slightest doubt of its present application. The word now means to so paint or screen objects that they either become invisible, or else are so disguised as to confuse the enemy."

Bement (ibid.:462): "To show how important camouflage has been in this war it is only necessary to state that one man out of every two hundred of the French forces is engaged in its practice."

•

Benefits of Camouflage
Its effectiveness (or not)

Harper Goff, interviewed in Naversen (1989:152): "We [camoufleurs] actually helped the soldiers' morale. They said they slept better at night knowing we had camouflaged the barracks. Civilians almost bestowed magical powers on camouflage; thinking it would protect them from harm. Many of the officers I worked for thought of us as dreamers with magic paint buckets and expected far more from us that we could possibly deliver."

Trevelyan (1957:113): "In these early days of the war [WWII] Industrial Camouflage was a perfunctory ritual that had very little to do with its proper function, which was, presumably, to protect [civilian] targets from being bombed by enemy aircraft. Much of it was done in an uncoordinated free-lance way; the little corner garage in a housing estate was quickly painted in wiggly lines, and so was the laundry and even the roof of the cinema. People seemed to feel that the green stripes were a charm that somehow bought them immunity from the unknown hazards of war, like the paper strips on the windows it made them feel that they had done their bit, had invested in their small piece of twentieth-century magic. In those early days it was easy to sell any kind of camouflage."

•

Howard S. Bennion
(1889-1971)
American engineer

Bennion attended West Point Academy, graduating in 1912 at the top of his class. Assigned to the US Army Corps of Engineers, he was sent to Europe during World War I, where he became Chief of the Camouflage Service and commanding officer of the 40th Regiment of Engineers. After World War I, he became assistant chief engineer of the Federal Power Commission, then district engineer in charge of flood control in New Orleans. Beginning in 1926, he worked for the National Electric Light Association as Director of Engineering, then as Managing Director of the Edison Electric Institute in New York. He was descended from a prominent Morman family. Faulkner (1973:105) speaks admiringly of him as "human, humorous, keen-witted and not over[ly] military."

•

Thomas (Duncan) Benrimo
(1887-1958)
American illustrator

Above Photograph of Thomas Hart Benton (1936) by Peter A. Juley & Son in the Photograph Archives, Smithsonian American Art Museum (J0001205). As reproduced in Stahl (1995), Dover Pictorial Archives.

Krass (2006:12): "Not only did these [WWI American] artist-soliders have no sense of military protocol, but their profession, by nature, was in direct conflict with military thinking: as artists they were accustomed to individual expression."

...

Guthrie (1993:49): "Small boys and professional soldiers know the magic of camouflage, and natural selection has produced highly camouflaged and practically invisible animals in every environment. A tawny, countershaded (darker above, lighter below) coyote walking on a tawny Colorado slope, for example, recently disappeared before my eyes merely by pausing and then reappeared simply by walking again."

Born in San Francisco, he studied in New York at the Art Students League, then worked as a theatrical designer. He designed the stage set for Shakespeare's *The Tempest*, as produced by John Corbin. Beginning in 1916, he served as a ship camouflage designer, working with William Andrew Mackay.

Witt (2002:72-73): "Benrimo became a lecturer in camouflage methods, training others in this art while at the same time gaining experience valuable to his later teaching career. Using tape measures, chalk lines, and rules, the camoufleur marked out the design on the ship and supervised quality control in the actual paint application... [Benrimo] apparently thought this design experience important because he kept his lectures notes and drawings." Reproduced in Witt (2002:73) is a page from Benrimo's camouflage notes, with drawings of dazzle camouflage schemes.

He taught at Pratt Institute from 1935 to 1939, where one of his students was (camoufleur) Gene Federico. Among his colleagues was his future wife, Dorothy, a jewelry designer who taught industrial design. They soon married and, in 1939, due to health problems, they moved to Taos NM, where they lived and worked for the rest of their lives. His Southwestern landscapes are often described as Surrealist. In Witt (2002:74), Benrimo is said to have been "well versed in the design principles of the German Bauhaus. His reputation as a design teacher continued even after he moved from New York. Laszlo Moholy-Nagy, founder of the New Bauhaus art school in Chicago, apparently invited him to take a position at the school some years after his move to Taos," but Benrimo declined.

In 1952, Dallas TX art dealer Donald S. Vogel mounted a solo exhibition of twenty Benrimo paintings, accompanied by a twelve-page catalog, with a foreword by Merle Armitage (Vogel 2000).

•

Frank W(eston) Benson
(1862-1951)
American painter

Born in Salem MA, Benson studied initially at the school of the Boston Museum of Art, then at the Académie Julian in Paris. After returning to the US, he joined the faculty of his Boston alma mater under the leadership of Edmund Tarbell, where one of his students was (camoufleur) Richard Meryman. He was a member of the alliance of artists known as The Ten, and was also associated with Childe Hassam and Abbott H. Thayer, both of whom were interested in protective coloration. Benson's own fascination with camouflage is documented in Bedford (2000), as is his correspondence with Thayer, to whom he wrote "I know absolutely that, if you are given a free hand [as a camouflage consultant], you could do more than any man living to make invisible everybody and everything used in the war." Benson's depictions of camouflaged birds in their natural setting are often remarkable, as is, for example, his watercolor of a ruffled grouse, titled *Rising*, as reproduced in Bedford (2000).

•

George C(ourtney) Benson
(1886-1960)
Australian painter

Born in Melbourne, Victoria, he produced cartoons for the *Sydney Bulletin* and the *Melbourne Punch*. He is mentioned in Kerr (2008), as having been an official war artist during World War I, and a camouflage artist in World War II.

•

Thomas Hart Benton
(April 15, 1889-January 19, 1975)
American painter

Born in Neosho MO, he studied at the Art Institute of Chicago, then at the Académie Julian in Paris. He returned to the US in 1913. During World War I, he enlisted in the US Navy and was stationed at Norfolk VA, where he was assigned to make drawings of camouflaged ships in the harbor.

Benton, in a letter to a friend, as quoted by Adams (1989:86): "I am now officially listed as [a] 'camoufleur' and have a nice quiet room in Norfolk and an office in which to work. There are two more 'camoufleurs' in the office with me, a photographer and a young would-be artist. Twice a week I leave the office at Norfolk with the fellow who takes the photographs. We go on board a 40-foot motor boat and cruise around the bay making sketches and photographs of newly arrived camouflaged ships. The sketches are finished back at the office (the colors of the camouflage are put in) and along with reports giving name, type, tonnage, etc., of each ship are sent to Washington to be filed. This is done so that if the ship should be torpedoed or lost in any way all the facts concerning her appearance etc. can easily be found."

He is also quoted in Cork (1994:193): "[Working as a camoufleur] was the most important thing that, so far, I had ever done for myself as an artist. The mechanical contrivances of buildings, the new airplanes, the blimps, the dredges, the ships of the base, because they were so interesting in themselves, tore me away from all my grooved habits, from my play with colored cubes and classic attentuations, from my aesthetic drivelings and morbid self-concerns."

Also see Louis Bouché, who had the same assignment and may have been the person who Benton describes as "a young would-be artist."

Above WWI American camouflaged artillery. Public domain.

Henry Berry (1978:208): "...I knew I had a good thing going, so I decided to play it for all it was worth. I had our company commander detach me from my squad and make me the company's camouflage expert. Each day I'd pick up a few men from the company and give them a detail. Oh, I'd ask them to rig up a dead horse or hide an artillery piece, something like that. I once even had them build a fake hill we could hide something behind. Then, at the end of the day, I'd examine what they'd accomplished."

...

British General James Grierson, when asked how a British Army unit had hidden so successfully, as quoted in Edmonds (1932:82): "The men put their waterproof sheets over their heads, and made noises like mushrooms."

Eugene Beringuier

(September 17, 1874-January 5, 1949)
French painter

Born in Toulouse, he was a student of Benjamin Constant and Jean-Paul Laurens. He is listed in Coutin (1988) as a World War I French camoufleur.

•

Oliver (Percy) Bernard
(April 8, 1881-April 15, 1939)
British architect, designer

Bernard was born in London, where his father was a theatre manager and his mother an actress. In his early life, he worked as a stage hand, but he was also interested in boxing, billiards and painting. He volunteered for military service in World War I, but was rejected because of impaired hearing. In 1915, while returning from a trip to the US, he was on the *Lusitania* when it was torpedoed by U-boats. He survived the sinking and even made eyewitness drawings that were published in the *Illustrated London News*. As an architectural designer, his work is characterized as Art Deco, specifically Streamline Moderne.

Wikipedia online biography: "...in 1916, he was commissioned into the Royal Engineers as a camouflage officer, serving in France, Italy and Belgium, reaching the rank of Captain, and being awarded the Military Cross and OBE."

•

Carroll T(hayer) Berry
(1886-1978)
American painter, printmaker

Born in Wicasset, Maine, he studied at the Pennsylvania Academy, then served as an officer in the American Camouflage Corps during World War I. He arrived at the American University encampment at the same time as Laurance Hitt and Victor White. See Rickard (1942:190).

•

Henry Berry
(dates undetermined)
American painter, writer

Henry Berry was an army camoufleur in World War I, then served in the US Marine Corps in World War II. He wrote two wartime memoirs: *Making the Kaiser Dance* (Berry 1978) and *Semper Fi, Mac: Living Memories of the US Marines in WWII* (Berry 1996).

In the latter, he writes: "Each day a German plane would fly over our base and check things out. We did our best to camouflage things, but I don't think we did much of a job.

Anyway, one day the German plane came over and dropped a little parachute with a letter attached. We finally got nerve enough to open it up:

'Dear Marines,' it read, 'if you will camouflage your position in this way'—they had enclosed a diagram showing us what we should do—'it will be much more difficult for us to see it'" (Berry 1996:201).

René-Paul Bertaux

(June 22, 1878-June 19, 1917)
French painter

Born in Paris, he studied at the École des Beaux Arts Décoratif. He worked as a fashion designer, while also teaching himself to paint. He is listed in Coutin (1988) as a World War I French camoufleur.

•

Above One of a series of paintings made by Charles Bittinger (pictured below) of the first atomic explosion at the Bikini Atoll in 1946. Courtesy the Naval Historical Center.

Below Bittinger painting a total solar eclipse in 1937, when he was part of a US eclipse expedition to Canton Island. Photo by Theodor Horyczak. Courtesy US Library of Congress, Prints and Photographs.

Charles (Désiré) Berthold-Mahn

(1893-April 1, 1975)
French painter, lithographer

He is listed in Coutin (1988) and Kahn (1984) as a World War I French camoufleur. Initially a stretcher bearer in the front lines, Berthold-Main was shifted to a camouflage research center located at the Paris Opera. There, according to Kahn (1984:30), he began to feel guilty about having it too easy while his comrades were in danger. Bt his own request, in June 1916, he was reassigned to a regional camouflage facility at Amiens, closer to the front lines. It was one of three such units, the other two being at Chantilly and Noyon. At Amiens, Berthold-Mann was placed in charge of about 1,500 workers.

•

Max Bill

(December 22, 1908-1994)
Swiss painter, designer

Born in Winterthur, Bill served in the Swiss Army in 1939-45, during which, according to Fleischmann (2001:53), "he was assigned to camouflage activities." As cofounder and director of the Hochschule für Gestaltung in Ulm, Germany (a school that was intended as a successor to the Bauhaus), he was a leading contributor to the development of Modern-era Swiss design and abstract geometric art.

•

Ross Birbeck
(dates undetermined)
Australian seaman

During World War II, Ross Birbeck was a sailor on the Australian cruiser *Perth*, c1939-42, which was painted by its crew in a camouflage scheme that somewhat resembled a rainbow, which Birbeck had come up with. Pfenningwerth (2007:87-88), quoting an eyewitness crew member: "'Our first camouflage, the rainbow, was the result of a competition held while passing through the Red Sea. This was won by my good friend, O.D. [Ordinary Seaman] Ross Birbeck. His prize was two bottles of beer. On return to Australia Ross was asked to become a Camouflage 'expert' by the Dept of Navy but he declined. One question he was asked was "How would you camouflage a battleship?" He replied, "Paint a destroyer on the side."' See mimicry.

•

Charles Bittinger
(June 27, 1879-December 18, 1970)
American painter, scientist

Originally from Washington DC, Bittinger studied for two years at MIT, with the intent of becoming a scientist. He then switched to art, and went off to Paris to study at the École des Beaux-Arts. There he married a concert singer named Edith Gay, and moved back to the US to study at the Art Students League. During World War I, he worked at Eastman Kodak Laboratories with physicist Loyd A. Jones on the scientific development of ship camouflage.

Between the wars, Bittinger relocated to Washington DC, and began experimenting with the use of scientific means in creating works of art. In 1929, he painted three murals for the Franklin Institute, depicting stages in the life of Benjamin Franklin. When observed in incandescent light, each painting appeared to be a certain picture, but when viewed under ultraviolet light, they appeared to be a different picture. In 1935, Bittinger exhibited another work in which Charles Lindbergh's famous transatlantic flight appeared under normal viewing conditions, but the same painting became the Mona Lisa when viewed through an optical instrument that Bittinger had invented.

Literary Digest ("Two Paintings in One" 1921): "...Charles Bittinger, a scientist who is primarily an artist, has hit upon the idea of utilizing these differences, increasing them where possible and making them serve his purpose. During the war Mr. Bittinger served in the department of camouflage of the United States Navy, conducting experiments in reflection and transmission of light-waves. By means of the spectro-photometer, he established the reflective powers of a number of pigments and dyes that had invisible spectral differences, and, with a palette set with paintings similar in color when seen in a white light, but contrasting sharply in degrees of light and dark when seen under a red light, painted his two-fold pictures, using round brushes for one series of paints and triangular ones for the other, to avoid confusion in the work....

In Mr. Bittinger's New York studio is a miniature stage set with a scene on the Riviera, which immediately changes to Madison Square in winter when the red light is switched on. Costumes, too, can be handled in endless effective ways by applying the principle to the dyes used and to the patterns in which the colors are put on. A chorus might come dancing on in dresses with horizontal stripes. The light changes—and instantly the stripes are vertical...

Mr. Bittinger has painted an airplane wing with the German cross upon it, which when viewed by our army through binoculars equipped with a red filter, discloses itself to be not the German cross, but the red, white and blue of the Allies. Thus an airplane could fly unscathed over the German lines and return home again without being fired upon."

Above Photograph of the camouflage modeling room at Fort Belvoir, near Richmond VA, during World War II.

Gertrude Stein (1945:17): "There was nothing more interesting in the nineteenth century than little by little realizing the detail of natural selection in insects flowers and birds and butterflies and comparing things and animals and noticing protective coloring nothing more interesting, and this made the nineteenth century what it is, the white man's burden, the gradual domination of the globe as piece by piece it became known and became all of a piece, and the hope of Esperanto or a universal language."

In 1937, Bittinger was invited jointly by the US Navy and the National Geographic Society to travel to Canton Island in the Pacific to paint a total solar eclipse. Nearly a decade later, he was the official artist for Operations Crossroads, for which he was one of the artists to paint the first atomic explosion at the Bikini Atoll in 1946. The paintings he made for the latter are posted online by the US Naval Historical Center at <www.history.navy.mil/ac/bikini/bikini1.htm>.

Clark (1939): "Some observations by Bittinger may be mentioned in connection with the separation by photographic means of two colors which are visually identical. He selected paints having predetermined and known reflection characteristics and a spectral difference which was not apparent to the eye. Scenes were painted in these colors, and illuminated with light of one color to produce a certain visual effect. By changing the spectral quality of the light in accordance with the known invisible spectral difference in the paints, he was able to produce an entirely different visual effect. For instance, in one example the painting shows a summer scene when viewed by white light, and an entirely different winter scene when illuminated by red light."

Washington Post ("Charles Bittinger" 1970): "Mr Bittinger served in the Navy during both world wars, receiving the Legion of Merit in 1946 for his work in the camouflage section of the Bureau of Ships."

See also *New York Times* ("First 'Invisible' Murals 1934; and "By Any Other Light" 1935) and Bittinger (1940).

Patents by Charles Bittinger:
US Patent No. 1,342,247 [June 1, 1920]: Combining Reflected and Transmitted Light Waves of Varying Lengths to Produce Subjective Changes in Scenic Effects.
US Patent No. 1,629,250 [May 17, 1927]: Production and Utilization of Diachronic Inks.
US Patent No. 1,781,999 [March 16, 1929]: Rear View Mirror.
US Patent No. 1,934,310 [with E.O. Hulburt, November 7, 1933]: Visibility Meter and Method of Measuring Visibility.

•

Morris (Atkinson) Blackburn
(October 13, 1902-1979)
American painter, printmaker

Born in Philadelphia, he studied at the Pennsylvania Academy of the Fine Arts, where he was a student of (camoufleur) Arthur B. Carles. He also worked as a WPA artist, and received a Guggenheim Fellowship. His papers are on file at the Archives of American Art, where they are described as including "outlines and notes for lectures, notes on camouflage, poetry, and writings on the nature of art."

•

Stan Blackford
(May 26, 1920-2007)
Indian-born Australian military officer

During World War II, while serving in the 13th Frontier Force Rifles of the Indian Army, he was assigned to teach a camouflage course in Urdu. In his memoir, there is a chapter titled "The Camouflage Expert" (Blackford 2002:186-193).

•

Above Photograph of Edwin Blashfield in his studio (1932) by Peter A. Juley & Son, in the Photograph Archives, Smithsonian American Art Museum (J0044087). Detail as reproduced in Cirker (1967), Dover Pictorial Archives.

Edwin (Howland) Blashfield

(December 5, 1848-October 12, 1936)
American mural painter

Born in New York, he attended the Boston Latin School (which prepared young men for Harvard), then lived abroad for thirteen years, studying principally in Paris with the French portrait painter Léon Joseph Florentin Bonnat. Returning to the US, he became a member of the National Academy of Design, for which he served as president for many years. Initially an easel painter, he turned instead to mural painting, writing, and lecturing in the 1890s. He was commissioned to paint a major mural for the Columbian Exposition in Chicago, as well as important projects for the Main Reading Room at the Library of Congress, the Detroit Public Library, the state capitols of Iowa, Minnesota and Wisconsin, and other important locations. See *New York Times* ("Edwin Blashfield" 1936).

New York Times ("Faking As An Art" 1917): "The American Association for Camouflage was organized some time ago to remedy this defect [the lack of US military camouflage], with Edwin H. Blashfield as Chairman, S[herry] E[dmundson] Fry as Secretary, and a membership including nearly every prominent member of the [National Academy of Design]. The association already has several hundred men enrolled for duty in the Camouflage Corps as soon as the War Department organizes it, and Mr. Fry is confident of his ability, when the call comes, to enlist any number that may be required."

•

Bill (William Ralph) Blass

(June 22, 1922-June 12, 2002)
American fashion designer

Born in Fort Wayne IN, he left the Midwest at age 17 to study fashion in New York. Three years later, he enlisted in the US Army and was assigned to the clandestine camouflage unit known as the Ghost Army. Following World War II, he returned to New

Below The dust jacket of the one-volume edition of Hardy Blechman's massive book on DPM or *Disruptive Pattern Material* (Blechman 2004). It is also available in a boxed, two-volume edition in which the second volume is entirely devoted to camouflage uniforms throughout the world.

York, and began a career as a fashion designer. His autobiography, *Bare Blass* (2002), including a chapter on his war experiences, was completed just before his death.

Blass (2002:38): "I enlisted in the [US] army's camouflage unit in the fall of 1942, and went straight to basic training at Fort Meade, Maryland. The 603rd was one of four noncombat units that were part of a phantom division called the Twenty-third Headquarters Special Troops [aka the Ghost Army]. I say phantom because throughout the war—indeed, for years after—only a handful of brass and the eleven hundred enlisted men involved knew of its existence. Our identity was kept secret for the simple reason that we were posing as other Allied troops in order to fool the enemy... We did this sort of thing in twenty-one engagements, often simulating troops ten and fifteen times our size, right up to the crossing of the Rhine."

Also see Ghost Army.

•

Hardy Blechman
(1968-)
British fashion designer, author

Blechman is founder and proprietor of Maharishi, a London-based clothing firm that is well-known for recycling surplus military clothing (particularly camouflage patterns), by re-dying and re-building it as fashionable Post-modern street clothing.

In 2004, he produced a massive, two-volume, 900-page book on camouflage, titled *DPM: An Encyclopedia of Camouflage* (Blechman 2004). Illustrated by scores of full-color images, it is an ambitiously singular view of contemporary culture in relation to camouflage.

•

Blending camouflage
The matching of figures with backgrounds

The use of blending (or background matching) in camouflage is contingent on our inherent tendency to see similar things as belonging together, and unlike things as belonging apart, known in perceptual psychology as *similarity grouping*. A figure might blend with a background, one figure with another, and so on. The adoption by the British in the 1880s of dust-colored (or khaki) field service uniforms was done with the intention of making soldiers blend in with such settings as India or South Africa. With the start of World War

I, American artist Abbott H. Thayer appealed to the British Army to abandon monochrome khaki because, in his view, it made the soldiers easier to see. Instead, he advocated highly disruptive, high contrast designs that would visually break up the figure.

•

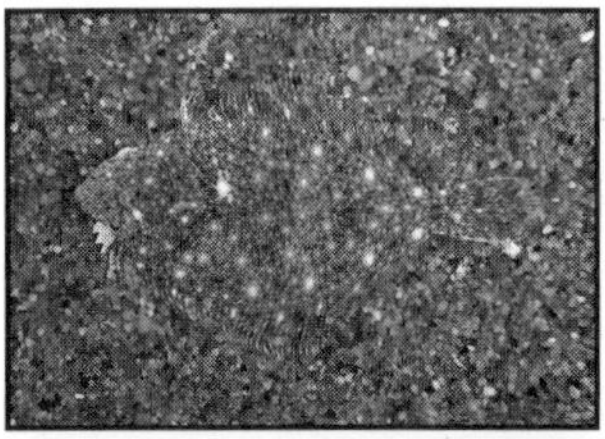

Above Certain organisms have the amazing ability to change the colors and patterns on their bodies in order to blend more convincingly with a range of backgrounds. Pictured here are two stone flounders or flatfish (*Kareius bicoloratus*) in two different settings. Public domain photos. In recent years, Roger Hanlon (Marine Biological Laboratory at Woods Hole MA) and other marine biologists have studied the phenomenal ability of cuttlefish to change their appearance in response to changes in their surrounding. They will even do their best to match a black and white checkerboard. See the documentary film NOVA: *Kings of Camouflage* (PBS).

Facing page US Army training exercise in which a camouflaged soldier emerges from a foxhole, c1919. Courtesy US National Archives (115-WW-71B-4).

Max Blondat
(November 30, 1872-1926)
French sculptor

He was born in Crain, France. Listed in Coutin (1988) as a World War I French camoufleur, he is best known for his bas reliefs, sculptures and public monuments, most of which are considered to be Art Deco in style. He was awarded the Legion of Honor in 1925.

•

Alighiero Boetti
(1940-1994)
Italian artist

Boetti was a conceptual artist who produced a series of camouflage-related artworks in the 1960s, as a result of his thrift shop discovery of rolls of Italian camouflage cloth from the 1930s. He did not alter the fabric, but simply mounted it as if it were painted canvas, and titled it *Mimetico* (1966). Further commentary and a reproduction of this work is in Blechman (2004).

•

Stephen Bone
(November 13, 1904-September 15, 1958)
British printmaker

Born in London, he studied with Henry Tonks at Slade School of Art. During World War II, he served first as a civilian camouflage officer (1939-43), then an Official War Artist for the British Navy. In later years, he was a newspaper critic, a radio broadcaster and panelist.

Adolphe Borie
(January 5, 1877-1954)
American artist

Born to a prominent family in Philadelphia, Borie studied with William Merrit Chase at the Pennsylvania Academy of Fine Arts, and with Carl Marr at the Royal Academy in Munich, Germany. He is listed in Wolanin (1983:64) as having worked with Arthur B. Carles as a camoufleur in the Philadelphia Navy Yard during World War I. He is also described by Dearinger (2004) as a ship camoufleur.

•

Henri-Louis Bouchard
(December 13, 1875-November 30, 1960)
French sculptor

Born in Dijon, France, he is listed in Coutin (1988) as a World War I French camoufleur, for which he was awarded the Legion of Honor. He was an associate of Paul Landowski. See Osborne (1981).

•

Louis Bouché

(1896-1969)
American painter

Above Photograph of Louis Bouché (n.d.) by Peter A. Juley & Son in the Photograph Archives, Smithsonian American Art Museum (J0001232). As reproduced in Stahl (1995), Dover Pictorial Archives.

Rudolf Arnheim (Arnheim 1989:91): "Our black cat lay on the window sill, against the black night outside. When his eyes were open, his body was visible, dimly outlined; but as soon as he closed them, the whole cat vanished, leaving only the unbroken darkness of the window."

Above David Bower, *Sheep Have No Fear Because of Their Whiteness* (1980). Mixed media shelf environment. Courtesy the artist.

Bouché was born in New York. He studied in Paris at the École des Beaux Arts with Jean-Paul Laurens, then returned to study at the Art Students League. During World War I, he served in the US Naval Reserve, where, like Thomas Hart Benton, he was assigned to making drawings of camouflaged ships in the harbor. He talks about this in various documents in his papers in the Archives of American Art. Here is an excerpt from a brief typewritten memoir:

"My war days, however, were not all smooth ones. Several times I was ordered to go to Long Island and make watercolor drawings from the shore of camouflaged destroyers. On a few of these assignments, when I had just started my sketch, I would get a semaphore signal from the destroyer's bridge, and unmistakably directed at me. I could not read the semaphore and was too lazy and indifferent to learn it. A launch would leave the ship and head in my direction. This happened several times, and after a while I knew when it was time to pack up my sketching materials, put them in my nice new leather brief case and wait for the inevitable. Onto the launch I would climb, and then be taken out to the destroyer. Being a real salt, I would forget to salute the quarter deck as I stepped aboard. This really made the situation seem more suspicious.

'Go down and come up again and salute properly,' the officer of the deck would bark at me. After doing so, I would be locked up in the brig, until the skipper or the highest ranking officer on board came to interview me, examine my papers (and I had many), and finally convince himself that I was a harmless seaman second class, in the pay of the Navy Department, and not a spy. After all the formalities, they were very nice. They didn't make me swim ashore but taxied me back to the wharf where I continued my art work in peace."

See also Thomas Hart Benton.

•

Jean-Louis Boussingault
(March 8, 1883-May 17, 1943)
French painter, printmaker

His grandfather was Jean-Baptiste Boussingault, a prominent French chemist. He was initially interested in the decorative arts and worked as an advertising artist. Beginning in 1902, he studied at the Académie Julian and Académie de la Palette, where he became friends with André Dunoyer de Segonzac, Charles Dufresné, and others who would later serve as camoufleurs. See Osborne (1981). He is also listed in Coutin (1988) as a World War I French camoufleur.

•

David Bower
(1936-)
American artist

Originally from Chicago, Bower's family moved to a rural setting when he was twelve. There he was influenced by the mysterious aspects of farmhouse interiors. In the 1970s, he began to invent precisely crafted "shelf environments," comprised of portions of miniature rooms. In time, Bower began to use military camouflage patterns in these, as a way of dissolving distinctions between the interior walls and the objects in front of them. The titles of these artworks often suggest connections between camouflage and psychoanalysis (Behrens 2002). Prior to his retirement, he was on the art faculty at Northern Illinois University.

•

Harold T. Bowler
(1903-July 22, 1965)
Portrait painter

Born and raised in Syracuse NY, he was commissioned to paint portraits of US Vice-President John Nance Garner, Secretary of State Cordell Hull, Vincent Price, Ruth St. Denis, Richard Haliburton and others.

New York Times ("Harold Bowler" 1965): "During World War II he served with a camouflage unit of the 999th Engineers of the First Air Force. After a 25-mile hike with the unit, he painted what is perhaps his best-known work, *After the Hike*, a battered pair of combat boots."

•

Albert Brabo

(April 13, 1894-August 20, 1964)
French painter

Farrer (1918:67): "The real thing about the human side of the war is the sheer fun of it."

Born in Ales (Gard), France, he is listed in Coutin (1988) as a World War I French camoufleur. He was assigned to serve in Italy in a French camouflage unit that was stationed there. In 1939, he was called up to serve again.

•

Above WWI railroad mounted howitzer, heavily camouflaged with paint.

Alex Danchev (2005:139-140): "The persistent irony of war is that it is worse than expected. [Georges] Braque had not forgotten the ironies of this one [WWI]: 'Our soldiers, in 1914, charged in red trousers! They came back from the war with pigs' snouts [gas masks on their faces].'...After the brightly lit battle pictures came a drab contest of restricted palettes, in which color was deliberately used to disrupt form—the word they used was *denature*—painting familiar objects in strange ways until they became unrecognizable. Camouflage was the Cubists' revenge, as Jean Paulham pointed out. The very art that resembled nothing and deformed everything, the bizarrerie in all its glory, was precisely what was now required."

Eveline (Warner) Brainerd

(1870-July 31, 1948)
Editor, feminist

Born in Haddam CT, she was the daughter of American lawyer Cephas Brainerd, who was head of the Young Men's Republican Club when Abraham Lincoln was US President. She was a member of the editorial staff of *The Independent*, and campaigned for women's civil rights, including the right to serve on a jury.

New York Times ("Eveline Brainerd" 1948): "During the first World War, Miss Brainerd served in France with the YMCA and ran a 'hut' at the US Central Camouflage Works at Dijon. On her return to New York she became editor of *The Bulletin*, published by the League of Women Voters, and continued in that office until 1938." Accompanying that article is a detail of a portrait of Brainerd by camoufleur Wilford S. Conrow, made for the New York YWCA.

In 1942, there appeared in the *New York Times* (Brainerd, 1942) a letter to the editor titled "Our Own Camoufleurs," unsigned except for the label "Overseas Woman" and the address of Haddam CT (Eveline Brainerd's hometown). She describes the efforts Americans made in producing camouflage in France, and mentions in particular "Wilford S. Conrow, Greville Rickard, Homer Saint-Gaudens, and Alfred [sic] Herter, Jr."

•

Georges Braque
(May 13, 1882-August 31, 1963)
French painter

Braque, the cofounder with Pablo Picasso of Cubism, is listed in Coutin (1988) as having been a World War I French camoufleur. He did serve as a lieutenant in the infantry, in the process of which he was wounded at Carency on May 11, 1915. But was he actually in the camouflage section? His close friend was Othon Friesz (they were from the same hometown) who camouflaged airplanes during the war, and other sources also list Braque as having been a camoufleur, for example, Ducasse (1962) and Wolinski (1998). But it is odd that he and his biographers have had so little to say about the experience.

Nonetheless, there is a well-known, provocative statement by Braque from an interview, in which he said: "I was happy when, in 1914, I realized that the Army had used the principles of my Cubist paintings for camouflage. 'Cubism and camouflage,' I once said to someone. He answered that it was all a coincidence. 'No, no,' I said, 'it is you who are wrong. Before Cubism we had impressionism, and the Army used pale blue uniforms, horizon blue, atmospheric camouflage'" (quoted in Liberman 1968:143).

•

Robert P(limley) Breckenridge
(May 8, 1907-1964)
American military officer

He was the son of George P. Breckenridge, a lawyer from St Louis, who was descended from John Caleb Breckenridge (sometimes spelled Breckinridge), a prominent Kentucky politician, US Vice-President under President James Buchanan, and Secretary of War for the Confederacy. In 1942, as a major in the US Corps of Engineers, Robert Breckenridge wrote a book titled *Modern Camouflage: The New Science of Protective Concealment* (Breckenridge 1942), with a foreword by General Ulysses S. Grant III. During World War I, he was a member of the Engineer Board at Fort Belvoir.

•

James A. Brennan
(1885-August 25, 1956)
American songwriter, musician

Born in Boston, he attended the Massachusetts Normal Art School, with the plan of becoming an advertising artist. In 1913, he instead began performing as a pianist for Maurice, a well-known dancer. Brennan was a song writer, and composed *Rose of No Man's Land* ("Mid the war's great curse stands the Red Cross nurse. She's the Rose of No Man's Land"), a popular song during

World War I. He also composed *In the Little Red Schoolhouse, Down at the Old Swimming Hole, Happy Go Lucky Days, Dreams of Mother, When Will I Know?* and *Are You in There O'Brien?*

Above Once you become aware of it, coincident disruption is ubiquitous. This is a brush and ink painting by American artist Leonard Ragouzeos titled *Draped Chair with Stripes* (n.d.). Sumi ink. Courtesy the artist.

During that war, he was a camouflage artist for the US Shipping Board. In 1927, he joined the vaudeville troupe of Billie Jones and Ernie Hare as a radio entertainer. In 1940, he again served as a camoufleur at the Fore River Shipyard in Maine. See *New York Times* ("James A. Brennan" 1956).

•

(John) Dwight Bridge
(dates undetermined)
American artist

Born in St Louis MO, he studied at the Art Students League with Albert Herter (the father of camoufleur Everett Herter). During World War I, he was the top enlisted officer (Sergeant Major) of the American Camouflage Corps in France. Faulkner (1973:88) speaks very favorably of him, calling him "just, sympathetic, and good-humored." On the web site of Alpha Phi Omega, a service fraternity, he is listed as having died in World War II.

Hiram Williams (1984:119): "Also, I think Picasso is often guilty of camouflage. Frequently Picasso makes a little bite look like a big chew."

Mrs. Henry Adams (1881): "It's not that he [novelist Henry James] 'bites off more than he can chaw,' as T.G. Appleton said of Nathan, but he chaws more than he bites off."

...

Beston (1919:120-121): "And now the mocker is sure to ask the pro-camouflager in the most serious of tones is he ever saw the ship disguised as a sunset which the Germans unhappily discovered on a rainy day."

Alec (Geoffrey) Brierley
(1911-?)
British-born Australian cartoonist, illustrator, writer

Born in England, he grew up in Australia, where he studied drawing at Melbourne Technical College. He worked for Australia Paper Mills as a timber selector. Skilled at landscape gardening, during World War II, "he was planting advisor to the State Camouflage Committee." In 1978, he published *An Illustrated History of the Kelly Gang*. See Kerr ("Alec Brierley" n.d.).

•

Nicholas Britsky
(1914-June 7, 2005)
American artist and educator

Britsky was a professor of art at the University of Illinois for 39 years, retiring in 1976. He taught camouflage for the US Army during World War II. *College Art Journal* ("Wartime Activities":363): "Lieutenant Nicholas Britsky is still stationed at Fort Belvoir as an instructor in camouflage."

•

Edward Brodney
(April 15, 1910-August 3, 2002)
American painter

Born in Boston, he studied art at the School of the Museum of Fine Arts, the Massachusetts College of Art, and Yale. In 1936, he was selected to paint a WPA mural for the Massachusetts State House. He was drafted into the US Army in 1940. Martin (2002): "Mr. Brodney was among the first American troops to land on New Caledonia. Though officially a medic, he carried sketchbooks, watercolors and oil paints. When a general saw how well he had camouflaged the tents of his medical compound, he was made chief of camouflage for all of New Caledonia."

•

Horace Brodzky
(January 30, 1885-February 11, 1969)
Australian-born painter

Born in Melbourne, Brodzky was the son of a prominent journalist and publisher. He studied art at the school of the National Gallery of Victoria, then later moved to London, where he was associated with the Vorticists. He is mentioned briefly in Cork (1994:232) in reference to his woodcut titled *Two Camouflaged Ships*.

•

Alfred (John) Brown
(October 8, 1893-October 21, 1976)
New Zealand architect, town planner

Brown was born in Auckland, New Zealand. Proudfoot (1993): "Enlisting in the New Zealand Expeditionary Force on 19 October 1915, Brown served with the 3rd Field Company, New Zealand Engineers; he was wounded in France in May 1917 and invalided to England... In 1940-45 he was employed by the [Australian] Commonwealth government (to devise camouflage)..."

Dayton R(eginald) E(vans) Brown

(1895-1971)
American artist

Born in Chicago, he studied at the Art Institute of Chicago, University of Illinois, Los Angeles Art Center, and San Diego State University. During World War I, he served in the US Army.

In 1940, he was commissioned as a lieutenant commander in the US Navy. He worked under Everett L. Warner in the US Navy Camouflage Section during World War II. In the years following World War II, he did extensive research in the camouflage of submarines.

Robert Hays (1999), who served with Brown, recalls: "By late 1944, I had covered a series of projects which had a Cdr. Dayton R.E. Brown, USNR, who arrived in Washington DC from the Pacific. He was covering aircraft camouflage and assisting Cdr. Charles Bittinger and Everett Warner, [and] had been a technical observer in the Pacific on ship camouflage problems/projects. Brown and I worked on many visual/visibility tests using subs and ships, subs and aircraft, lectures at MIT...

...You would have found 'Brownie' or 'DREB' very, very interesting. Ha! One evening after a quick nine-hole golf game, a delicious meal preceded by 2 or 3 double marguerites, by midnight we were in his study and he dashed off an oil painting of my head in an hour's time—I still prize it."

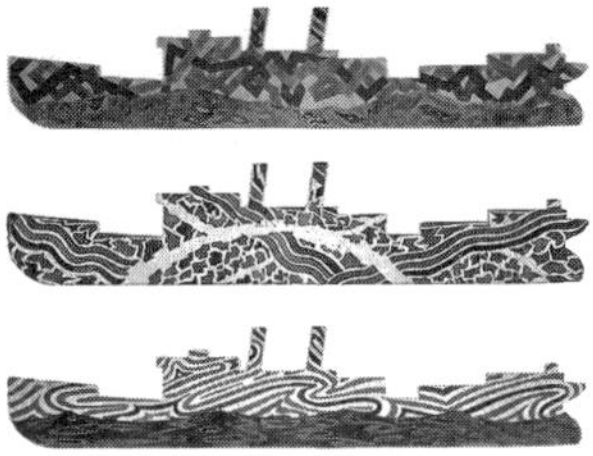

Above Ship camouflage proposals, painted with watercolor on cut-out boards, each about 30 inches wide. These appear to be ship silhouettes that were tested on the shore of Lake Ontario by Eastman Kodak physicist Loyd A. Jones during World War I. Artist unknown. Courtesy US National Archives, Records of the Bureau of Ships.

Lewis R. Melson (quoted in Gould 1992:227): "Many years ago, I was summarily ordered to assume the responsibility of directing the efforts of the US Navy's Ship Concealment and Camouflage Division, relieving the genius who had guided this effort throughout World War II, Commander Dayton Reginald Evans Brown. Dayton had perfected the camouflage patterns employed on all naval ships and aircraft throughout the war. In his briefing of what I could expect in directing the continuation of his work, I found his theories and designs were based upon Abbott H. Thayer's earlier work in the field of concealment and camouflage..."

•

Paul F. Brown

(1871-December 9, 1944)
Newspaper illustrator

Born in Concord NH, he began his career as a newspaper artist at the *Boston Record* in 1898, making chalk plates for the reproduction of drawings. He also worked for the *Boston Herald*, where he was a staff artist for forty years. He served as president of the Boston Press Club. *New York*

Times ("Paul F. Brown" 1944): "...Mr. Brown during the first World War was in charge of the Navy's camouflage division at New Orleans, where he directed the camouflaging of ships."

•

John Gaitha Browning
(1913-1992)
American artist

Brown was born in Oxford MS, then moved with his parents to Texas. While living in the Southwest, he was an Indian trader on the Navajo and Zuni reservations. At age 30, he joined the US Army, and was sent to the South Pacific for two years, where he served as a translator among Native American soldiers. He was also a camoufleur, as is documented in his wartime diary, *An Artist at War: The Journal of John Gaitha Browning* (Toliver, 1994).

Browning (Toliver 1994:41-42) on April 13, 1943: "Captain Gudgeon came by and discussed camouflage work with me this morning...Our executive officer suggested that if the captain wants me to direct his camouflage, he better do something about getting me a rating. The captain replied that if it were possible, he would like to have me on his staff. Of course, I would like the work, but you just can't depend on anything. Tomorrow they may stick someone into the place who has never done any of the work, for no reason other than that someone else happened along and said, 'I want John Jones to be your camouflage,' and the captain wouldn't hesitate for a minute."

•

Charles F. Brunckhorst
(dates undetermined)
American architect

Listed in Embury (1917:138) as "from Schmidt, Garden & Martin's office in Chicago." He painted six murals for the interior of the Fox Theatre in Fullerton CA.

•

George de Forest Brush
(September 28, 1855-April 24, 1941)
American painter

Above Photograph of George de Forest Brush (1936) by Peter A. Juley & Son. Courtesy Photograph Archives, Smithsonian American Art Museum (J0001300). As reproduced in Stahl (1995), Dover Pictorial Archives.

In Sumrall (1971), it was stated (based on US Navy records) that, in 1899, "Robert DeForest [sic] Brush, a prominent artist known for his studies of protective coloration in nature, suggested a scheme" for applying comparable principles to the concealment of ships. In a subsequent letter to the editor, a Brush relative pointed out that the person referred to was George de Forest Brush. Sumrall (1972) replied that the records clearly indicate "Robert," which they did. But I have seen at least one US Navy memorandum ("Marine Camouflage" n.d.) with handwritten notations by Everett L. Warner. In part B, page 1 of that document, Warner has circled "Robert" with a link to the note "correction *George*."

Born in Shelbyville TN, Brush studied in Paris under Jean-Léon Gerome at the École des Beaux-Arts, where one of his classmates was Abbott H. Thayer, with whom he became lifelong friends (although they had their quarrels). In the 1880s, Brush was primarily known for his idyllic paintings of Native Americans, among whom he had lived at various times in his life. He and his family moved to Dublin NH, where the Thayer family had resettled. There, he turned to painting portraits of his wife (sculptor, inventor and aviator Mittie Taylor Brush) and their children. But he also collaborated with Thayer on camouflage experiments.*

In Bowditch (1970:188) there is considerable mention of the Brush family's interest in camouflage. She describes how Thayer and he (presumably during the Spanish-American war) experimented with the camouflage of toy battleships (on Dublin Lake) that they covered with tin foil and then countershaded. Later, coincident with World War I, Brush bought a small monoplane that he named *The Moran Sonnier*, then collaborated with a Mr. Harrison (possibly Dublin painter Birge Harrison) in devising a transparent fuselage and wings, by covering the frame with an all but transparent silk fabric, which, although they strengthened it with varnish, was not sufficiently durable (p. 152). According to Bowditch, her mother was even more deeply involved in aviation and airplane camouflage.

•

*In addition to being called Robert in US government records, George de Forest Brush is often mistaken for his son, Gerome, who also contributed to camouflage, and who, in 1902, was listed as the co-applicant with Abbott H. Thayer for a ship camouflage patent. Oddly, Gerome Brush was only fourteen at the time. As an indication of the persistence of the confusion between father and son, there is mention in Williams (2001:37) of "George de Forest Brush (1855-1941), also referred to as Jerome Brush."

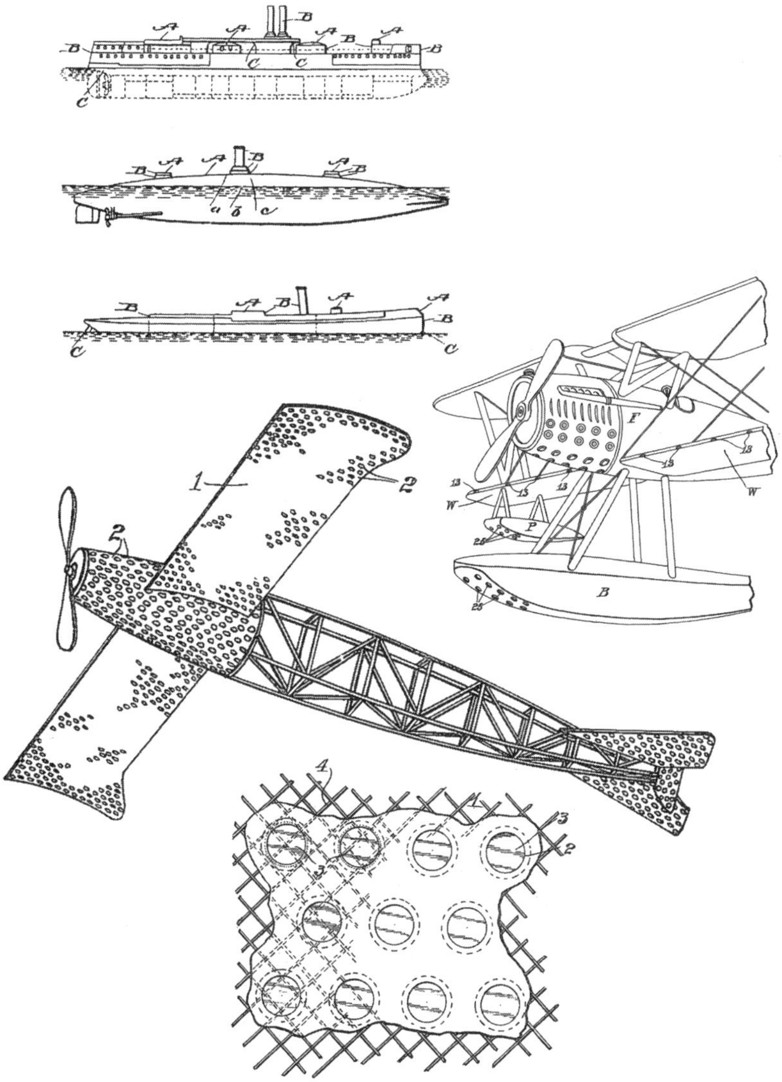

Above Members of the Brush family (parents and children) were amazingly resourceful. The father, his wife Mittie Taylor Brush, and their son were artists as well as inventors. Their patents, some of which pertain to camouflage, are on file at the US Patent Office. At the top of this page are three drawings that Gerome Brush and Abbott H. Thayer submitted in 1902 for US Patent No. 715013, which uses countershading for ship camouflage. The remaining drawings are for aviation patents by Mrs. Brush. One of their daughters was Nancy Douglas Bowditch, a painter and theatrical designer, who also wrote a biography of her artist-father (Bowditch 1970).

Gerome Brush

(1888-September 13, 1954)
Painter and sculptor

Below (top) A portrait photograph of Gerome Brush (n.d.) by Peter A. Juley & Son. Courtesy Photograph Archives, Smithsonian American Art Museum (J0001301). Below that is a watercolor drawing from the US National Archives, labeled "Camouflage for *USS Mount Vernon*" and (in lower right corner) "by Gerome Brush." Perhaps the design was rejected or simply used on a different ship, because to the right of that is a photograph of the *USS Mount Vernon* and, while camouflaged, it is not the pattern that Brush had proposed. Courtesy US Naval Historical Center (NH 45747).

Born in New York, he was the son of painter George de Forest Brush and sculptor/inventor Mittie Taylor Brush. He was named for his father's teacher, Jean-Léon Gérome. He studied art with his father, and with other artists in the US and Italy. Among his friends were (camoufleur) Barry Faulkner, of whom he created a plaster bust, and (camoufleur) Sherry Edmundson Fry, who also made a bust of Brush that is now in the collection of the National Academy of Design. One of Brush's own sculptures is at the Gettysburg Monument.

According to the *New York Times* ("Caruso Buys Painting" 1914), a portrait of his [Gerome Brush's] wife, Louise, was purchased by Italian opera singer Enrico Caruso at Knoedler Galleries on November

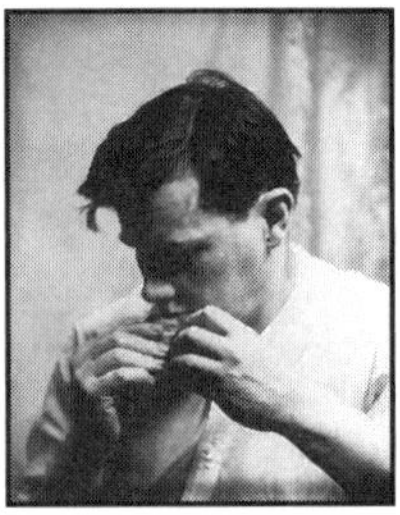

24, 1914. In the weeks prior, the Brush family had returned from Europe on a passenger ship on which Caruso also sailed. For an amusing account of this voyage, see Bowditch (1970:144).

Bowditch (1970:151) also mentions the ship camouflage patent that was jointly filed by Abbott H. Thayer and her brother, as well as the fact that, during World War I, Gerome was put in charge of "the painting of merchant ships all along the eastern seaboard. He worked at Norfolk, Virginia, Boston Harbor, New York Harbor, and many other places."

John Gaitha Browning (Toliver, 1994:34): "...[Sergeant] Wilson would not have thought of concealing his whereabouts, being a person who considers himself a 'tough' soldier. He's also dull enough to overlook the fact that boasting to show—or try to show—how 'army' he is makes him much less a man."

Toch (1919): "The first workers on ship camouflage were Messrs. [William Andrew] Mackay, Gerome Brush, Louis [sic] [Lewis] Herzog, Lt. [Everett L.] Warner and myself, and of course in those days we didn't have the guidance of the brilliant scientific research of Messrs. [Harold] Van Buskirk, Warner and [Loyd A.] Jones."

Patents by Gerome Brush:
US Patent No. 715013 [co-filed with Abbott H. Thayer, December 2, 1902]: Process of Treating the Outside of Ships, etc., For Making Them Less Visible
US Patent No. 1296753 [March 11, 1919]: System for Reducing the Visibility of Objects
US Patent No. 2414632 [January 30, 1943]: Method for Preventing Wake Formation

•

[Mary] Mittie Taylor [Whelpley] Brush
(1866?-July 29, 1949)
American sculptor, inventor and aviator

Mittie Brush, who was both an artist and a pioneering aviator, was the wife of George de Forest Brush, and the mother of Gerome Brush. From all indications, she was an extraordinary person. *New York Times* ("Mrs. George Brush," 1949): "She was an inventor, credited with a number of devices for camouflage and lighting effects for airplanes." According to Bowditch (1970), with her husband's purchase of a monoplane, Mittie Brush also began to experiment with airplane camouflage. Taking up her husband's failed idea of a transparent airplane, she came up with the alternative strategies for enhancing its invisibility, such as installing lights within the fuselage. A contemporary of Amelia Earhart (whom she was surely acquainted with, since Earhart had a summer home in Dublin), she trained as a pilot and became the second woman (Earhart being the first) to be admitted to the Aeronautical Chamber of Commerce.

Patents by Mittie Taylor Brush:
US Patent No. 1293688 [1919]: Aircraft
US Patent No. 1442323 [1919]: Reinforced Fabric for Aircraft
US Patent No. 1367294 [1921]: Aircraft
US Patent No. 1619100 [1922]: Aircraft
US Patent No. 1686649 [1927]: Reenforced Transparent Sheet and Process of Making the Same

•

Richard (McLean) Buckley

(dates undetermined)
British military officer

According to Goodden (2007), Buckley served as a camoufleur under Solomon J. Solomon during World War I (presumably in the same unit as Alister Hardy), and was chosen by Frederick Beddington to be the chief camouflage instructor at the Royal Engineers' Camouflage Development and Training Center during World War II. Trevelyan (1957), Gardner (1993) and Lewis (1994:57) all attest to what Lewis describes as Buckley's "appalling temper," the outbreaks of which would be followed by bursts of tears. Nevertheless, Lewis recalls, "We all grew very fond of this intelligent, unstable commandant."

Above US President Theodore Roosevelt (Abbott H. Thayer's outpoken opponent) dressed in buckskins. Photo by James Suydam (1885). Courtesy US Library of Congress, Prints and Photographs.

Gardner (1993:96-97): "I also discovered that, despite appearance to the contrary, his odd mind had a quixotic grasp of the situation. When ordered to provide the army with an issue of camouflage officers, he had decided to select civilians who had reached the top by using their eyes. This being a rare enough occurrence in Britain, he had collected a mixed bag—Victor Stiebel the couturier, Jasper Maskelyne the magician, Oliver Messel the set designer, Talbot Kelly the bird watcher, a bunch of arty artists, and two honorables (probably included to give the unit 'a touch of class')."

Popular Mechanics (1918:3): "The [US] War Department has announced that camouflage camps, or classes for training women in camouflage, are not authorized by it. Work of this character is not fitted for women, it is pointed out."

•

Buckskins

Fringed woodland hunting attire

Facing page This 1918 news photograph shows a camouflage training exercise for members of the Women's Reserve Camouflage Corps of the National League for Women's Service. Their work is described in detail in Smith (1921). Courtesy US National Archives (165-WW-599H-1). See Cagoules.

Wigginton (1979:411), reporting on an interview of hunters Dale Black and Hawk Boughton by Jeff Reeves and Kirk Patterson: "When asked about their buckskins, Hawk said, 'The story is that it [the fringe] helped water drain off in a downpour instead of soaking into the cloth, but I don't go along with that...But I believe that one thing that it did do, whether they realize it or not: It helped break up their silhouette when they were in the woods. I mean, if you were dripping with fringes and things, you would blend in with the woods easier than if you were just a hard silhouette."

Mauricio Bueno
(1939-)
South American artist, architect

Bueno was born in Quito, Ecuador, and studied architecture at the Universidad Nacional de Bogotá in Columbia. While living in the US, he worked for the National Endowment for the Arts, and the Graham Foundation. He also studied at the Center for Advanced Visual Studies at MIT, where he worked with György Kepes, who had taught camouflage at the New Bauhaus in Chicago in the 1940s. In one phase of his experiments with technology in relation to art, Bueno used military camouflage in combination with aerial views of urban landscapes.

•

Walter (Ralston) Bunning
(May 19, 1912-October 13, 1977)
Australian architect, town planner

Born in South Brisbane, Australia, he studied art in Sydney before turning to architecture. After living and working in London, he returned to establish a practice in housing design and town planning. Spearritt (1993): "During World War II Bunning worked as a specialist in camouflage..."

Charles Burchfield

(April 9, 1893-January 10, 1967)
American artist

Above Photograph of Charles Burchfield (n.d.) by Peter A. Juley & Son in the Photograph Archives, Smithsonian American Art Museum (J0001324). As reproduced in Stahl (1995), Dover Pictorial Archives.

Krass (2006:18): "[During World War I, in traveling across the Atlantic] A moonlight night was 'an excellent night for Fritz'—a good night for getting 'a tin fish in you'—for U-boats stayed hidden among the diamond shapes shimmering on the water's surface. On day nine at sea, [Ernest] Peixotto and [Wallace] Morgan's convoy was met by a dozen destroyers that were to escort them into port. The coloring of these warships caught Peixotto's eye; they were 'brilliantly camouflaged like wasps, queerly striped with black and white, with spots between of yellow, gray-blue, and water-green'—the softened tone of [Claude] Monet's paintings. 'Like wasps too they darted around us,' he wrote, 'zigzagging across our bows, dropping astern, watchful, then, with a burst of speed, forging up ahead again.'"

*There is a reproduction of a collage by Byron Burford, titled *Camouflage* (1996), in Behrens (1999:58).

Burchfield was born in Ashtabula OH, and attended the Cleveland Art Institute. *New York Times* ("Charles Burchfield Dies" 1967:25): "In July, 1918, he was inducted into the [US] Army, ending up in the Camouflage Section."

Baur (1982:91): "In July 1918, Burchfield was called for military service and sent for his training to Camp Jackson in South Carolina...At first, he wrote, 'The new life threw my mind into utter chaos,' but later he enjoyed the army, particularly after he was transferred to the Camouflage Corps with the rank of sergeant. His commanding officer, Lieutenant William Yarrow, was an artist—a Philadelphia portrait painter who later did murals at Princeton. Burchfield told him he admired Cézanne, which created an immediate bond between them. Yarrow assigned him to camouflaging artillery pieces and gave him time off—once a whole week's leave. The work was not without interest. Burchfield made numerous notes on camouflage techniques ('Utmost importance to avoid straight lines and angles'), and the small watercolor *Camouflaged Gun Pit* [1918] is probably a record of his own labors."

•

Byron (Leslie) Burford (Jr.)

(1920-)
American artist

Born in Jacksonville MS, Burford* was for many years a professor of art at the University of Iowa. For much of his life, he was both a performing circus drummer and a painter whose subject matter pertained at times to camouflage. During World War II, he served in the US Army Air Corps. He recalls, "I tried to get into camouflage, but they needed radio operators at that time. Because I was a drummer, I passed the Morse Code aptitude test 100%. As it turned out, my entire platoon were jazz musicians, for their scores on the aptitude test were all very good. A less military group I never saw" (Burford 1998).

Joseph E. Burgess
(1890-January 16, 1961)
American artist and portrait painter

Born in Dansville NY, Burgess earned a BA degree at Syracuse University in 1914. He also studied with George Bridgman and Harvey Dunn at the Art Students League in New York. He taught briefly at the University of Illinois, then joined the US Army as a second lieutenant.

New York Times ("Joseph Burgess" 1961): "He later received citations from President Woodrow Wilson and the British high command for his work with camouflage units." He painted a portrait of American baseball player Christy Mathewson, and was the studio manager for the American Lithography Company from 1928 to 1932.

Above Participants in a camouflage class at New York University in March 1943. They are camouflaging features of a landscape model prepared from aerial photographs. Photo by Marjory Collins. Courtesy US Library of Congress, Prints and Photographs.

Leslie Thomas Burrage
(1909-1983)
South African artist

Journal of the South African Military History Society ("With the Sappers" n.d.): "In November 1943, L.T. Burrage was appointed as one of seven official South African war artists...he served with the Concealment and Camouflage Section, his role being to illustrate instructional pamphlets and posters on camouflage..."

•

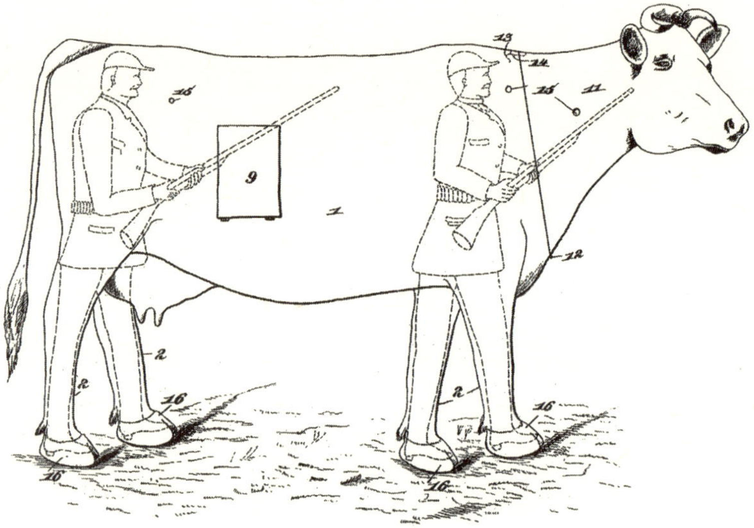
13
14
15
11
15
9
1
12
2
2
2
16
16
16
16

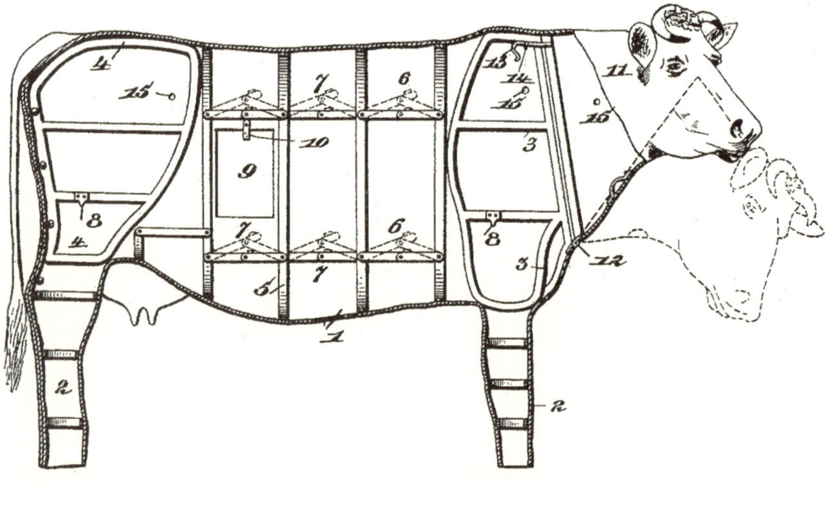
4
15
7
6
13
14
16
11
16
10
9
3
8
4
7
6
8
3
12
5
7
1
2
2

C

Cagoules
World War I hooded battlefield outfits

Cagoules were paint-splattered hooded cloaks devised by the first French camoufleurs, to prevent themselves from being seen by aerial observers as they manned long-range artillery. A wonderful painting of this was published on the cover of *Scientific American*, on September 29, 1917 (see following page). Some of these cloaks still exist, including prototypes designed by French camoufleur Eugene Corbin, and are reproduced in Blechman (2004:126-127) and Newark (2007:83). They are also described in Kahn (1984:148), who states the following: "Hidden beneath camouflaged *cagoules*, hooded uniforms...was a ghoulish image of the modern soldier, whose finely fitted and brilliant red and blue clothing was replaced by an amorphous costume of drab greens and browns that turned the individual into a frightening form."

•

Roger Caillois
(March 3, 1913-December 21, 1978)
French writer, philosopher

Caillois' writings are characterized by a mixture of sociology, philosophy and literary criticism. With regard to camouflage, he is best-known for his essay on "Mimicry and Legendary Psychasthenia," first published in a French

Facing page Drawings submitted for US Patent No. 586,145 titled "Hunting Decoy," invented by J. Sievers, Jr. (1897).

Martha Banta (1987:222): "Camouflage conceals by the *obliteration* of visual signs. It invites observers to forget that there is anything to look at."

...

C.H.R. Chesney (1943:64): "Camouflage is the art of concealing the fact that you are concealing."

Above right and facing page: Various examples, from WWI-era publications, of disruptively-painted camouflage cloaks called *cagoules*.

C.H. Rowe (Rowe 1940: 168):"Leonardo da Vinci, a great artist and scientist, pointed out Nature's use of protective coloration before [Abbott H.] Thayer was born. Artists before and since have been fully aware of the principles of 'countershading,' 'coincident pattern,' 'disruption,' and 'deflection,' only they have been called in art schools since the time of Rembrandt such other names as 'counterchange,' 'discord and harmony,' 'atmospheric and solid perspective.' Indeed, it is impossible to paint a picture without using these principles."

intellectual journal in 1935, then translated into English fifty years later and published in the US in *October* magazine (Caillois 1984). *Psychasthenia* (which translates as "descent into hell") is a now rarely used label for psychological states characterized by paranoia, anxiety and obsessive-compulsive disorders. In his essay, Caillois chose to focus on mimicry, particularly as it occurs in various exotic insects, which he compares to certain aspects of human behavior in relation to cultural settings. He coined the term *legendary psychasthenia*, by which he refers to the inability of people (or other organisms) to distinguish themselves from their surroundings, social or otherwise. Owing to space limitations, in his own words, he "had to leave aside certain related questions, for example those of obliterating coloration and dazzling coloration" (p. 32), both of which were terms proposed by Abbott H. Thayer, who, oddly, he fails to mention. For a Caillois-inspired discussion of camouflage in relation to aspects of culture, particularly architecture, see Leach (2006).

•

Henri-Eugene Callot
(December 20, 1875-December 22, 1956)
French painter

Born in La Rochelle, France, he was a student of Jules Lebebvre. He is listed in Coutin (1988) as a World War I French camoufleur. He was awarded the Legion of Honor.

•

Charles Camoin
(September 23, 1879-May 20, 1965)
French painter

Camoin came from a family of paint manufacturers. He studied at the École des Beaux-Arts with Gustave Moreau, and is listed in Coutin (1988) as a World War I French camoufleur. In Clement (1994), there is an account of his affiliations with Paul Cézanne and Henri Matisse, as well as other artists, including some who later served as World War I camoufleurs, such as Jean Puy and Othon Friesz. Clement (1994:585): "Camoin was mobilized in 1914 and sent to the trenches. In 1916 he was attached to a blackout and camouflage unit with Charles Vildrac and Dunoyer de Segonzac."

•

Camouflage and Camoufleur

WWI-era French camouflage terms

In both World Wars, various names were applied to camouflage and camoufleurs. Here are a few:

According to Kahn (1984:27) camouflage was considered, during World War I, to be "*un bon filon* ('a good, cushy job')."

Elias (2003): "The term 'camoufleur' presented problems in itself [for the Australian infantry in WWII], being too feminine. Acknowledging this, Frank Hinder ironically referred to himself and his colleagues as 'camopansies'."

Sykes (1990:73): "Geoffrey Barkas was the force behind a quite unusually welcome collection of bumph which was sent to all outlying camouflage people—'Camoufluers' as he called us. This publication was called *The Fortnightly Fluer...*"

Bond (1918-1919:499): "A new name was coined for these warriors of the paint brush—*camoufleurs* they were called, and their work was known as *camouflage*."

Harper Goff, interviewed in Naversen (1989:150): "'P.C.' we called it [camouflage]. P.C. for 'protective concealment.' They figured 'P.C.' would be unclear to any spies."

•

Camouflage Society of Professional Stage Designers

World War II US civilian camouflage group

Founded at the start of World War II, the society's members included Boris Aronson, Bradford Ashworth, Lemuel Ayres, Robert Barnhardt, Stewart Chaney, Adrian Cracraft, Manuel Esman, Fredrick Fox, Edward Gelbert, Phillip Gelbert, Mordecai Gorelik, Harry Horner, Carl Kent, David Langworthy, Johannes Larson, Thomas Lee, Sam

Ad Reinhardt (1975:175-176): "When the issue of how artists can help in this war [World War II] is raised...the answer must be direct and clear that the fine artist can be important to the war effort as fine artist, not as soldier, draughtsman, camoufleur, cartoonist, poster designer, or work-relief recipient (WPA artist)."

...

Rudolf Arnheim (1989:90): "[British painter J.M.W.] Turner had to search nature for subjects resembling the swirls of color congenial to his temperament. He found them in storms, waves, clouds, flames, smoke, and in the camouflaging hazes."

Above Robert Gibbings, *Fowey Harbor* (c1919). Wood engraving. See fade away.

Leve, Jo Mielziner (President), Donald Oenslager (Treasurer), and Howard Wisler (Naversen, 1989:178-185).

Fredrick Fox, quoted in Naverson (1989:145): "The Camouflage Society [of Professional Stage Designers] was founded by several of the members who attracted the rest to join the group. If we were going to go in the army we might as well do the things we knew how to do. We each paid $50.00 for membership so we could rent an office in which to hold classes and to pay the guys [Herbert Kniffen of the American School of Design and Baron Nicholas Cherkassoff] who would be teaching us camouflage. A Russian officer, the Baron, was the main teacher, and he sold us copies of his camouflage book for $25.00 each."

Of the twenty pre-war members of the Camouflage Society, it appears that only two, Jo Mielziner and Donald Oenslager, served as commissioned officers. According to Henderson (2001:132), "Several of the members were rejected for medical reasons, while others were accepted as enlisted men." See also Naversen (1989:43),

•

Gordon Campbell
(January 6, 1886-July 3, 1953)
British naval officer

"Gordon Campbell" (n.d.): "Campbell was instrumental in the development of Q-ships, specially-outfitted and armed merchant ships designed to present easy targets to U-boats. When a U-boat surfaced, the Q-ship dropped the camouflage hiding its armament and opened fire."

•

Arthur B(eecher) Carles
(March 9, 1882-June 18, 1952)
American painter

Carles was born in Philadelphia PA, where he studied at the Pennsylvania Academy of the Fine Arts (1900-07) with Cecilia Beaux, William Merrit Chase and others. While living in France in 1907-10, he became friends with John Marin and Edward Steichen. In the US, his first solo painting show was at Alfred Steiglitz's gallery *291* (1912). His work was included in the Armory Show (1913). He also taught at the Pennsylvania Academy in 1917-25. His daughter, Mercedes Carles, was the wife of graphic designer Herbert Matter.

Wolanin (1983:64): "His [Carles'] greatest contribution to the [WWI] war effort, however, was as a camoufleur in the Philadelphia Navy Yard where he worked in 1918 with many of his friends and students, including Franklin Watkins, Adolphe Borie, Jean Knox, Waldo Pierce, and Carroll Tyson. Watkins devised some of the camouflage designs, but it was Carles who actually oversaw the application of the paint on the vessels, supervising the work from various angles and distances."

Above The *USS President Grant* in dazzle camouflage during WWI. Courtesy US Naval Historical Center (NH 41747).

Gerald H. Thayer (1923:162): "[By the end of WWI] The trick painting, closely parallelling or even imitating nature's protective coloration, was largely abandoned. Dust and mud and wreckage, say the French and Yankee camoufleurs, work such rapid havoc with these elaborate 'stunts' of camouflage painting (as on ambulances, tanks, railroad cars, etc.), that in the long run the game proves not worth the candle."

Philadelphia Museum of Art (*Arthur B. Carles* 1970): "Carles spent the years during World War I in Philadelphia. The United States Navy set up a department to devise camouflage design, technique, and patterns to confuse the German U-boat commanders who were sinking so many ships with their torpedoes. A great deal of this activity was carried on at the Philadelphia Navy Yard and many local artists found work there. Young Franklin Watkins was hired by the Navy to help design the camouflage schemes and in turn got Carles employed to oversee their execution on the ships. Carles worked for a year marking vessels and became a project director in camouflage."

•

Richard Carline
(1896-1980)
British painter

Sydney Carline
(August 14, 1888-February 14, 1929)
British artist

The Carline brothers were sons of another British painter, George Carline. They were associated with Stanley Spencer, who married their sister Hilda, and with the Canadian-born camoufleur Percyval Tudor-Hart. In 1917, they were both serving in the Royal Flying Corps when Tudor-Hart, who had received government approval for his experiments in ship camouflage, requested that they be reassigned to his project, along with Hugh de Poix and James Wood. See Hartcup (1980:41). During World War II, according to Goodden (2007:24), Richard Carline was appointed to the British Civil Camouflage Section.

•

Sturgis Carne
(dates undetermined)
MGM film designer

Carne served as a US Army camoufleur in World Wars I and II. According to George Gibson, in an interview by Naversen (1989:147): "Sturgis Carne worked in camouflage for the Army, both World War I and II. He was the brother of [Hollywood film director] John Sturgis."

•

Cromwell H. Case
(dates undetermined)
American architect

He is listed in Embury (1917) as a World War I American camoufleur.

•

Hugh (Maxwell) Casson
(May 23, 1910-August 15, 1999)
British architect, designer and writer

Born in London, he was educated at Cambridge University, and University College, London. During World War II, he was an officer in the Camouflage Branch of the Air Ministry, 1940-1944. A distinguished architect and designer, he taught interior design at the Royal College of Art for twenty years, and served as its Provost. He was also President of the Royal Academy.

Goodden (2007) provides information on Casson's camouflage service, along with reproductions of his watercolor paintings and drawings of buildings camouflaged by planned cultivation, the disguise of other buildings by the painted addition of windows and doors, and the distortion of a building's shape using forced perspective (much like Adelbert Ames' "architect's room" or Everett L. Warner's illusory wallpaper). See anamorphosis.

•

Chaff

Metal foil strips used to confuse radar detection

Known as "window" by the British and *Düppel* by the Germans, it was a clever means of subverting the detection of airplanes by radar by releasing out the window small strips of metal foil. These were picked up as airplanes on the radar screen, thereby concealing the location of the real plane. See Jones (1973, 1978) and Reit (1978:217).

•

Chameleon

Mascot of the World War I French camouflage unit

Peixotto (1917:3): "[According to a friend] The sign of the aviator is a wing on the arm; that of the telegraphic corps is the braided forked lightning. The sign of the *camoufleur* used to be the chameleon, but this has lately been removed. There could be no better insignia for the camouflage than this—a disappearing chameleon." There is a photograph of the World War I French camouflage unit's red and white armband, with a chameleon in metallic thread, in Blechman (2004: 42).

Above Lithographic illustration of an African chameleon by Karl Joseph Brodtmann, from *Naturgeschichte und Abbildungen der Reptilien* (1833). Public domain.

Liberman (1968:189): "[French painter Fernand] Léger wore a checked shirt, and the violent pattern of his clothes against the violent pattern of his paintings made him seem like a chameleon."

...

Eric Hoffer: "When people are free to do as they please, they usually imitate each other."

•

Leslie Cheek, Jr.

(1908-1992)
American museum director

Cheek was born in Nashville TN, where his family made their fortune from Maxwell House coffee. After studying art at Harvard, he turned to architecture and studied at both Yale (where he took courses from theatrical designer and camoufleur Donald Oenslager) and Columbia University. Throughout his life, he served as an arts administrator with various institutions, including the College of William and Mary, the Baltimore Museum of Fine Art, and the Virginia Museum of Fine Art. In 1942, he joined the US Army and was assigned to camouflage at Fort Belvoir in Richmond VA, with the 936th Engineer Camouflage Battalion at the Army Air Base. According to Henderson (2001), he became close friends with the unit's executive officer, theatrical designer and camoufleur Jo Mielziner. He also revised the Army's camouflage training manual. His papers are at the Museum of Modern Art and the Archives of American Art, Smithsonian Institution.

•

Nicholas Cherkassoff
(dates undetermined)
Russian military officer

Baron Nicholas Cherkasoff describes himself on the cover of his 16-page booklet titled *Basic Principles of Camouflage* (1941) as "Formerly Colonel of the Russian Imperial Army, Captain of the British Army, and Member of the Council of Civilian Camouflage." It was Cherkassoff (and Herbert Kniffen, author of a book called *Masks*) who gave instructions to the members of the Camouflage Society of Professional Stage Designers in New York.

Scenic designer Fredrick Fox, (not to be confused with Frederic E. Fox of the Ghost Army), quoted in Naverson (1989:145): "As I recall, the Baron was a very impressive fellow—small, wiry with a thin black moustache—very pallid, slicked down black hair, and wore glasses rimmed in pale gold... The training was basically lectures and theories of camouflage and we worked on model projects which Cherkassoff would critique. The class met one night per week and lasted six weeks. After that, we marched *en masse* to the induction center at Church Street."

•

Clement H(ope) R(awdon) Chesney
(dates undetermined)
British military officer

Lieutenant Colonel Chesney was a British camoufleur stationed at Amiens during World War I. He later published *The Art of Camouflage* (Chesney 1941), in which he was openly critical of the use of artists as camoufleurs.

Chesney (1941:89): "Camouflage by mid-1917, had begun to interest a big field and the factory was visited by people in high places of every nation allied with us."

•

Above A flag-like British dazzle design, as applied to the *HMS Argus*, c1918. Courtesy US Naval Historical Center (NH 63030).

Everett L. Warner (quoted in White 1951:136): "It was inevitable that countershading should have been revived during World War II and the result proved most unfortunate. An area normally in shadow would be painted pure white, and the roll of the ship would throw it into glaring illumination. Strongly criticized, it was discarded after a short time, but it may well have been one of the factors which led to abandonment of all [disruptive] pattern camouflage during the closing months of the war. I am certain that countershading will come back to haunt us during the next war, because the 'invisible ship' is the dream of the inexperienced. So, also, the novices will start 'measuring the sky' all over again. Much money was spent on this in the First World War, and in the last war even more of the taxpayers' money was spent to determine the best average color for ships. As Captain [Charles] Bittinger pointed out once, an artist with a trained eye would not need to spend a quarter of a million dollars. After a few days observation of ships he would just mix the best color up on on his palette."

Thomas Chimes

(1921-)
American artist

Chimes, who was born in Philadelphia, is of Greek descent. In 1939, he enrolled at the Pennsylvania Academy of Fine Arts, but withdrew two months later. Soon after, he completed aircraft mechanics training in order to enlist in the US Army Air Force. During basic training, he was stationed at Greensboro NC, where he met artist Michael Lekakis, who taught camouflage there. Following the war, he studied philosophy at Columbia University, and art at the Art Students League in New York, where Lekakis was also a student. In 2007, the Philadelphia Museum of Art mounted a retrospective of about 100 of his paintings and works on paper, with a time span of nearly 50 years.

•

Reginald (Hughes) Church

(1892-November 6, 1942)
British-born church organist and choirmaster

Church was born in Brighton, England, then immigrated to the US. He made his living as an organist and choir director at various churches in the New York area. *New York Times* ("Reginald H. Church" 1942:53): "In the first World War Mr. Church was employed as a camouflage expert by DeVoe and Raynolds" (the oldest paint manufacturing firm in the US).

•

Emilius R. Ciampa

(1896-1996)
Italian-born American sculptor

Born in Taurasi, Italy, Ciampa immigrated to the US at age 10. He grew up in Boston, where he attended the Belle Arti School, Copley Art Studio, Butera Art School and the Normal Art School (now Massachusetts College of Art). He is mostly known for his sculpture and plaques for various Boston-area memorials. Online Wikipedia biography: "In 1918 (at age 22) he became a soldier in World War I, stationed in France. He was put in the camouflage unit because of his artwork. Ciampa kept a detailed notebook of the camouflage techniques and also made many sketches for the soldier's newspaper and for various French periodicals."

•

Phyllis (Margaret) Cilento
(December 23, 1923-2006)
Australian painter and printmaker

Born in Sydney, Australia, she was the daughter of a diplomat who was later assigned to the United Nations. In 1944, in conjunction with World War II, she contributed to the camouflage of the city of Ipswich, Queensland. When her family moved to New York (because of her father's role in the United Nations), she became determined to study with British printmaker (and camoufleur) Stanley William Hayter (who was living in the US at the time), as well as at the workshop of Mexican muralist Rufino Tamayo. Around the same time, she began interested in Abstract Expressionism, and is sometimes credited with introducing that movement to Brisbane, when she moved back to Australia in 1965. Prior to that, she also lived in Europe, where she studied with Hayter at his Atelier 17, and at the École des Beaux-Arts. See Callaway (n.d.).

•

Cameron Clark
(March 24, 1887-March 23, 1957)
American Architect

Born in Holyoke MA, he studied at MIT and the American Academy in Rome. He designed model homes for *House Beautiful*, *Ladies Home Journal* and other prominent magazines, and the Easton CT home of Helen Keller. According to the *New York Times* ("Cameron Clark" 1957:86), "During World War I he served with the [US] Navy's camouflage section," camouflaging airplanes.

•

Sir George Sydenham [Baron Sydenham] Clarke
(July 4, 1848-February 7, 1933)
Australian statesman

According to Lettice (1981): "As a truly professional military officer his [Clarke's] ideas were ahead of his times and many—especially on fortifications, submarines, camouflage and army reform—proved remarkably prescient."

Robert A. Clifford
(dates undetermined)
American architect

He is listed in Embury (1917:138) as a World War I American camoufleur: He "has been in charge of the decorations of several big New York hotels and did some planting of the Far East Gardens in the Vanderbilt, the Japanese Gardens at the Ritz, and also work at Newport."

•

Graham (Barry) Clilverd
(April 6, 1883-1959)
British architectural painter, printmaker

Clilverd studied at the Central Arts School in London. He was best known for his drypoint engravings and etchings of architectural settings. Art of the Print website: "During the First World War, Clilverd served as a camouflage artist (1916-1918)." Portsmouth Museums website: "[During World War I, Clilverd] was chief camouflage officer to the [British] Admiralty."

•

Griffith (Baily) Coale
(1890-1950)
American painter

Born in Baltimore MD, he studied at Maryland Institute of Art, and the Art Students League in New York. He also studied in Europe. During World War I, he designed ship camouflage for the US Shipping Board (1917-1918). After the war, he worked as a mural painter, producing public murals for major New York buildings, among them the Metropolitan Life Insurance Company, the Criminal Courts Building, and the City Bank Farmers Trust Building. In 1941 to 1947, he served in the US Navy as a Combat Artist for the Office of Public Affairs. See "Griffith Baily Coale" (n.d.), and Archives of American Art (1995:39).

Above An extraordinary example of coincident disruption in the West African Gaboon Viper (*Bitis gabonica rhinoceros*). Public domain photo by Tom Vickers.

Eric Sloane (1942:44): "Nature weaves fantastic designs but she avoids round or angular patterns, or monotonously regular curves. Irregularity is what the camoufleur should seek at all times when designing disruptive concealment."

...

Tom Czarnopys, in Kamholtz (1988:12): "...[while bow hunting with his father] we were covered with camouflage. You do everything to eliminate the boundary between yourself and the natural world. More than concealment, camouflage is eliminating a boundary. It's like the [New] Guinea mud-men covering themselves with mud. They would, in effect, eliminate the boundary between them and the earth."

Alvin Langdon Coburn
(January 11, 1882-November 23, 1966)
American-born British photographer

Coburn was associated with the British Modern artists called Vorticists, one of whom was the camoufleur Edward Wadsworth. Some of his experimental photographs were made with a mirrored device called a *vortoscope*. The resulting *vortographs*, which are often said to resemble Cubism, are reminiscent of dazzle painting as well.

•

Coincident Disruption
Disruption and background blending combined

The term was coined c1940 by British zoologist Hugh B. Cott. It refers to aspects of camouflage that make tandem use of blending (high similarity) and disruption (high difference). In other words, not only is the camouflaged object broken up, some of its fragments may also connect (visually) with portions of the background, or, in many cases, with adjacent portions of the object itself. Although it is the least well-known, it is the most prevalent kind of camouflage in nature as well as in the military.

•

Thomas Casilear Cole
(1888-1976)
American artist

Cole studied with Edmund Tarbell and Frank Benson, and in Paris with Jean-Paul Laurens. According to the inventory of the William Rutherford Savage Papers at the University of North Carolina Manuscripts: "In the 1900s, William [Rutherford Savage] began a close correspondence with his nephew, Thomas Casilear Cole, who attended Riverview Military Academy in Massachusetts and the School of the Museum of Fine Arts in Boston. Thomas Casilear Cole began to make a name for himself as a portrait painter in the 1910s and designed camouflage patterns for the Navy in 1918."

•

Colorblindness
Its relevance to camouflage

The experimental use of colored filters for the detection of camouflage began during World War I. Inevitably, it led to the notion that colorblindness might actually be an advantage in detecting camouflage. Here are excerpts from that discussion:

Luckiesh (1922:214-215): "Some time before the Great War began, it occurred to the writer that colored filters could be utilized in aiding vision by increasing the contrast of the object to be viewed against its surroundings. Studies were made of various filters, made with the object of the experiment in mind, in viewing the uniforms of various armies. Further developments were made by applying the same principles to colored lights and painted pictures...With the development of the science of camouflage, filters came into use for the detection of camouflage."

Nature ("Colorblindness and Camouflage" 1940:226): "[With regard to camouflage detection] certain colorblind observers might see more than the ordinary person. For example, in a building camouflaged with large irregular patches of color, the actual outline of the building may be lost in the jumble of these patterns. But the colorblind person may be scarely conscious of the variegated colors, so that to him the outline of the building may be almost unaffected by the camouflage."

Scientific American ("Colorblind Flyers" 1941:103): "[That colorblind pilots can see through camouflage was discovered while] the Army was testing the effectiveness of camouflaging heavy guns so they could not be spotted from the air. A regular air corps observer picked out 10 of the 40 guns which had been disguised. An observer from the field artillery picked out all 40—and confessed, after landing, that he could not distinguish one color from another. The answer was of course that he was colorblind."

•

David C. Comstock
(dates undetermined)
American architect

A graduate of the architectural program at Cornell University in 1912, he is listed in Embury (1917:138) as having "left a professorship in Cornell and

Above An example of coincident disruption in a Swallowtail butterfly (*Iphiclides podalirius*), in which the visual connections occur between patterns on physically separate wings. Public domain photo by Tim Bekaert.

Abbott H. Thayer (in a letter to Franklin D. Roosevelt, dated September 3, 1917): "Gerome Brush's scheme, which by countershading reduces a ship to a flat gray, is one for which he and I took out a patent fifteen years ago, before I had learned by experiment that it took the brightest white on vertical surfaces to match the horizon sky in cloudy weather, which predominates 95% to 90% of the time in the 'war zone.'"

...

Jungmann (1918:346): "This book [*Concealing Coloration in the Animal Kingdom*] is also known to have played a leading part in the development of camouflage in Germany!"

has done splendidly as a soldier." He is also mentioned in *Literary Digest* ("Company A" 1917) as having been a camoufleur.

•

Wilford (Seymour) Conrow
(July 14, 1980-November 24, 1957)
American artist

Born in East Orange NJ, he studied at Princeton University, at the Académie Julian with Jean-Paul Laurens, and later with Percyval Tudor-Hart and Jay Hambidge.

New York Times ("Wilford Conrow" 1957:30): "Mr. Conrow, a first lieutenant in World War I, is credited with having been the first American officer to call the military's attention to the use of camouflage in warfare.

He helped to organize the first camouflage unit in the Fortieth Engineers, and served as chief instructor in the art to the American Expeditionary Forces."

Scientific American ("Seeing But Not Being Seen" 1918:451): "Lieutenant Wilfred [sic] Conrow is credited with the invention of an invisible helmet..."

MacGregor (1961:96) [re Tudor-Hart]: "One day Wilford S. Conrow, a pupil from New York, was asked by another pupil whether he could recommend a good authority on color. 'What a stupid thing to ask!' Conrow replied. 'Why not consult our own teacher—the Darwin of Color? ' Conrow, I believe, was director of American Camouflage during the Second World War [sic]."

•

(Jean-Baptiste) Eugene Corbin
(1867-1952)
WWI French camoufleur

According to Kahn (1984:166), there is an early memo in which Corbin is listed as one of the first French camoufleurs in World War I, along with five others. As illustrated in Blechman (2004:28 and 127) and Newark (2007:82-83), Corbin was largely responsible for the design of hooded camouflage robes, called *cagoules*.

A wealthy amateur artist, art collector and businessman, he was the son of the founder of the Magasins Réunis, a large, lucrative department store in Nancy, France. Today, his former residence is the Musée de l'École de Nancy, where a major part of the holdings consist of Corbin's own collection of Art Nouveau artifacts. He founded the magazine *Art et Industrie*.

Blechman (2004:128): "[In August 1914] Corbin went to his friend Louis Guingot, a painter/decorator who was experimenting with ways of applying paint to fabric. On 14 September, with his brother Henri, Guingot produced a large piece of canvas and five uniforms, all hand-painted to match local terrain. During a demonstration to French Army personnel, the canvas was suspended on poles above a gun operated by five men clad in the new uniforms. Impressed by this innovation, the army ordered a group of painters to produce more of the camouflage fabric, under the direction of Corbin and Guingot."

•

Hugh B. Cott (Cott 1938:506): "It will be noted that in countershading we have a system of coloration the exact opposite of that upon which an artist depends when painting a picture. The artist by skillful use of light and shade, creates upon a flat surface the illusionary appearance of solidity: Nature, on the other hand, by the precise use of countershading, creates upon a rounded surface the illusionary appearance of flatness."

...

Abraham Rattner, as quoted by Culkin (1980): "The camoufleur reverses the principle used by the artist who paints a picture representing an object in nature. The artist follows the principle of making a three-dimensional object appear as a three-dimensional object and he builds this appearance on the flat surface of the canvas. The camoufleur uses the plastic means of the artist in an opposite way because he is working on a three-dimensional form already existing—a cannon, a soldier, a pile of ammunition, an airplane, etc. and his object is to make these three-dimensional objects appear flat. The camoufleur must destroy, dissemble, make disappear a form."

...

Kahn (1984:19): "To the traditional eye a Cubist image appeared deformed and was extremely difficult to recognize. For the camoufleur, Cubist theory was relevant in that the final goal of camouflage was to disguise or prevent a coherent reading of an object."

Kate (Thompson) Cory

(February 8, 1861-June 12, 1958)
American painter, illustrator

Born in Waukegan IL, she studied at Cooper Union and the Art Students League in New York. In 1905, she met artist Louis Akin, who persuaded her to move with him to Arizona. For the next seven years, she lived on the Oraibi and Walpi mesas with the Hopi Indians, whom she painted, photographed and wrote about. For most of her remaining life, she lived in Prescott AZ, where, according to Lippard (2002:154), she was "poor and eccentric, still visited by Hopi friends."

Fran Elliot (n.d.): "During World War I, she [Kate Cory] returned east to join the Women's Land Army on Long Island, New York, in growing food in 'victory gardens' to stave off wartime shortages... she also worked for Standard Aircraft in Newark, creating experimental camouflage designs for British airplanes."

In Lippard (2002:154), there is a photograph taken by Cory in 1906 of two fighting Hopi men, one of whom is being attacked for being friendly with non-Indians. Other Cory photographs from the same period can be found in Gaede (1986).

•

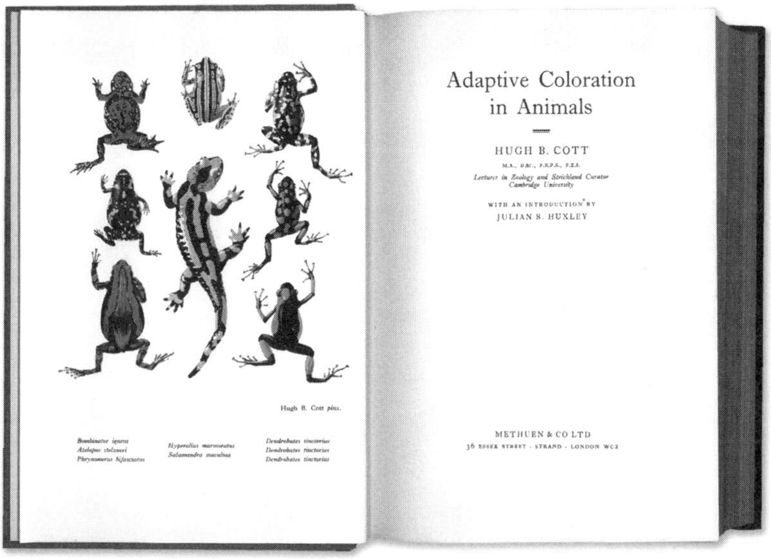

Above British camoufleur Hugh B. Cott was not only a zoologist; he was also a very capable illustrator and photographer. Shown here is the title spread for his widely admired book on animal coloration (Cott 1940), in which his own painting appears here as the frontispiece, with scores of pen and ink drawings throughout. Author's collection.

Jean-Louis Marcel Cosson
(1878-1956)
French painter

Born in Bordeaux, he is primarily known for his paintings of the opera, the ballet and the circus. He is listed in Ducasse (1962) and Coutin (1988) as a World War I French camoufleur.

•

Hugh B(amford) Cott
(July 6, 1900-April 18, 1987)
British zoologist, scientific illustrator

Born in Leicestershire, England, he studied at Cambridge University, then at the University of Glasgow with John Graham Kerr, a prominent zoologist who had long been interested in camouflage. Cott served in the British Army during World War I, then taught army camouflage during World War II. A zoology professor for many years, perhaps his greatest accomplishment was *Adaptive Coloration in Animals* (1940), a book that is widely regarded as the most authoritative source on biological camouflage. A skilled artist and photographer, he illustrated his own publications.

Abbott H. Thayer [in a typewritten autobiographical note in the Thayer family papers at the Archives of American Art]: "Around 1890 a new element came into my life in the shape of the discovery of the Protective Coloration of the Animal Kingdom; being in the field of Art and never before, apparently, having come under an artist's attention it had remained a sealed book."

...

Gladys Thayer, regarding (her father) Abbott H. Thayer's experiments with countershading, as quoted in White (1951:112): "...he set up in the Dublin [NH] town hall a large wooden box wired for operating two sets of electric lights to demonstrate the disappearing act of a small countershaded Venus de Milo. It was the delight of the school children to press the buttons and make her come and go." [This set-up appears to have been similar to that of the disappearing tiger installed at Harvard University, as seen on the facing page.]

...

White (1952:112-113): "As late as the fall of 1905 his [Thayer's] idea of making money from the principle of concealing coloration continued to distract Thayer, and we find him still experimenting with life models in tights which he tried to make appear and disappear in lighted boxes as he had done with the small statue [of Venus de Milo]."

Trevelyan (1957:118): "We laughed at him [Cott] for his passionate addiction to countershading, the trick by which, for instance, the white belly and dark back of a gazelle, when seen at a distance in strong light, seem to flatten out and destroy its form. In fact, countershading was used in camouflage in many different contexts."

Hartcup (1980:98) reports that "in his spare moments [Cott] could be found attending to the snakes, beetles and lizards which he kept in petrol cans."

•

Countershading

Obliterative shading or Thayer's law

American artist Abbott H. Thayer began to realize the survival function of countershading in the early 1890s. It began with his discovery that a large number of animals have light-colored undersides, with darker coloring on the top. His son Gerald H. Thayer (Thayer 1923:158) described the effect as follows: "If a rounded object, say a ball or cylinder, 'in plain sight,' is to be made to disappear, it has, in the first place, to be countershaded. That is, its shadowed parts must be made lighter in color, must be painted lighter until the shadow no longer shows; and the portions facing toward the source of light must be just proportionately darkened. In this way a rounded, solid form can be made to look perfectly flat. Solidity is known chiefly by light-and-shade; and this has here been 'cancelled,' neutralized, by trick painting."

In order to prove what the elder Thayer called a "universal principle," he began to demonstrate it to others, using hand-carved wooden ducks or painted sweet potatoes. Barry Faulkner (1973:18)

Above (top row) Four stages in the demonstration of countershading, from left to right: (a) a flat expanse of paper; (b) the artistic technique of shading, by which a flat surface appears to be dimensional; (c) countershading, in which the undersides of animals are lighter than the top surfaces; and (d) the flat expanse of tone that results from countershading. (second row) A countershaded mouse. (third row) Thayer's demonstration of countershading in a stuffed tiger, as displayed at Harvard University. The animal is clearly visible on the left, when lighted from below, but all but disappears on the right, when lighted from above. (bottom) Thayer's standard demonstration, using two wooden duck decoys (or sometimes sweet potatoes). The one on the left (clearly visible) is the color of the ground but has not been countershaded, while the one on the right (all but invisible) has been carefully countershaded.

recalled that his first vivid memory of his cousin Abbott Thayer was in the school yard, where he was demonstrating countershading to the drawing teacher, using two wooden ducks.

J. Huddlestone (in Chesney 1943:25-26): "To demonstrate his theory, [Abbott] Thayer made models which can be seen in the British Museum of Natural History, and also at Oxford and Cambridge. They certainly support his demonstration in a most realistic and convincing manner. The models are contained in a glass-fronted box lined with gray flannel, and carefully lighted from above. A rod which can be turned by means of a handle runs through the box, and on the rod are perched two models of ducks, covered with exactly the same shade of gray flannel as that lining the interior of the case. One of these model ducks is left bare, and being lighted from over head, its upper parts contrast strongly in tone with the lower, which are in shadow. The second model has these 'contrasts' obliterated, by stippling the upper parts with a dark gray which negates the strong lighting effect. The shadow of the underparts is painted out with a white paint. Consequently when the models are scrutinized from a distance of about two yards, the painted or 'countershaded' bird disappears or at the most, appears as a faint grayish indefinable shade, and that only after the keenest observation. When the handle of the rod is turned and the models are upsidedown, the bare flannel duck is still conspicuous, for now the back is in shadow, and the underparts lighted. The painted bird is more conspicuous still, for the darkened back is darker than ever with the shadow on it, and the white breast now turned to the light is really dazzling. When the handle is again turned, the shaded bird once more loses itself."

He also used wooden ducks when he demonstrated countershading to Frank Chapman (editor of *The Auk*, the American journal of ornithology) in the spring of 1896. Chapman was so persuaded that he published Thayer's first article on the subject ("The Law Which Underlies Protective Coloration") in the next issue of the journal, and invited him to present the same demonstration at the annual meeting of the Ornithological Congress in November. As described by John Hall Sage in the society's minutes (Sage 1917), this is what took place at the meeting: "The feature of the afternoon session was the open-air talk by Mr. Abbott H. Thayer, demonstrating his theory of the principles of protective coloration. Mr. Thayer placed three sweet potatoes, or objects of corresponding shape and size, horizontally on a wire a few inches above the ground. They were covered with some sticky material, and dry earth from the road on which they stood was sprinkled over them so that they would be the same color as the background. The two end ones were then painted white on the under side, and the white color was shaded up and mixed with the brown of the sides. When viewed from a little distance these two end ones, which were white below, disappeared, while the middle one stood out in strong relief and appeared much darker than it really was. Mr. Thayer explained that terrestrial birds and mammals which are protectively colored have the under parts white or very light in color, and that the color of the under parts usually shades gradually into that of the upper parts. This is essential in order to counteract the effect of the shadow, which otherwise, as shown by the middle potato, makes the object abnormally conspicuous and causes it to appear much darker than it really is...The effect was almost magical."

Homer Saint Gaudens (1919:23-24) [referring (perhaps sardonically) to the followers of Abbott H. Thayer]: "The faithful understood that if we painted the bottom of a potato white and graded it up to brown on top, they could not see it on the road."

Gardiner Cox
(1906-1988)
American painter

Born Holyoke MA, he studied at the Art Students League in New York with Charles Hawthorne. He is mentioned by Reit (1978) as having worked on camouflage during World War II with William Pahlmann at Jefferson Barracks, St Louis MO (p. 79). His portrait of Dean Acheson (1974) is in the National Portrait Gallery.

•

Ralston Crawford
September 5, 1906-April 27, 1978)
Canadian-born American painter

Crawford was born in St. Catherines, Ontario, but grew up in Buffalo NY. He studied art at the Otis Art Institute in California, and worked at the Walt Disney Studios. He also studied at the Pennsylvania Academy of Fine Arts and the Barnes Foundation.

According to Haskell (1985:62), when Crawford received a draft notice in 1942, he responded "by applying for an appointment as a photographer in the Navy. Failing to pass the Navy physical, however, he enlisted in the camouflage division of the Army Engineers Corps. His six months in basic training were psychologically disastrous. For 'the enemy was not Hitler, or Mussolini' but what he considered 'those miserable, stupid and sometimes vicious people' who conducted the recruits through basic training 'in modern assassination techniques.'" As a result, Crawford never served as a camoufleur, but instead requested transfer to Washington DC, where he was assigned to the Weather Division of the Visual Presentation Unit . There, "His task was to translate the meteorological data necessary in planning military operations into a form quickly and accurately comprehensible by Army Air Force personnel."

•

Cubism in relation to camouflage
Art style that began in France in 1907

Cubism is usually said to have started with the completion of *Les Demoiselles d'Avignon* by Pablo Picasso in 1907. Throughout World War I and in the years since, there has been endless speculation about the purported resemblance of Cubism to camouflage. Here are a few examples of that:

Gertrude Stein (1938:11): "I very well remember at the beginning of the war being with Picasso on the boulevard Raspail when the first camouflaged truck passed. It was at night, we had heard of camouflage but we had not yet seen it and Picasso amazed looked at it and then cried out, yes it we who made it, that is Cubism."

Rosenblum (1961): "...the complex, concealed identies of Cubist art found a surprisingly parallel in the military art of camouflage, developed largely by the French during the First World War."

Richardson (1971:114): "Although it affords a definite illusion of depth the space of Cubism is an uncanny sort of space on which the viewer can never put his finger or rest his mind."

Sypher (1962:85-86): "[T.S. Eliot's Prufrock] camouflages himself, at teas, on long foggy afternoons, where women come and go. It is said that Cubism invented camouflage, and indeed the First World War used camouflage in a Cubist way.

Cubism broke up the object into fragmented patterns of changing appearance; eventually it deprived the object of identity entirely. Under the Cubist attack the object first disintegrated into uncertain planes, then disappeared into the illusion of an object. The technique of camouflage assumed there was an object to camouflage, that there was a self to take on a protective coloration."

Gaunt (1949:173-174): "The tremendous [World War I frontline] bombardments which left behind a stark tree-stump, a criss-cross of trenches on a barren shell-pocked plain from which all color had gone, created a Cubist landscape. The science of camouflage, devised by the most correctly academic painters, arrived in some strange fashion at the abstract forms which had seemed too revolutionary a few years before."

•

Rogelio Lopez Cuenca
(1959-)
Spanish artist

The work of this artist is discussed and illustrated in Méndez Baiges (2007:106-109). He is described as intending to challenge the "hidden agenda" of capitalist societies, in which artistic expression

André Schwab [World War I French art critic] quoted in Kahn (1984:62): "Just like the bouillon kub, Cubist painting is a product of 'boche' [slang for German] exportation...Cubism alias tubism, thomboedrism, etc., grotesque and unhealthy conception, emerged one evening from the square head of an esthetic who probably also had a hangover."

...

Ernest Hemingway (1922): "The the plane began to move along the ground, bumping like a motorcycle, and then slowly rose into the air. We headed almost straight east of Paris, rising in the air as though we were sitting inside a boat that was being lifted by some giant, and the ground began to flatten out beneath us. It looked cut into brown squares, yellow squares, green squares and big flat blotches of green where there was a forest. I began to understand Cubist painting."

...

Charles DeKay (1918:106): "That this kind of Cubist painting on a colossal scale [dazzle painting] should have proved useful in the world war is only one example more of the fact—that you can never tell! What could be duller, more trivial and tiresome—one is tempted to say imbecile—than the pictures so-called of the cubists, with their broken lines, ugly corners, wretched colors, and long-winded explanations that signify nothing? Yet some of their extravagances can be made use of, it seems, in such marine and moving deceits. If indeed it cannot be said that the practitioners of *Cubisterie* have suffered a sea change into something rich and strange, at any rate they have found some place to stand upon."

Cubism

I have laughed, but seen it,—under Ditchling Down,
Blue cubes, yellow cubes, crimson cubes and brown.
I have laughed, but seen it,—shouting at the sky,
Crazy as a crazy quilt, over Telscombe Tye:

Cubes of russet ploughland, graying in the sun,
Cubes of honeyed clover, red as blood could run,
Cubes of yellow mustard, clean as hammered gold,
Bleating cubes of clouds or sheep, crammed into a fold.

Clinging to the Sussex down,—did we crawl like flies?
Ask the proud Antipodes towering in their skies.
I have laughed and seen it, solid in the sun,
All the myriad planes of earth, blocked and wedged in one;

Solid as your flesh and bones, blocked with bits
of sea,
Squared with dusky semi-tones, and cubed
with mystery,
Planes of Anglo-Saxon art, planes of modern
mirth,
From an aeroplane above—or below—the
earth.

Butting through the solid blue like a submarine;
While my eyelids clung to cubes of blue and
gold and green,
Till the level meadows rose, upright to the sky,
And we looped the loop again, over
Telscombe Tye.

—Alfred Noyes (1920)

is relegated to remote and harmless arenas, where it can easily be ignored. His art is more subversive: He installs it not in art galleries but on the street in the form of signs, billboards and posters.

•

Gladys (Maeva) Cumpston
(May 31, 1887-June 18, 1975)
Australian horticulturalist, Braille transcriber

Born in Rosedale, Victoria, she became an expert gardener, as well as an active contributor to the Society for Growing Native Plants, the Australian Red Cross, and the Canberra Nursery Kindergarten Society. In the final two decades of her life, she transcribed fifty books into Braille. Waterhouse (1993): "[During World War II, she] made camouflage nets, draping them across the fireplace in her home and inviting visitors to add to them."

•

Robert (Emerson) Curtis
(October 4, 1898-March 23, 1996))
British-born painter, illustrator, cartoonist

Born in Croydon, England, he attended school in Chile. In 1914, he moved to Australia where he worked as an advertising arstist and cartoonist. He traveled to the US in 1922 (accompanied by filmmaker Charles Chauvel) and studied at the Art Institutes of San Francisco and Chicago. While residing in Chicago, he worked as a draftsman in the architectural firm of Daniel H. Burnham. One of his books of architectural illustrations, titled *A Vision Takes Form* (1967), documents the various stages of the construction of the Sydney Opera House.

Holder (n.d.): "During World War II Curtis recorded working life in the Commonwealth Munition factories (1939-41), worked as a Camouflage Officer in Australia and with the RAAF in New Guinea (1941-43), until he was finally appointed an official war artist to record the nation's industrial wartime production." He was a member of the Australian Camouflage Unit, part of the Department of Home Security, and an associate of Max Dupain, Joshua Smith, Frank Hinder, William Dobell and Clem Seale.

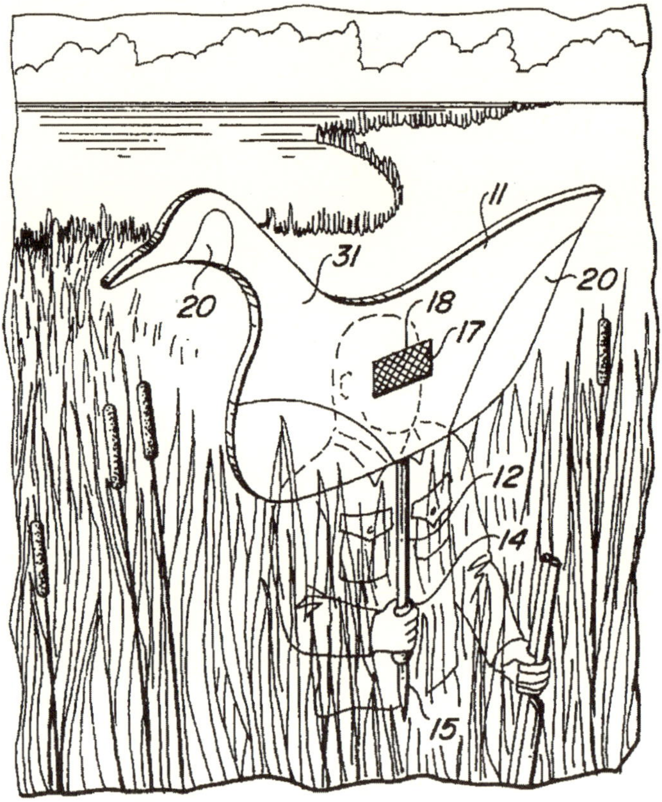
11
31
20
20
18
17
12
14
15

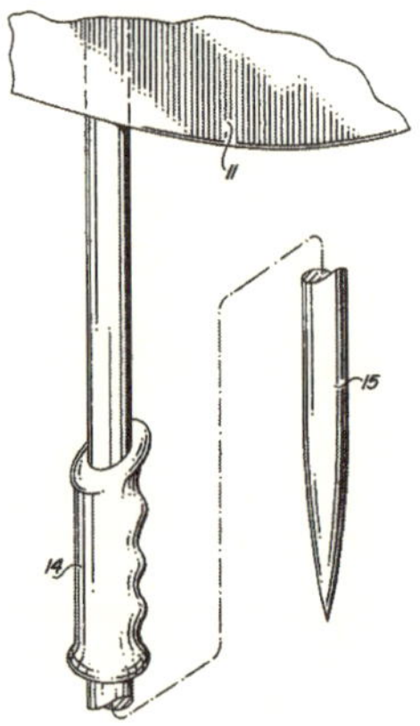
11
15
14

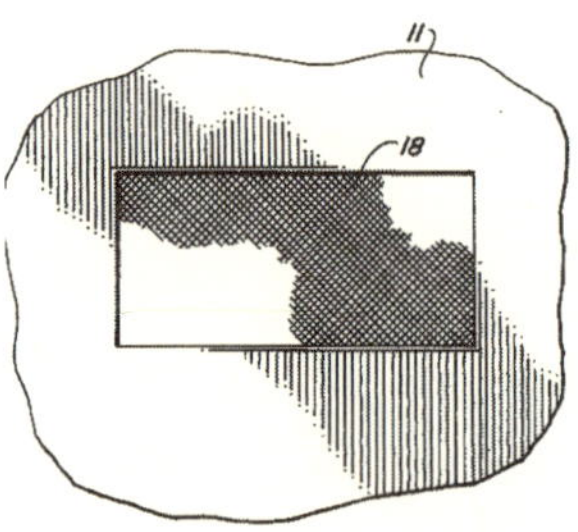
11
18

D

William (John) Dakin
(April 23, 1883-April 2, 1950)
British-born Australian zoologist

Dakin was born in Liverpool, England, where he graduated with honors in zoology from the University of Liverpool. In 1929, he moved to the University of Sydney, Australia, where he taught zoology until his retirement in 1948. Elias (2003, 2008) has written extensively on his curious preoccupations with camouflage, sex hygiene, and eugenics.

Bygott (1981): "During World War II Dakin in 1941 became technical director of camouflage for the Ministry of Home Security and built up a research institute and museum in his department at the university. His publication, *Art of Camouflage* (1941), which drew heavily on his marine experience, summarized much of his work in this field."

Elias (2008:256): "Dakin's camouflage manuals for soldiers in jungle-warfare instruct them to cover their bodies, their belongings, and their immediate surroundings, with bitumen, mud, soot, oil sludge, dyes, stains, blackened seaweed, and grease. Blackness is given almost fetishistic attention. One of the most noticeable inversions in Dakin's writing on warfare and his writing on health is the recognition that in war the dirty and the dark are important for the nation's salvation, but in civilized life they represent the nation's ruination. Dankin's camouflage manuals make frequent reference to 'primitive' man."

Facing page Drawings submitted for US Patent No. 5,572,823 titled "Hand Held Decoy and Hunter Shield," invented by J.R. Savaria (1996).

Roy (Frederick Leslie) Dalgarno

(December 2, 1910-February 1, 2001)
Australian artist and designer

Born in Melbourne, Australia, he worked as an apprentice to a lithographer. He became associated with left-wing artists and intellectuals, and joined the Communist party. While working for an advertising firm, he was fired when he provided editorial cartoons to striking union workers, some of whom were employed by the same corporate clients who Dalgarno was designing for. Kerr (n.d.) "Roy Frederick Leslie Dalgarno": "[During WWII] he began work as a camouflage artist for the RAAF based in Darwin, serving in North Queensland and the offshore islands." He moved to Europe in 1949, where he studied lithography and etching with British printmaker (and camoufleur) William Stanley Hayter in 1951-53 at Atelier 17 in Paris. He associated with other expatriates, including Dadaist Tristan Tzara. In 1956, he moved to India, where he lived for twenty years, working as an art director and lecturing at the Bombay School of Fine Arts. He resettled in Auckland, New Zealand, in 1976.

•

Salvador Dali

(May 11, 1904-January 23, 1989)
Spanish surrealist artist

Nearly all of Dali's work is relevant to camouflage, if sometimes indirectly. There is a lengthly section in *The Secret Life of Salvador Dali* (Dali 1993:69, first published in 1942) in which he recalls his childhood use of a leaf mimic insect in a sleight of hand magic performance. He writes: "When, much later, at the outbreak of the war of 1914, I saw the first camouflaged ships across the horizon of Cadaques, I jotted down in my notebook of personal impressions and reminiscences something like the following— 'Today I found the explanation of my *morros de con*, (for this was what I called my leaf insect) when I saw a melancholy convoy of camouflaged ships pass by. Against what was my insect protecting himself in adopting this camouflage, this disguise?'"

Ann Elias (2008:253): "Drawing on Darwin's theory of the survival of the fittest, [Australian camoufleur William] Dakin's 1918 work on sex hygiene and sex education referred to citizens who were morally and physically strong and lacking disease as 'the fit.' He had a second category for those who were morally and physically weak and diseased, referring to them as 'the unfit' because he believed they were not sufficiently healthy in mind and body to lead the British Empire forward to world dominance and supremacy. By the Second World War Dakin added a third category of fitness to his vocabulary and this was the 'misfit,' a term he used in his capacity as the Technical Director of Camouflage for Australia to refer to those in war who did not blend into their background. The misfit in war was someone of the wrong size, shape or color for their environment and who therefore posed a threat to the welfare of other soldiers."

...

Time magazine ("Camouflage" 1939:42): "Camouflage in the last war meant whirls, blotches, stripes and curlycues with which 'experts' made common objects look like a Futurist's bad dream."

...

Marston (1942:1): "A good camoufleur usually has a tough road to travel. He must expect, on numerous occasions, to be treated as some sort of harmless lunatic by his friends. He must expect, more often than not and probably for very good reason, that cold water will be poured on many of his pet schemes and ideas, and that they will be labeled fantastic and impracticable."

In the same year that *Secret Life* was released, Dali also published a 3-page article in *Esquire* magazine titled "Total Camouflage for Total War" (Dali 1942). In it, he credits Picasso and Cubism with having influenced World War I military camouflage, and proposes that now it is *his* turn to influence a new approach to camouflage. He continues: "And just as the camouflage of 1914 was Cubist and Picassian, so the camouflage of 1942 should be Surrealist and Dalistic. For this time, the discovery is mine—namely the secret of total invisibility and psychological camouflage."

Later in the same article, he offers still more detail about his idea of camouflage: "For ten years I have made a systematic study of the problems of vision, and I have come to the conclusion that we have had but the merest glimpse of the psychological significance of such phemomena. We see what we have some reason for seeing, above all what we believe we are going to see. If the reason or the belief is upset—we see something else. In this connection, visual reactions can be controlled. To borrow the radio terms, they can be beamed, or even jammed, by purely psychological effects. My long investigation leads me to believe therefore that psychological camouflage is no idle dream."

•

Marcel Damboise
(August 8, 1903-1992)
French sculptor

Born in Marseille, Damboise began to study art at the École des Beaux-Arts in that city, but withdrew to become an apprentice to a stone cutter. After moving to Paris in 1926, he became associated with various people, among them (World War I camoufleurs) Charles Despiau (his teacher) and Charles Malfray. He was also associated with comedian Jean-Louis Barrault (who he met while both were serving in a French Army camouflage unit during World War II) and with writer Albert Camus.

•

Fred H(amilton) Daniels
(1872-?)
American art educator, painter

He was originally from Worcester MA. In 1889, he enrolled at Massachusetts Normal Art School. Nine years later, he was hired as the art supervisor for Buffalo NY. He also worked at Prang Educational Company, an art materials manufacturer. In 1901, he founded the *Applied Arts Book*, now known as *School Arts* magazine, which he edited for two years. He is mentioned by Faulkner (1957) in his radio talks as a World War I American camoufleur. He was still painting as late as 1947.

Robin (Vere) Darwin

(May 7, 1910-January 30, 1974)
British painter and educator

He was the grandson of British naturalist Charles Darwin. Educated at Eton College and the Slade School of Fine Art, he was associated with the British Camouflage Directorate during World War II as secretary to its Camouflage Committee. In 1943, he submitted a statement to the Ministry of Information, contending that artists and designers are of undoubted value in the development of camouflage, but that they should work side by side with engineers, architects and scientists. In 1967, he became the first rector and vice-provost of the Royal College of Art. Apparently colorful, he is mentioned in various sources, among them Hartcup (1980), Gardner (1993), and Goodden (2007).

Graves ("Adventures in Prohibition" 1920:581): "The subject of one of the most popular jokes in a certain part of Luzerne Country, for many weeks during the summer, was an 'apple truck' that left the town of Glen Lyon every night between twelve and three. At each end were bags and crates of apples, for purpose of camouflage, but in the middle were barrels of whiskey."

Above This astonishing dazzle camouflage scheme, applied to the *USS Andra* during WWI, was designed by US naval camoufleur Frederick Waugh. It is virtually identical to its (apparent) earlier application to the *SS Congaree*. US Navy photograph. See Frederick Waugh.

•

James (Henry) Daugherty

(June 1, 1887-February 21, 1974)
American author and illustrator

Born in Asheville NC, he spent his early years in Indiana and Ohio. In 1896, his family moved to Washington DC, where he studied art at the Corcoran School of Art, then at the Philadelphia Art Academy. While living in England for two years, he studied with Frank Brangwyn. An accomplished book illustrator, he won the Newberry Prize for illustration in 1940, and was twice a runner-up for the Caldecott Medal (in 1939 and 1957).

Lawton (n.d. "Heroic America"): "Upon America's entry into World War I, Daugherty volunteered to work as a camouflage artist in the Navy. What could have been a routine year of monotonous work, became instead a liberating experience. Daugherty treated the ship diagrams as if they were enormous expanses to be covered with an arrangement of harmonious colors. In some designs, he experimented with bright hues to observe the structure of color relations within the outline of a battleship." As a civilian camouflage artist, Daugherty worked at the Newport News

Cruickshank (1979:17-18): "[During World War II, Winston Churchill was a vocal advocate of camouflage.] In late 1941, Churchill asked the War Office expert on camouflage to listen to his nephews, John and Peregrine Churchill on the subject. 'The first...executes camouflage, the second is concerned with its technical aspects: The first is an artist, the second an engineer.'"

and Baltimore shipyards. Also see Daugherty (n.d.) and an online Wikipedia biography.

•

Henry Davenport
(April 1, 1881-1965)
American painter

According to Love (1999), "In the New York area, the artist in charge of camouflaging ships in military service during World War I was Henry Davenport, an accomplished painter who had worked in Europe." Originally from Brookline MA, Davenport attended Harvard University. He also studied architecture at the École des Beaux-Arts in Paris until he decided that his first interest was in painting, at which time he enrolled at the Académie Julian. Returning to Boston in 1914, he continued his study of art under Charles Hawthorne and George Elmer Browne at Provincetown MA. In 1916, he founded his own school in Paris, called the Clonet School of Art. Later, having returned to the US, he taught studio art and art history at Yale University.

•

Nat (Natalia) Davies
(January 25, 1907-April 29, 1951)
Welsh-born Australian women's rights advocate

Born in Lampeter, Cardiganshire, Wales, she worked as a public school teacher, during which she fought for women teachers' rights. In advance of World War II, she founded the Defense Society (most of whose members were women) through which she lectured and taught courses on first aid, home nursing, electrical maintenance, pistol shooting, map reading, self-defense, and precautions with regard to fire, air raids and poison gas.

According to Jones (1993:583): "The sole female civil defense area officer in Adelaide, Davies advised the Education Department; military authorities accepted her camouflage designs; and at Keswick Barracks society members made sniper's suits from hessian." After World War II, Davies wrote educational radio plays, and, in 1947, became a radio quiz champion.

•

Montague Dawson
(1895-1973)
British marine painter

Born in Chiswick, England, his grandfather was landscape painter Henry Dawson. He is mentioned in Williams (1989) and Atterbury (1975) as having served with Norman Wilkinson in the British dazzle painting section in World War I. During World War II, he was an Official War Artist. Some of his commissions were for "portraits" of boats for their owners.

•

Irwin (1917): "Early this summer, Allied airmen began to encounter German planes 'painted like harlequins and spotted like circus ponies,' said a French aviator. One would have a blue wing and a yellow wing, a green body; one would be polka-dotted. But always the colors met in wavy uncertain lines, and always they were blended where they joined."

Above Nothing is quite as dramatic as a series of dazzle camouflage woodcuts created c1918 by Edward Wadsworth, a British Vorticist artist who also served in World War I as a dock supervisor for the dazzle painting of ships. This one, titled *Drydocked for Scaling and Painting* (1918), was reproduced on the poster for an *Englishe Graphik* exhibition in 1923. His masterpiece is a large oil painting titled *Dazzle Ships in Drydock at Liverpool* (1919), in the National Gallery of Canada.

Etchells (1919): "[Dazzle-painting] was one of the bright spots of the war and it is not too much to say that it would probably never have developed as it did had it not been for the experiments in abstract design made by a few modern artists during the years immediately prior to 1914."

Dazzle and/or Dazzle Painting
Disruptive confusion forms

Perhaps no approach to camouflage is of wider appeal than dazzle painting. The term pertains specifically to the use of brightly colored, high contrast disruptive shapes, a method first applied to ships during World War I. Its originator is almost always listed as a marine painter named Norman Wilkinson, who proposed it to the British Admiralty in 1917. It was he who named it dazzle. As he explained many times, its purpose was never to prevent a ship from being seen (that was simply not possible, if only because the smoke from its smoke stacks could easily be detected). Rather, its purpose was to reduce the likelihood that ships would be struck by torpedoes from German U-boats.

Warner (1919:215): "The sole purpose of this method of painting [dazzle painting] was to guard vessels against torpedo attack. The torpedo has only a speed of about 35 knots, and the submarine commander will rarely be in a position to fire point blank at a ship. He fires at some point that he expects the ship to reach in a given time [i.e., he leads the target]. Now if his estimate of the direction in which the ship is steering is incorrect it is obvious that the torpedo will pass harmlessly ahead or astern of its target."

Cott (1940:47): "The word 'dazzle,' which came into popular use during the War [WWI] in connection with camouflage painting, is an American term applied by Thayer in his work on *Concealing Coloration* to disruptive coloring in animals. It is based on the American slang 'razzle dazzle,' which has a meaning—expressive of active confusion—

quite different from the correct English use for partial blinding by brilliant lights."

Warner (1919:216): "[Another purpose of dazzle painting was to prevent a ship from being identified because, if it were recognized] the enemy would at once acquire a valuable aid in judging the speed....[as pointed out in a WWI German submarine training manual] at 1,000 meters, even if the course of the ship has been correctly judged, an error of 2 knots in the estimate of her speed will cause the torpedo to pass 63 meters ahead or astern."

It is possible that the earliest use of the word "dazzle" in relation to biological camouflage is in Thayer (1909:151-152): "[Certain animals] mammals and birds alike, have a few sharp markings in their mainly immaculate white costumes. These evidently serve as what may be called 'distractive' or 'fixed dazzling' marks. They are, in most cases, too small to show except in a very near view—when, by their sharp but isolated and noncommittal conspicuousness, they tend to draw and hold the eye's attention—in a sense, to dazzle it, so that it less readily discerns the faintly shown snow-white body of their wearer... "

Later in the same book, Gerald Thayer also notes (Thayer 1909:198-199): "Somewhat more brightly colored, as a rule, are the many species which habitually perch on or amidst terrestrial vegetation... These [their wings] are often brightly and boldly colored, without minute patterns, and when expanded in flight, against sun-variegated backgrounds, they lessen the conspicuousness of the moving insect just as do the bold 'ruptive' markings of many butterflies (and moths). Some of them, again, must under certain conditions act as 'dazzling'-marks by their sudden display and equally abrupt eclipse."

Warner (1919:216): "The term 'dazzle painting,' which was used by the British to describe the ultimate type of camouflage for course distortion, was a very misleading one, and at one time the [US] Navy Department considered the question of finding a new name. A name was desired which would be at once more descriptive, and at the same time differentiate the American development of camouflage from the British. While no name that proved worthy of acceptance was brought forward, a good deal of merriment was caused by the suggestion of 'Jazz Painting.' It was generally agreed that this epitomized the popular impression about ship camouflage."

Cork (1994:232): "The explosive diagonal and zigzag bands [used in dazzle painting] had a wild, ragtime impact. Hulls and funnels alike were transformed into crazily festooned appartions, intended to disorientate any U-boat commander scrutinizing them through his telescope."

US Secretary of the Navy (Taylor 1918:516): "Up to October 1, 1918, 1,127 [American] vessels had been specially painted with [camouflage] designs prepared by the Bureau [of Construction and Repair]."

Right An American World War I facility in which camoufleurs are preparing camouflaged ship models for testing (1918). Photo courtesy US Naval Historical Center (NH 41721).

Below A disruptively patterned Mandarin Duck. Public domain photograph by Adrian Pingstone (2005).

Gerald H. Thayer (1923:161): "The moving object cannot, as a rule, be *hidden*, but it *can* be made less definite, more puzzling, a more 'tricky' and difficult target, by certain arrangements of color and pattern. This my father and I pointed out in 1909 in our book *Concealing Coloration in the Animal Kingdom*; and we there used the terms 'dazzle' and 'dazzling' very much as they have since been used in connection with the camouflage of ships."

New York Times ("Shows New Ship Painting" 1917:20): "In the British Navy the name of the new war color is 'eccentric painting.'" See also baffle painting.

•

Albertus De Friest
(1889-July 26, 1953)
American civil engineer, bridge designer

Born In Nashua NH, he graduated from Yale University. He was the chief engineer for the Union Pacific Railroad for sixteen years. *New York Times* ("Albertus De Friest" 1953): "During World War II, he planned and was responsible for the camouflage applied to Los Angeles aircraft plants." He was also an important contributor to the installation of the 200-inch telescope at Mount Palomar Observatory. Also see John Detlie and John Ohmer.

•

Louis Dejean
(June 9, 1872-?)
French sculptor

Born in Paris, he began his career as an assistant to Auguste Rodin, and is listed in Coutin (1988) as a World War I French camoufleur. He was awarded the Legion of Honor.

•

Left A dazzle painted British troopship, the *SS Empress of Russia*, traveling in a convoy with other dazzle-painted ships. Public domain photograph by Stewart Bale (1918).

Sonia Delaunay
(1885-1979)
Russian painter, designer

Wolfram (1975:25-26): "Sonia Delaunay was more interested in color and light than anything else, and her use of found materials grew quite naturally out of her fascination and experimentation with juxtaposing bits of colored materials to see what sort of compositional possibilities they might offer. When her son was born in 1911, she set about making a blanket for his cradle out of scraps of cloth, sewing them together as she had seen peasants do in Russia [as in a crazy quilt]. The result, a configuration of rough squares and triangles, seemed to both the Delaunays to look quite a lot like the Cubists' compositions, and so she set about modernizing their home, making lampshades and cushions in similar striking patchworks of colors..."

•

Georges Delaw
(1874-December 8, 1929)
French painter, designer

Born in Sedan, France, Delaw was primarily an illustrator for books and magazines. He was particularly interested in folklore, and was also known for having created interior mural-like panels for various passenger ships. He is listed in Coutin (1988) as a World War I French camoufleur.

•

Above As early as 1913, Russian-born painter and fashion designer Sonia Delaunay was experimenting with dazzle-like patterns in clothing design, as in her famous "simultaneous dress." Twelve years later, she designed an outfit using multi-colored rectangles, then painted a car with a matching design. See cagoules and harlequin.

Paul Atterbury (1975:26): "Up to the end of October 1918, 2,719 British merchant ships were dazzle painted. Designs were also produced for a number of warships. In addition, large numbers of French, Belgian, Italian, Russian and Japanese ships were painted according to dazzle principles."

...

J. Bowyer Bell (1991:257): "Modern artist Christo began his career with a Bulgarian army camouflage unit..."

Gabriel Deluc
(October 1, 1883-September 15, 1916)
French painter

A student of Léon Bonnat at the École des Beaux Arts, he is listed in Coutin (1988) as a World War I French camoufleur. He died in the war.

•

Henri Deluermoz
(December 9, 1876-1943)
French painter, printmaker

Deluermoz was a student of Gustave Moreau at the École des Beaux-Arts. Primarily an animal painter, he illustrated an edition of Rudyard Kipling's *Jungle Book*. He is listed in Coutin (1988) as a World War I French camoufleur.

•

Edwin W(illard) Deming
(August 26, 1860-October 15, 1942)
American sculptor, painter

Born in Ashland OH, Deming had a great interest in Native American culture, starting from an early age. During periods of his life, he lived with various Indian tribes, including Crows, Apaches and Pueblos. Although he was urged by his father to study law, he chose instead to become an artist, studying first at the Art Students League in New York, then at the Académie Julian in Paris, with Jules Lefebvre and Gustave Boulanger. His papers, including his correspondence with Theodore Roosevelt (c1913-1917), are in the collection of the New York Public Library. *New York Times* ("Edwin W. Deming" (1942:19): "During the World War, Captain Deming went to Washington, where he instructed the Fortieth Engineers in marksmanship. He was commissioned a captain in the National Army the next year. He was also active in camouflage work during the war years, painting the designated targets that were lithographed and now are used in all Army and National Guard schools in this country."

Above WWI-era photographs of ships in the process of being dazzle painted. At the top right are two views of the same US ship, before and after being camouflaged; the chalk lines are faintly visible in the top photo. On the left in the second row, the ship is being marked with chalk to guide the painters. The remaining photos are views of the painting in progress.

Everett L. Warner (n.d., Part 2, p. 5) [paraphrased from his lecture notes]: "Of how much value was World War I ship camouflage? Take the period from March 1 to November 11, 1918. Ninety-six American ships were sunk during those months, of which 2500 tons or over eighteen of them were camouflaged, but of that number four were sunk through collisions and three by striking mines. So German U-boats sunk only eleven out of about 1250 ships that we camouflaged, which is less than one percent. We never lost a single camouflaged troop transport, although three monochrome gray troop transports (*Antilles*, *President Lincoln*, and *Covington*) were all sunk by U-boats. We had issued camouflage designs for the last two, but owing to delay they were never painted. It is easy to claim that camouflage would have saved them, but I don't feel absolutely certain of that. Some loss is inevitable."

Charles Despiau

(November 4, 1874-October 28, 1946)
French sculptor, draughtsman

Despiau was associated with Auguste Rodin for a number of years, during which he managed the marble store at Rodin's studio. He is listed in Coutin (1988) as a World War I French camoufleur. He made portrait busts of, among others, the wife of (camoufleur) Othon Friesz, and of (camoufleur) André Dunoyer de Segonzac. He was awarded the Legion of Honor.

•

Arthur Preuss (*Fortnightly Review*, October 15, 1917:313-314): "Another new and fortunately less disreputable word than 'jazz' is 'camouflage.' It is defined by a writer in the London *Saturday Review* as 'concealment of the presence of a person or thing by causing his or its color to blend with his or its surrounding.' Like all arts, it is an imitation of nature, which is the greatest of camouflagists."

John (Stewart) Detlie

(1908-November 30, 2005)
American architect and set designer

Detlie was born in Sioux Falls SD. He earned degrees in architecture at the University of Pennsylvania, then moved to Hollywood in the early 1930s to work as a set designer. He married film star Veronica Lake, whom he later divorced. He received an Oscar nomination for the production design of *Bitter Sweet* (1940), starring Jeannette MacDonald and Nelson Eddy.

Above An American freighter, the *USS West Mahomet*, painted in dazzle camouflage (1918). Courtesy US National Archives (19-N-1733).

In 1942, he left his position at MGM to oversee the camouflage of the Boeing Aircraft plant in Seattle WA, for the US Army Corps of Engineers. *Washington Post* ("John S. Detlie" 2005): "To confuse enemy bombers, Boeing Aircraft camouflaged nearly 26 acres of the plant in Seattle, where the B-17 and, later, the B-29 were built. Boeing's Plant 2 was covered with a three-dimensional wire, plywood and canvas structure that was made to look like a town, including trees, houses and schools, instead of a wartime airplane factory.

After the war, Mr. Detlie joined the architecture firm that hired him to do the camouflage project. He eventually became a partner there." Also see Albertus De Friest and John Ohmer.

Anon (an American journalist, quoted in Wilkinson 1969:92): "You should see our fleet! It's camouflaged so, it looks like a flock of sea-going Easter Eggs. If you shut your eyes good and tight, and stand behind a wall, you can't see a ship a cable's length away. It was an English guy [Norman Wilkinson] thought of it first, and his name's the first toast now at all the paintmakers' social reunions."

•

SEA CAMOUFLAGE AGAINST U-BOATS: THE ART OF "DAZZLE-PAINTING."

"DAZZLE-PAINTING": A BRITISH ART OF CAMOUFLAGE BECOME UNIVERSAL.

Above This two-page pictorial feature on dazzle camouflage appeared in the *London Illustrated News* on January 4, 1919.

André-Victor-Édouard Devambez
(May 26, 1867-March 18, 1943)
French painter

Born in Paris, he studied with Jules Lefebvre and Jean-Joseph Benjamin Constant. He is listed in Coutin (1988) as a World War I French camoufleur. He was awarded the Legion of Honor.

•

George Diestel
(dates undetermined)
American film set designer

He was a member of the Ghost Army, as mentioned in Park (1985). He was also interviewed in Naversen (1989).

•

Walt(er Elias) Disney
(December 5, 1901-December 15, 1966)
American artist and entrepreneur

Henning (*The Walt Disney Story*): "In the summer of 1918, Walt [Disney] was 16—too young for the military. When he heard that the Red Cross Ambulance

Corps would accept 17-year-olds, he lied about his age, joined, and began training. All the same, he almost missed his chance when he came down with influenza in an epidemic that killed about 20 million people worldwide. The war ended. But the Ambulance Corps still needed 50 more men, and Walt was the fiftieth selected. He was on his way to France. For the next year, Walt drove an ambulance, chauffeured officers, played poker, started smoking, and wrote letters... He made money with another young man painting helmets with camouflage colors, banging them up to look battle-scarred, and then selling them to Americans in search of realistic souvenirs." See also Michel Aubrey.

Archibald Wavell (World War II British general), quoted in Reit (1978:82): "...the elementary principle of all deception is to attract the enemy's attention to what you wish him to see and to distract his attention from what you do not wish him to see."

...

In Marston (1942:26), there are four "photographs of models illustrating camouflage schemes prepared by the Walt Disney Studio unit for the 604th Engineer Battalion."

JustDisney.com (*Walt Disney: Short Biography*): "During the fall of 1918, Disney attempted to enlist for military service. Rejected because he was under age, only sixteen years old at the time. Instead, Walt joined the Red Cross and was sent overseas to France, where he spent a year driving an ambulance and chauffeuring Red Cross officials. His ambulance was covered from stem to stern, not with stock camouflage, but with Disney cartoons."

During World War II, the Disney Studio produced at least two animated cartoons on camouflage: A Donald Duck short titled *The Vanishing Private*, (1942), and one in the Private Snafu series, titled *Camouflage* (1944).

Above Two American dazzle-painted ships during World War I, the *USS Saint Louis* and the *USS Kimberly*. Courtesy the US Naval Historical Center (NH 90775 and NH 51080).

•

William Dobell
(September 24, 1899-May 13, 1970)
Australian painter

Born in Newcastle, New South Wales, Australia, Dobell studied in London at the Slade School of Art with Henry Tonks and Philip Wilson Steer, and privately with William Orpen. In Eagle (1996), he is described by a companion, Eric Wilson, as being so poor that he might go out in the evening "holding a newspaper under his right arm to cover the tear in his overcoat. Whenever his socks have holes...he simply paints his leg to match." He served as a camouflage artist during World War II, along with

Above A dazzle-painted American troop ship, the *USS DeKalb* in late 1918. Courtesy US Naval Historical Center (NH 68750).

Douglas Annand, James Cook, Adrian Feint, Donald Friend, Robert Emerson Curtis and Joshua Smith. He later became an official war artist with the Allied Works Council (1942-1944).

Eagle (1996) provides a fascinating account of a scandal involving Dobell and two other camoufleurs.

•

Jay Doblin
(December 10, 1920-1989)
IIT professor, designer

Born in New York, Doblin studied industrial design under Alexander Kostellow, Rowena Reed, Donald Dohner and Gordon Lippincott at Pratt Institute in Brooklyn NY from 1939 to 1942, during which he received camouflage training certification. Following graduation, he became executive designer for Raymond Loewy Associates in New York, as part of which he was a "designer for camouflage and other government projects, 1942-49." See Morgan (1984:158-159).

Fort (2007): "Jay Doblin went to Pratt Institute and his degree was in camouflage, which will surprise some modern ID students. But, remember, it was the war years and people were trying to do things that were relevant. I can remember more than one vivid conversation with Jay where he was teaching me principles of camouflage. What do you do to disguise a tank, or a factory,

Mingo White (a former Alabama slave, paraphrased from an interview by Levi D. Shelby, Jr, in 1937 as part of the Federal Writers Project, now in the Library of Congress): "[Confederacy President] Jeff[erson] Davis was as smart a man as you ever want to see. During the [American Civil] war he sheared his horse in such a way that he looked like he was going one way when he'd be going the other."

Above During World War II, aircraft manufacturers throughout the US hired civilian camoufleurs to conceal their factories from the air. It was usually done by covering the factory buildings (including the parking areas) with dummy structures that appeared to be suburban neighborhoods, with simulated streets, houses and vegetation. This was done not only at the Douglas Aircraft Factory, but at those of Lockheed (shown above) and Boeing as well. Also see Albertus De Friest, John Ohmer, and John Detlie.

or ammunitions plant, or ammunitions storage facilities so that anybody on a bombing strafe will go past it and tend to drop their bombs in a way that misses the target? It is not a question of making it completely disappear. It is a question of actually trying to get it to appear somewhere else.

It may also interest modern-day design students to know that Jay's favorite color palette, both for the clothes that he wore and for virtually all of the things he designed, went towards military draft colors. He loved khakis and olives and a color palette that was decidedly muted and in line with theories and practices of camouflage."

Nathan Lerner (1978): "I believe most of the [camouflage] material you are inquiring about was lost when the school made its move from the building on Ontario and Dearborn to the IIT campus. I was told that after the new director Jay Doblin came, whatever was left in Moholy's file was thrown out."

•

Douglas Aircraft Factory
Camouflaged in 1942

Rasmussen (2002): "With World War II raging in Europe, [Donald] Douglas realised well before Pearl Harbor that his plant was a sitting duck for an air attack. He didn't wait for the government to protect him; he took the controls. Douglas asked his chief engineer and test pilot, Frank Collbohm, and a renowned architect, H. Roy Kelley to devise a way to camouflage the plant. (Later, Collbohm would found the Rand Corporation and Kelley would design its headquarters.)" See also Stanley (1998:171-176).

•

Keith (Castellain) Douglas
(January 24, 1920-June 9, 1944)
British poet

Born at Tunbridge Wells, Kent, England, he studied briefly at Merton College, Oxford, and then, in 1939, enlisted in the British Army. However, his activation was deferred until 1940, at which time he was assigned to cavalry officer's training, then sent to the Middle East.

Soldier Poets on BBC Channel 4 ("Keith Douglas" n.d.): "While the rest of the regiment was gearing up for action, Douglas was sent to Alexandria to teach camouflage and then to HQ at Wadi Natrun. On 23 October 1942, the battle of El Alamein began; after four days, Douglas could bear it no longer. Against orders, he left HQ, commandeered a lorry and joined his regiment in the desert—an action that, in less desperate times, could have led to court martial. But he went unpunished: the regiment had lost 16 of its 22 officers and needed reinforcements."

Unfortunately, only three days after surviving the invasion of Normandy, he was killed by mortar fire.

•

(Stephen) Olin Dows
(1904-1981)
American mural painter

Born in Irvington-on-Hudson NY, he studied at Harvard and Yale. Between the wars, in 1933, he became director of the US Treasury Relief Art Project. In 1942, he enlisted in the US Army, and served in Europe with the Ghost Army. See Goossen (1973:103).

•

Jacques Dresa
(Janury 11, 1869-1929)
French designer

Born in Versailles, his primary achievements were in the area of textile design. He was also a set designer for the Paris Opera. He is listed in Coutin (1988) as a World War I French camoufleur.

•

Douglas Duer
American illustrator, poet
(1887-1964)

Born in Baltimore MD, he studied under Howard Pyle. He was a friend of (camoufleur) Leslie Thrasher, who had also studied with Pyle, and the two of them served together in the American Camouflage Corps during World War I. See Duer (1916).

Charles-Georges Dufresné

(November 23, 1876-August 8, 1938)
French painter

Dufresné studied at the École des Beaux-Arts. He was especially influenced by the work of Henri de Toulouse-Lautrec and (camoufleur) Jean-Louis Forain. He is listed in Coutin (1988) as a World War I French camoufleur. His later work is characterized as Art Deco, including his murals for an interior at the 1925 Exposition Internationale des Arts Décoratifs, from which Art Deco got its name.

•

Right A dazzle painted American ship, the *USS Mercury* and another US transport in a French harbor, c1918. Courtesy Naval Historical Center (NH 98655).

Jay Raymond (*The Golf-father*): "A thought struck him. 'Surely,' he said, 'the Chinese could easily spot the new golf courses from the air.'

Krulenko grinned and said something to his interpreter.

'The minister congratulates you on your perspicacity. However, it is our intention to camouflage the golf courses, until we think the correct time has come to officially break the news that the Russians have taken up golf.'

'How on earth will you be able to camouflage them? Golf courses are pretty conspicuous.'

'It is simple,' Krulenko said with a smile through his interpreter. 'We shall camouflage them so that they resemble missile sites.'"

...

Alister Mackenzie (1920:128-129): "There is an extraordinary resemblance between what is now known as the camouflage of military earthworks and golf course construction...Successful golf course construction and successful camouflage are almost entirely due to the utilization of natural features to the fullest extent and to the construction of artificial ones indistinguishable from nature."

Raoul Dufy

(June 3, 1877-March 23, 1953)
French painter

Although there is no indication that Dufy was actually a camoufleur, he is mentioned by Cork (1994:60) in the following context: "Camouflage grew so sophisticated during the war [WWI], for artillery and navy alike, that it eventually inspired the formation of [an organization called] Les Peintres de la Guerre au Camouflage...One of the French artists who subsequently became involved with Les Peintres de la Guerre au Camouflage was [Raoul] Dufy."

The following entry appears in Clement (1994:148): "1917 Dufy supervises painting and drawings at the Musée de la Guerre, under Camille Bloch, and purchases work by [camoufleurs] André Dunoyer de Segonzac and Luc-Albert Moreau. Les Peintres de la guerre au camouflage shows their work at the Galeries Barbazanges."

•

Augustus (William) Dunbier
(January 1, 1888-September 11, 1977)
American painter

Born in Shelby NE, his immigrant family moved back to Germany, their homeland, in 1903. In Cologne, he became a friend of the Surrealist painter Max Ernst. He studied art at the Dusseldorf Royal Academy, and then, when World War I began, returned to the US and enrolled at the Art Institute of Chicago. In 1918, he joined the US Army Camouflage Corps, Company A, 97th Engineers. See Dunbier (2004).

•

George (Bernard) Duncan
(January 7, 1904-May 8, 1974)
New Zealand artist

Born in Auckland, New Zealand, he studied in England and Europe in the late 1930s. He then served as a camoufleur during World War II. See Power (2002).

•

André Dunoyer de Segonzac
(July 7, 1884-September 17, 1974)
French painter

Born in Bossy-Saint-Antoine in France, he studied at the Académie de la Palette, but dropped out, deciding instead to work on his own. He shared a studio with Jean-Louis Boussingault, who, like himself, would later be a camoufleur. He participated in both the Salon d'Automne and the Salon des Indépendants. From 1916 to 1918, he was in charge of the Section de Camouflage for the North, where he worked with Boussingault, Charles Dufresné, Charles Camoin, Jacques Villon, Jean Puy and others. After the war, he continued to paint, while also working extensively in theatre design (for the Russian Ballet, and Isadora Duncan) and book illustration. His recollections of his wartime duties as a camoufleur were published in a statement in a book of his drawings (Dunoyer de Segonzac 1970:163-166). Also see Osborne (1981). He was awarded the Croix de Guerre.

Above Selection of natural patterns by French designer E.A. Seguy (originally published c1920), from *Seguy's Decorative Butterflies and Insects in Full Color* (New York: Dover Publications,1977) from the Dover Pictorial Archives.

In Blechman (2004:132), there is a photograph of members of the French Section de Camouflage at its first anniversary banquet on February 12, 1916. According to the caption, among those pictured are André Dunoyer de Ségonzac, Louis Guingot, André Mare, Abel Truchet, Jean-Louis Forain, and Lucien-Victor Guirand de Scévola.

•

Max(well Spencer) Dupain
Australian photographer
(April 22, 1911-July 27, 1992)

Born in Sydney. GroveArt.com: "During World War II Dupain worked first in a camouflage unit and then for the Department of Information (until 1947). This experience and the influence of British filmmakers enhanced his interest in the creative potential of documentary photography. Returning to private practice, he concentrated on architecture and industrial work and became Australia's foremost architectural photographer."

•

Victor-Francois-Eugene Dupont
(July 12, 1875-?)
French painter

Born in Boulogne-Sur-Mer, he studied at the École des Beaux-Arts and the Académie de Boulogne. He was particularly interested in religious subjects. He is listed in Coutin (1988) as a World War I French camoufleur.

•

Jean-Charles Duval
(1880-1963)
French painter, designer

Duval was a landscape painter, but he also worked for the Paris Opera and the Russian Ballet. A friend of Edgar Dégas, he wrote a book titled *Dégas as I Knew Him*. He is listed in Coutin (1988) as a World War I French camoufleur.

•

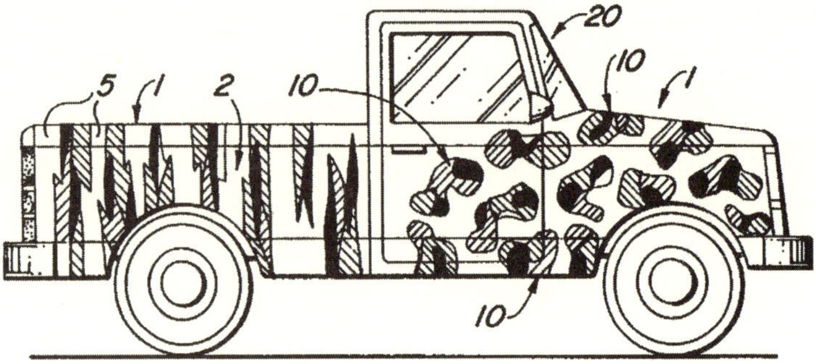
20
10
1
5
1
2
10
10

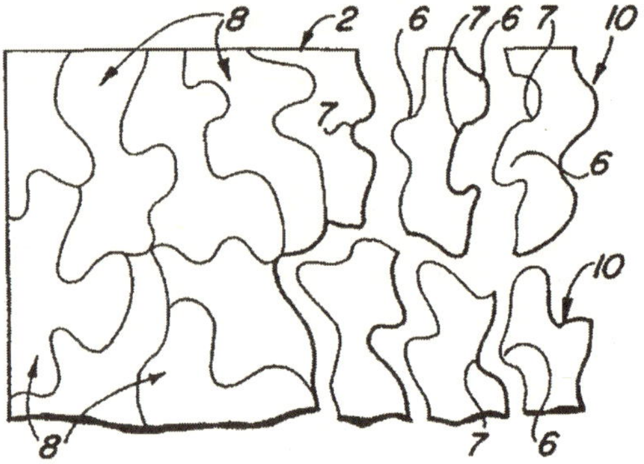
8
2
6
7
6
7
10
7
6
10
8
7
6

7
10
3
7

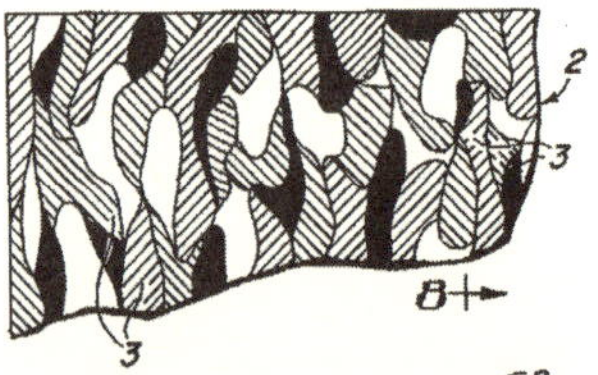
8
2
3
8
3

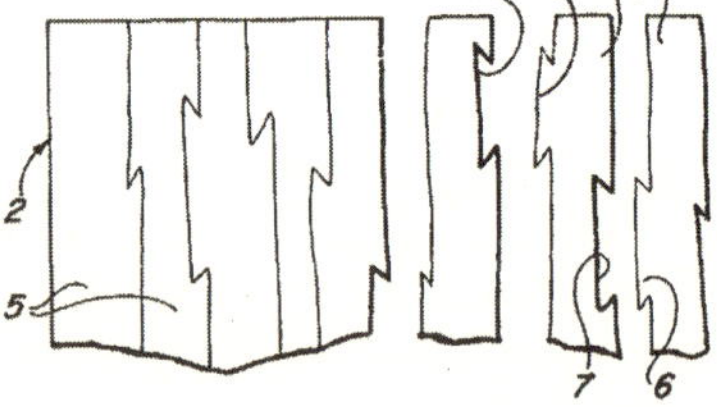
7
6
5
2
5
7
6

2
4

E

Harley Earl
(November 22, 1893-April 10, 1969)
American industrial designer

Born in Hollywood, he learned to custom build automobiles in his father's firm, Earl Automotive Works. In 1927, he was commissioned to design the 1927 Cadillac LaSalle. Soon after, General Motors president Alfred P. Sloan hired him to be the first director of GM's newly formed Art and Color Division (later called the Styling Division). Earl is said to have originated tail fins, wrap around windshields, two-tone body paint, clay modeling for automotive design, and the hardtop sedan. During World War II, he started a camouflage research and training component at General Motors. See Lauve (1942) and Earl (1954). Other camoufleurs who worked for General Motors were Ledyard Towle, David Jones and McClelland Barclay.

•

Kerr Eby
(October 19, 1889-1946)
Canadian-American artist

His parents were missionaries stationed in Tokyo when Eby was born. Throughout his childhood, he lived in various regions of Canada. He worked as a printer's devil, then moved to New York, intending to become an artist. He studied briefly at Pratt Institute and later at the Art Students League. Old

Facing page Drawings submitted for US Patent No. 5,549,938 titled "Removable Camouflage," invented by G.G. Nesbitt (1996). By this method, magnetic abstract shapes can be applied, rearranged and removed, without damaging the vehicle.

Print Shop: "He spent most of World War I on the front line camouflaging the big guns." His antiwar battlefield drawings were published as a book titled *War* (1936/1971). During World War II, he served in the Philippines, in the course of which he contracted a fatal tropical disease.

Eby (1936/1971): "I saw a great deal of America's part in the last war by an accident of service. In our army anything that was queer or out of the way was put into the Engineers, so we of the Camouflage working with and attached to the field artillery were engineers, the 40th Regiment of the regular army. There were comparatively few of us divided into groups of ten or twelve men to a regiment of artillery. We had few replacements and of necessity when our division of the moment was withdrawn from the line for rest and replacements we of the Camouflage were transferred to the incoming outfit and stayed with them in action. How many divisions had the doubtful blessing of my services I cannot remember nor the names of the places we went—all that is a haze—but what I do remember and brilliantly—is what it looked like and felt like."

•

Above These are before and after photographs of Thomas A. Edison's camouflage for the *SS Ockenfels*, a captured German freighter (later renamed the *USS Pequot*). Everett L. Warner (n.d.) [paraphrased from his lecture notes, in reference to the *Ockenfels*]: "The structural changes were the plan of Thomas Edison, and I painted her. It was not a success. The 'Hawaiian skirt' on the bow, as the longshoreman called it, was carried away before she got out of New York Harbor. The colors of the spots were purple and green, with the addition of a few small orange spots to warm up the resultant gray. Although the spots were fifteen feet long, they blended to a neutral gray at less than two miles, while the vessel remained perfectly visible. While the experiment was not a success, valuable lessons were learned from it" (Part 2, pp. 1-1).

There is an online essay by E. David Cronon on Edison's other WWI contributions, titled "Thomas A. Edison: Unorthodox Submarine Hunter" (1986) at www.worldwar1.com/sfedsub.htm.

Thomas (Alva) Edison
(February 11, 1847-October 18, 1931)
American inventor

During World War I, Edison (as the country's preeminent problem-solver) was invited by the US government to come up with solutions for the "submarine threat." According to Everett L. Warner, as quoted in White (1951:138): "[When solicited for ship camouflage ideas, Edison proposed a] scheme involving a big spread of canvas. But Edison was an inventor, so they let *him* try out his idea, and a very wild idea it turned out to be. I know because I had the job of doing the painting work on the vessel (*SS Ockenfels*). Part of the added camouflage structural work was so unseaworthy that it got carried away before the vessel got out of New York Harbor."

There are several accounts of Edison's role. In Crowell (1921), for example, it states (uncritically) that it was Warner's proposal, *not* Edison's, that was applied to the *SS Ockenfels*: "The pattern which he [Warner] applied made the ship's water line elusive. He cut down the funnels and masts and stretched a screen of canvas from bow to stern, the upper edge of the screen being on a level with the tops of the truncated masts" (p. 499). A few pages later, it also states that a Cunard passenger liner, the *Valeria*, "was turned over to Mr. Edison for experiment. Among other things that he did to the ship, he screened her upper works in canvas" (p. 502).

•

Édy-Legrand
(July 24, 1892-September 1970)
French painter, illustrator

Édy-Legrand was the pseudonym of Édouard-Léon-Louis Legrand. Born in Bordeux, he painted murals for major Paris stores and ocean liners. He was also a book illustrator. He is listed in Coutin (1988) as a World War I French camoufleur.

•

Albert Eiffel
(dates undetermined)
French painter

A seascape painter, born in Levallois-Perret, he was the son of Gustave Eiffel. He is listed in Coutin (1988) as a World War I French camoufleur.

•

Ann Elias
(1955-)
New Zealand-born Australian art historian

Elias is a senior lecturer and Director of Theoretical Enquiry at the Sydney College of Art, University of Sydney, Australia. Among her research interests are the involvement of Australians in World War II camouflage, the social implications of the beliefs of zoologist and camoufleur William Dakin, and the conceptual relations between Surrealism and camouflage. See Elias (2003 and 2008).

•

Frank Eliscu
(July 13, 1912-June 19, 1996)
American artist

Born in Washington Heights NY, he attended Pratt Institute and the New York Teacher's College. In 1935, he designed the Heisman Memorial Trophy. He taught at the School of Industrial Arts in New York for more than 25 years.

There is an online biography at (www.frankeliscu.com), where it states the following: "Frank was an Army veteran of WWII. Originally he was stationed in [Fort Belvoir] Virginia working on camouflage and maps for the Allied Invasion. Later he was moved to Pennsylvania where they learned that he was an artist in an unusual way. Soldiers [who] had been injured in war and had large burns or missing facial parts were given a new face by the doctors but the scars were made more realistic by the techniques developed by Frank. He was so successful at this that they were reluctant to release him from the Army."

•

Douglas D. Ellington
(June 26, 1886-August 27, 1960)
American architect

Ellington studied architecture at Drexel University, University of Pennsylvania, and the École des Beaux Arts in Paris. Returning to the US, he taught architecture at Columbia University, and at the Carnegie Institute of Technology, where Homer Saint-Gaudens and Everett L. Warner were on the faculty. An online Wikipedia biography states: "During World War I, Ellington worked with the United States Navy [in the design subsection, working under Warner], supervising the design of camouflage for ships."

•

Martha Banta (1987:223): "By the mid-1890s poster art and book design turned pictorial surfaces into a maze...Design blurred the outlines that commonly give recognizable shapes to unique objects or...that distinguish between multiple planes. Everything within the grip of the design became part of the same boldly flattened-out field: wonderful to look at, somewhat a puzzle to read."

Above American cattle thieves used embedded figures to conceal existing brands on stolen livestock. After stealing the cattle, the rustlers changed the animal's brand not by deleting the earlier mark (which wasn't possible), but by adding other shapes to convert it to a new logo. Tattoo artists do the same when deleting an old tattoo: if it can't be removed, they design a new, larger tattoo in which to conceal the earlier one. See Hough (1897:294-296) and Johnson (1973:236-237).

Regarding these examples, it may also be of interest to look at US Patent No. 6,634,644 titled *Concealment-Type Word Puzzle Game* (2003), invented by Jeffrey W. Falana.

Jean (Melicent Jane) Ellis
(March 23, 1887-July 6, 1974)
Australian women's rights advocate

Born in Brisbane, she was a co-founder (in 1937) of the Penguin Club, which had the purpose of training women to speak clearly and fluently in public. Rutledge (1996): "Early in 1941, she was asked to establish centers to make large, mesh camouflage nets from sisal twine for the army. She started the Central Netting Depot in the ballroom of Mark Foy's Ltd, soon transferring to Anthony Hordern and Sons Ltd. By mid-1945 she was running fifty centers which had supplied thousands of nets."

•

Paul Robert Ellsworth
(1923-?)
American painter

He studied at the Art Institute of Chicago and the Art Students League in New York, then joined the US Marine Corps in 1942. Crane (1944:226): "In the service he has been a tank driver, an intelligence map draughtsman and a camouflage expert."

•

Embedded Figure
Shapes concealed in other forms

During the Iraq War, it was commonly said of journalists that they were "embedded" with infantry units, partly for their own safety but as well to prevent interference with military operations. The term implied that news reporters, although attached to military units, were also independent.

The term may have first become popular in the 1920s, in relation to experiments by the Gestalt psychologists in Germany. One of them, Kurt Gottschaldt, conducted experiments having to do with the influence of past experience in which he showed his subjects a series of visual puzzles, known in psychology as "embedded figures" (Behrens 2000). This kind of figure was not at all new, because it had been used for centuries in children's puzzles (whether pictorial or abstract), sight gags, and erotic art.

There is mention of embedded figure puzzles in an article by Salvador Dali (Dali 1942): "Every Saturday I received a juvenile publication to which my father had subscribed for me. Its final page was always devoted to a puzzle picture. This would present, for instance, a forest and a hunter. In the tangled underbrush of the forest the artist had cleverly concealed a rabbit; the problem was to find it. Or, again, a doll must be discovered, lost by a child in an apparently empty room. My father would bring me the puzzle, and what was his astonishment to see me find, not one but two, three or four rabbits; not a single doll but several—and never the one which the artist had meant to conceal."

See also Arcimboldo and coincident disruption.

•

Aymar Embury II
(June 15, 1880-November 15, 1966)
American architect

Embury was born in New York. He studied architecture at Princeton University, where he subsequently taught, while also practicing architecture in New York. He was mainly known for designing residences, university buildings and social clubs, and for working under Robert Moses as a senior member of the architectural team that constructed and renovated New York City city parks. He himself may have supervised the design of more than six hundred public structures, among them the Central Park Zoo, the New York City Building at the 1939 New York World's Fair, and the Henry Hudson Bridge. He also published six books on architecture. Online Wikipedia biography: "Embury served for fourteen months during World War I as a Captain in the Fortieth Engineers, US Army Corps of Engineers, where he helped establish a unit of eight professional artists to document the activities of the American Expeditionary Force in France."

In Embury (1917:138), he states: "Since Major [Evarts] Tracy left, I have been honored by being given charge of the training, equipment, and personnel of camouflage troops." See also *New York Times* ("Aymar Embury" 1966:47).

•

Star (May 8, 1931): "Although the majority of women seem to prefer shoes with just two colors to match their frocks, 'dazzle' footwear are a good second."

Above In 2002, American conceptual artists Monica Duncan and Lara Odell collaborated on a video spoof of fashion and camouflage, titled *Winter Camouflage Catalog 2002*, from which this is a video still. Courtesy the artists.

London Times [reporting on the Chelsea Arts Club dazzle ball] (1919): "The very fact that this disruptive coloration broke up the usual lines of form gave to many costumes a grace and charm as delightful as they were unexpected..."

...

John Townsend [in *The Young Devils: Experience of a School-Teacher* (1958)]: "The boy...his yellow dazzle socks flashing like twin beacons."

...

Economist (January 11, 1958): "Girls in tight jeans and dazzle socks."

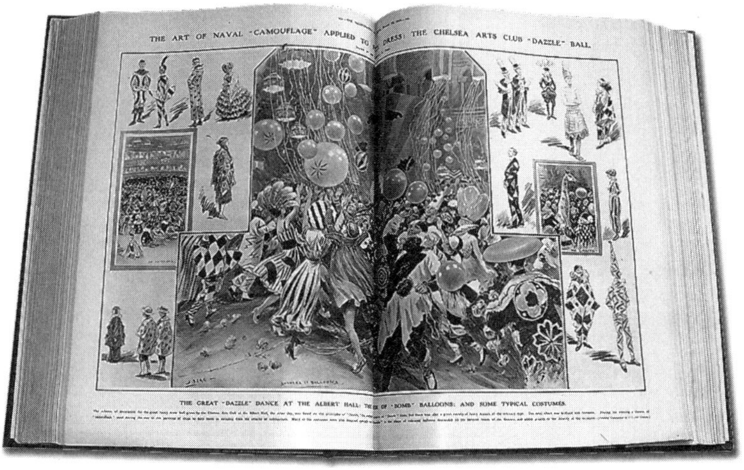

Above Examples of the post-WWI popularity of camouflage in British and American society, including (top left) dazzle-patterned bathing suits; (top right) dazzle-patterned stockings; (center) Louis Weinberg's diagrams (in Weinberg 1927) showing how clothing styles can "camouflage" the physical shapes of two different women (one too thin, one too stout), each modeling three outfits; and (bottom) a festive pictorial page spread, from the *Illustrated London News* (March 22, 1919:414-415), reporting on a Chelsea Arts Club "dazzle ball," that was held in Royal Albert Hall on March 12.

Georges d'Espagnat

(August 14, 1870-April 17, 1950)
French painter, illustrator, stage designer

Espagnat was born in Melun, Seine-et-Marne, France. In the 1890s, he exhibited at the Salon des Indépendants, and was affiliated with the Impressionists, including Auguste Renoir, Paul Signac, Maurice Denis, Pierre Bonnard and Édouard Vuillard. He was also one of the founders of the Salon d'Automne. He served as a French camoufleur during World War I.

US National Research Council, from the minutes of a meeting of the Executive Board of War Organization, on February 25, 1918: "Upon nomination of the chairman of the Division of Physics, Mathematics, Astronomy, and Geophysics, the appointment of a Camouflage Committee was approved with membership as follows: M. Luckiesh, Chairman, Bassett Jones, Lloyd [sic] A. Jones, G.H. Thayer."

•

[Herbert Barnard] John Everett

(1876-1949)
British marine painter

...

Gestalt psychologist Wolfgang Köhler (1947:92-93): "...sometimes we see from a distance a strange object which later, when we approach it, splits into a well-known thing and parts of other objects...The puzzle-pictures which years ago amused the readers of magazines were examples of this kind of thing. In modern wars it has become a real art to makes objects such as guns, cars, boats, etc., disappear by painting upon these things irregular designs, the parts of which are likely to form units with parts of their environment. In such cases the objects themselves no longer exist as visual entities, and in their place appear meaningless patches which do not arouse the enemy's suspicion; for similar patches are constantly produced by accidental combinations of parts which fuse, for instance, because of their similarity."

Everett was born in Dorset, England. He was a close friend of Marine painter WWI camoufleur Julius Olsson, and together they founded a school for seascape painting in 1895. He studied at the Slade School of Fine Art, then at the Académie Julian in Paris.

Yarker (n.d.): "In the spring of 1918 he was commissioned by the Ministry of Information to complete some drawings and paintings of wartime London docks and the Thames which were subsequently exhibited in America...In his paintings Everett also explored the visual effects of dazzle painting. This was a form of camouflage put on ships during the war. He experimented with the formal possibilities these designs offered and *A Convoy 1918* reveals the rhythmic relationship between the hull, funnel, waves and the sea. At the same time the attention to flag details indicates the work hovers uncertainly between a conventional and formalistic approach."

Wallace Stevens (in a letter to William S. Braithwaite, December 5, 1921): "From time immemorial the philosophers and other scene painters have daubed the sky with dazzle paint."

Above These are from a series of colored drawings of World War I dazzle-painted ships, completed by John Everett for the British Ministry of Information. The ship at the top right is the *SS Lepanto*, while the bottom one is *HMS Coronado*. From postcards in the author's collection. Gift of Les Coleman.

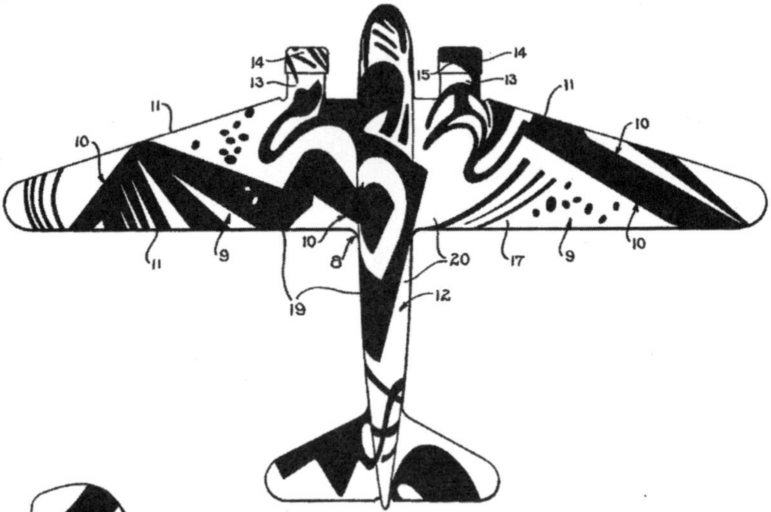
14
13
11
10
14
15
13
11
10
11
9
10
8
19
20
17
9
10
12
19

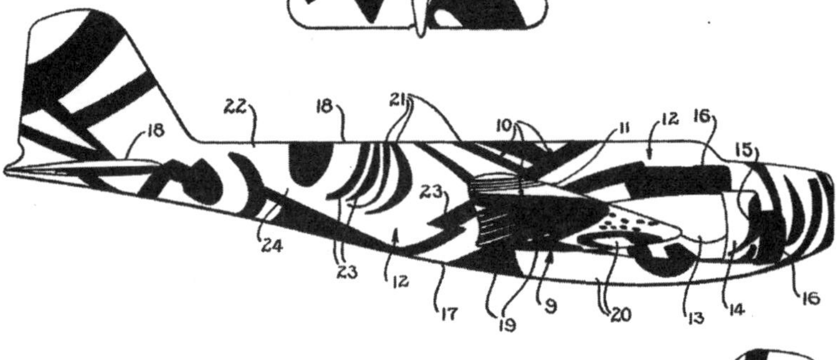
18
22
18
21
10
11
12
16
15
23
24
23
12
17
19
9
20
13
14
16

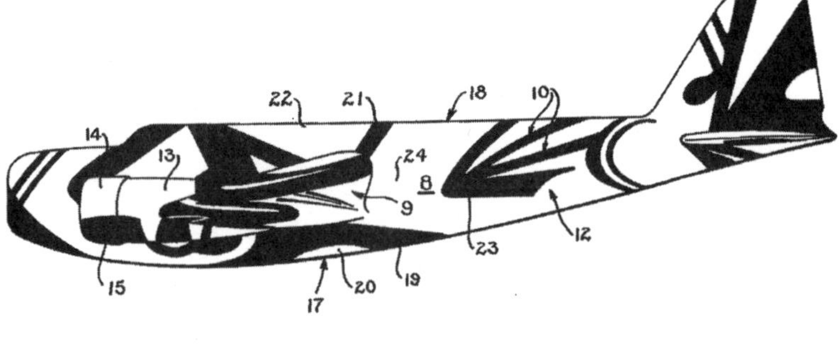
22
21
18
10
14
13
24
8
9
23
12
15
17
20
19

F

Fade Away
Deliberate omission of edges

The term "fade away" was in popular use around 1900 and later to refer to images in which some (not all) of the edges of a figure were deliberately omitted, allowing the partial merger of a figure with its background. Among the best known examples of these were advertising images of fashionable women, known as "Fade Away Girls," created by Coles Phillips and other illustrators for the covers of *Life*, *Good Housekeeping*, and *Century Magazine*, in addition to countless magazine ads. This was also then in common use among artists, designers and others, including the Beggarstaff Brothers, Ludwig Hohlwein, J.C. Leyendecker, and Robert Gibbings. Its purpose was to selectively withhold information, resulting in a puzzling form that would evoke audience participation through *closure*.

•

Jules-Abel Faivre
(March 30, 1867-August 1945)
French painter, illustrator

Faivre was born in Lyons, France, and studied at the École des Beaux-Arts with Jean-Joseph Benjamin Constant and Jules Lefebvre. A WWI French camoufleur, he was awarded the Legion of Honor.

Facing page Drawings submitted for US Patent No. 2190691, titled "Camouflaging," invented by McClelland Barclay (1940). See also Barclay.

Pierre Falké

(May 24, 1884-June 1947)
French painter, illustrator

He is listed in Coutin (1988) as a World War I French camoufleur. He illustrated the writings of Edgar Allan Poe, Jules Reynard, and an edition of Daniel Defoe's *Robinson Crusoe.*

Above Robert Gibbings, woodcut titled *The House Painters* (1921), an example of a "fade away" image.

Stephen Kern (Kern 1994): "The Cubists tried to blend the fruit into the background of their still lifes, just as camouflage attempted to blend soldiers into the terrain."

•

Barry Faulkner

(July 12, 1881-October 27, 1966)
American muralist

Barry Faulkner was born in Keene, NH, and studied with Abbott Thayer (his cousin), George de Forest Brush and Augustus Saint-Gaudens. Faulkner's formal training was at Phillips Exeter Academy, briefly at Harvard University (where his roommate was camoufleur Homer Saint-Gaudens), and at the American Academy in Rome. While in Europe he was commissioned to paint his first mural. Throughout his career he completed a number of important murals, including those at the National Archives Building in Washington DC, a bank in Keene NH, the RCA building at Rockefeller Center, the John Hancock Building in Boston, and the State House in Concord NH.

Above Photograph of a contemplative Barry Faulkner (n.d.) by Peter A. Juley & Son. Courtesy Photograph Archives, Smithsonian American Art Museum (J010874).

There is a book about Faulkner's murals (Rumrill 2007) that includes recordings of Faulkner on a CD-ROM, speaking about Abbott Thayer and other subjects.

In anticipation of the US entry into World War I, he collaborated with sculptor Sherry Fry (also a student of Augustus Saint-Gaudens) in forming the New York Camouflage Society, which eventually contributed to the formation of the Army's American Camouflage Corps. Faulkner and Fry and others were among the first US camoufleurs to serve in France. For extensive details, see Faulkner (1957) and (1973).

In his autobiography (Faulkner 1973:109-110), he mentions his completion, after World War I, of a large mural for the Cunard Building in New York. Two other veteran camoufleurs worked on that project as well: Ezra Winter painted the ceiling and John Gregory designed bronze medals that were installed in the floor.

Royal B(ailey) Farnum
(1884-August 28, 1967)
American art educator and administrator

Farnum was born in Somerville MA. He graduated from Massachusetts Normal Art School in 1906, then went on to study at the Cleveland School of Art, Brown University, and Rhode Island State College. At various times, he was head of art education for the New York State school system, President of the Massachusetts College of Art, head of art education for the State of Massachusetts, President of the National Association of Schools of Art and Design, President of the National Art Education Association, and Vice-President of the Rhode Island School of Design. *New York Times* ("Dr. Royal B. Farnum" 1967): "He had developed a color combination for military camouflage and was the author of *Learning More About Pictures*, published in 1958."

•

Gene Federico
(1918-October 23, 1999)
American graphic designer

Born in New York, he studied at Pratt Institute, graduating in 1939. He was widely known for his work as a graphic designer at *Fortune* magazine, Grey Advertising, and other firms. Antupit (2000): "World War II interrupted his career; Federico was drafted into the Army, where he served in the Camouflage Corps in the US, Europe, and North Africa."

•

Adrian (George) Feint
(June 28, 1894-April 25, 1971) Australian printmaker and painter

He is listed by Elias (2003) as a member of the Sydney Camouflage Group. For biographical details, see Butler (1996).

•

John Duncan Fergusson
(March 9, 1874-January 30, 1961)
British painter

Fergusson studied medicine at Edinburgh University, but then became influenced by Arthur Melville and others of the Glasgow School, which led him to study in Paris. He was a friend of Charles Rennie Mackintosh. In 1918, he joined the British Navy, in connection with which he made wartime paintings of the dockyards at Portsmouth. The foreword to his 1961 memorial exhibition was written by (camoufleur) André Dunoyer de Segonzac.

•

Keith Ferris

(1929-)
American artist

Ferris grew up on a military base in Texas, where his father was a Air Force pilot. He studied aeronautical engineering at Texas A&M University, with the plan of becoming a pilot. For health reasons, he turned instead to aircraft illustration, working as a civilian artist for an Air Force publications unit. Throughout his life, he has continued to be a prolific illustrator (see Ferris 1978), as well as a major contributor to US Air Force camouflage. Ferris is usually credited with the elimination of brightly colored insignia on combat planes, the use of non-reflective low contrast colors in airplane camouflage, radical asymmetry, and a method of disruption that he has described as "longitudinal," like a sausage sliced lengthwise. See Ferris (1976), Browne (1987) and DeMeiss (1987).

•

Ian Hamilton Finlay

(October 28, 1925-March 27, 2006)
Scottish artist, poet, gardener

Born in Nassau, the Bahamas, Finlay's work as an artist was notoriously diverse. He was in part a writer, sculptor, gardener, poet and philosopher. He served in the army during World War II. A number of his artworks refer to camouflage. Stephen Bann, in Abrioux (1992): "In *Strawberry Camouflage*, the very bold Fauvist impact introduces the element of force more typically associated with Finlay's Suprematism. Finlay explains in his *Detached Sentences on Camouflage* that 'The domestic strawberry disguises certain of its leaves as ripe fruit; this is the principle of dazzle camouflage.'"

Stuart Guthrie (1993:231): "The diverse kinds of camouflage that have evolved to hide the eye pattern also suggest that our inclination to see it is innate."

...

Kahn (1984:11): "By a ministerial order of 12 February 1915, just seven months after the outbreak of war, the French minister of war established *La Section Camouflage*, whose function he described in a memo of the same date to the military governor of Paris: 'To arrange on the ground masking devices to hide the works from aerial views and to disguise the equipment which could be spotted by the enemy.' By 1917, 3,000 men and women worked in this division."

...

Charles Burchfield (University of Arizona Art Gallery 1965:24): "[During World War I, he was appointed] as one of several designers of layouts for camouflaging (with paint) the various pieces of field artillery. The work was pleasant and I would set my drawing board under a young oak tree, more or less by myself. To make my designs more interesting [to himself] I would invent themes, which I kept secret, which embodied some previous experience in Nature (such as March wind, sunlight and rocks). To see these later, on say, a field gun, gave me quite a thrill."

Above US dazzle-camouflaged emergency ship, the *Zinkel*, as photographed in 1918. Photo courtesy US Naval Historical Center (NH65052).

•

William Flemer III
(1922-April 22, 2007)
American botanist, environmentalist

Fleming graduated from Lawrenceville School, then attended Yale University. *New York Times* (Dube 2001): "He served for three years with the [US] Army's 603rd Engineer Camouflage Battalion [the Ghost Army], part of that time in Europe."

Ibid: "At 79, William Flemer III, stands tall and erect, is soft-spoken and learned and has a memory sharp as a tack. Still active, he holds more than 50 patents to tree and shrub species. 'It's my avocation,' says the Yale-educated gentleman. 'I do it instead of tennis or golf.'" He is also mentioned in Fox (1978).

•

Louis Floutier
(dates undetermined)
French painter

Born in Toulouse, he was a student of Fernand Cormon. He is listed in Coutin (1988) as a World War I French camoufleur.

•

Jean-Louis Forain
(October 23, 1852-July 11, 1931)
French painter, illustrator

A native of Rheims, he studied at the École des Beaux Arts. Throughout his life, he was widely admired for his achievements as a caricaturist, satirist and painter, with close links to the Impressionists and Post-Impressionists, among them Henri de Toulouse Lautrec and Paul Cézanne. During World War I, he volunteered for active duty as a camoufleur, although he was sixty-two. He was awarded the Legion of Honor.

Peixotto (1917:2): "It was the artists also—Forain among the first—who, at the very beginning of the war, first throught of dissimulating war material by means of protective coloring."

Browse (1978:59-60): "Forain could not bear the seeming inactivity of remaining a civilian. Much to his pride, Jean-Loup [his son] was already with his regiment at Rheims, and at the age of sixty-two, Forain was also determined to enlist. Somehow or other he managed to enroll as a camouflage artist and was put in charge of concealing the viaduct at Chantilly."

•

W.D. Foster
(dates undetermined)
American architect

Foster is listed in Embury (1917:138) as "head draftsman in H. Van Buren Magonigle's office." He is also cited as a camoufleur in *Literary Digest* ("Company A" 1917).

•

William F(rederick) Foster
(1883-1953)
Illustrator

He studied with Frank Duvenek at the Art Academy of Cincinnati, and with Robert Henri and William Merrit Chase at the Art Students League. He also studied at the Académie Julian in Paris. He is mentioned as a camoufleur by Gallatin (1919) as well as by Love (1999). Primarily known as a magazine illustrator, his wartime drawings are in *Scribner's* (April 1919) as part of his article titled "A Day with a Sketch-block at the Front."

John Windsor (1996): "Camouflage is the creation of aesthetes, not military tacticians. In fact, ever since the first purpose-designed camouflage of the First World War, hardened fighting men have gone into battle wearing patterns dreamed up by artists who considered them rather gorgeous."

Above The *USS Rhode Island* in an experimental camouflage pattern, c1917-1918 (detail). Photo courtesy US Naval Historical Center (NH101192).

Paul K. Saint-Amour (2003:350): "By 1918, young British aviators were being trained to see an avant-garde exhibition unfurling beneath their cockpits: a First World War Air Force photo atlas for new pilots used 'FUTURIST country' and 'CUBIST country' in its taxonomy of aerial landscapes, alongside more everyday mneumonic headings such as 'FRUIT GROWING' and 'PATCHWORK QUILTING.'"

Frederic E. Fox
(August 19, 1917-1981)
American theologian

Fox graduated from Princeton in 1939, worked briefly for NBC, then enlisted in the US Army. He was a captain in the Signal Corps, and served in the Ghost Army, as a result of which he wrote an "Official History of the 23rd Headquarters Special Troops." After World War II, he attended Union Theological Seminary, and served as a minister in Congregational churches in various parts of the country. He was also speechwriter for US President Dwight D. Eisenhower. In 1964, he became the Recording Secretary at Princeton University, and in 1976, its Keeper of Princetoniana. See Goossen (1973:103).

He is quoted in Park (1985:142): "Since no one knew how a deception unit was supposed to operate, the training program was not easy to write. Officers, who had once commanded 32-ton tanks, felt frustrated and helpless with a battalion of rubber M4s [dummy tanks], 95 pounds fully inflated. The adjustment from man-of-action to man-of-wile was most difficult." See also Fox (1978).

•

Fredrick Fox
(1910-?)
American theatre designer

According to Naversen (1989:180), Fox was educated at Phillips Exeter Academy, Yale University, and the National Academy of Design. He was a member of the Camouflage Society of Professional Stage Designers, and the set designer for such films as *Darkness at Noon*, *King of Hearts* and *The Seven Year Itch*.

•

Milton S. Fox
(1904-1971)
American artist, editor

Fox was for many years an art book editor at Harry N. Abrams publishing company in New York. Prior to World War II, he was associated with the Cleveland Museum of Art and the WPA. During the war, he was head of the Civilian Camouflage Committee for northeastern Ohio. There is an interview with him on this and other subjects at the Archives of American Art, Smithsonian Institution (Fox 1964). For his own published writings on art and camouflage, see Fox (1942, 1943a and 1943b).

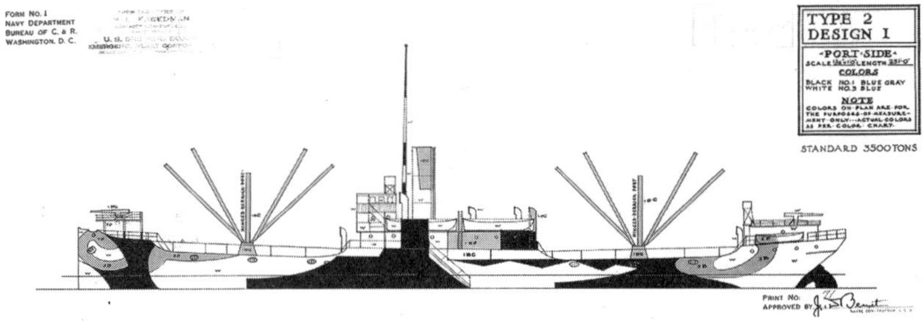

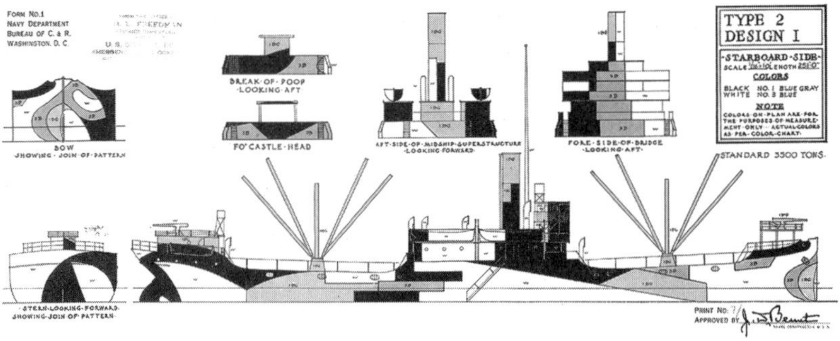

Above After World War I, American artist Maurice Freedman gave the Rhode Island School of Design (his alma mater) more than 450 plans for camouflaging merchant ships (top) and twenty black-and-white photographs of various camouflaged ships (bottom). The ship in the photograph is the *SS Everglades* at the Oscar Daniels Shipyard in Tampa FL (c1918). Courtesy the Fleet Library at Rhode Island School of Design.

Maurice L. Freedman
(July 15, 1898-December 1983)
American artist

Claudia Covert (2007:53): "During World War I, Maurice L. Freedman was a District Camoufleur in the 4th District of the US Shipping Board, in Jacksonville FL. In 1918 and 1919 he attended RISD [Rhode Island School of Design] to study decorative design as well as freehand drawing and painting. According to the Library Annual Report of 1919, he donated to the RISD Library 'charts and directions for camouflaging ships and samples of colors used.' Today the collection consists of nearly four hundred discontinued plans for the short-lived dazzle ships. The collection also includes twenty black-and-white photographs of World War I dazzled merchant ships. Unfortunately, the whereabouts of the 'samples of colors used' is unknown."

Following World War I, Freedman worked as an advertising artist in Providence RI, and, in the 1940s, developed one of the first versions of a game like *Battleship*, called *Warfare: Naval Combat*. In the 1950s, he was an assistant art director for Paramount Cards, a major greeting card company in nearby Pawtucket.

Left Cover of a game developed in the 1940s by Maurice Freedman, titled *Warfare: Naval Combat*. Courtesy of the Fleet Library at Rhode Island of Design.

French Section de Camouflage
First official camouflage unit

Peixotto (1917:1): "...I have among my papers a single list of three hundred and fifty of them [French artists] killed in action...
...The French Government awakened to the fact that the artists of its country should not be thus wantonly sacrificed; that there were very useful things that they could do. So it directed its efforts toward employing them in work for which they were specially equipped."

According to Kahn (1984:166), there is an early memo in French military archives about the formation of a "unit called camouflage," in which the first six camoufleurs are listed as [Lucien-Victor Guirand de] Scevola, [Eugene] Ronsin,[Jean-Baptiste Eugene] Corbin, [Louis] Roger, [Joseph] Pinchon and [Georges] Mouveau.

Milton Fox (1943): "[Camouflage is] one of those fusions of art, science, and technology, of which two have already provided the most popular forms of expressions in our time, namely, photography and the movies...

This hybrid art, at its best, spans many fields, amongst them art, photography, architecture, biology and nature study, physics, chemistry, psychology, and aviation. Obviously then, thorough training for camouflage is no matter for art classes alone. But equally obviously, camouflage is the manipulation, in one way or another, of visual effects and appeals to vision. And the basic preparation for such manipulation is to be found in the training of perception and the powers of observation such as one should expect in art classes."

...

Robert P. Watson (2004:241): "...President [Franklin D.] Roosevelt, to the credit of observers at the time and the years since, immediately rejected a porposal from Civil Defense officials to paint the White House camouflage for the duration of the war [World War II]."

...

H.G. Wells (117:112): "The French, being I think on the whole much better acquainted with the commonplaces of science than the senior British officers, have taken a tip from the coloration of animals in this matter. As every magazine reader knows nowadays, animal colors even when apparently conspicuous are arranged almost always so as to *break the outline*."

(Achille-Émile) Othon Friesz
(February 6, 1879-January 10, 1949)
French painter, printmaker

Friesz was born in Le Havre, France, as were Georges Braque and Raoul Dufy, with whom he was lifelong friends. After moving to Paris, he studied with Léon Bonnat at the École des Beaux-Arts. Initially drawn to Impressionism, he became influenced by Henri Matisse, and veered toward Fauvism, a style he later abandoned. He is listed in Coutin (1988) as a World War I French camoufleur, during which he specialized in airplane camouflage. He was awarded the Legion of Honor.

•

Sherry Edmundson Fry
(September 29, 1879-June 9, 1966)
American sculptor

Fry was born to a farm family in Creston IA. He studied with Lorado Taft at the Art Institute of Chicago, then in Paris at the Académie Julian and the École des Beaux-Arts. He also worked with Frederick MacMonnies in Giverny, France. In 1909, he received an award at the Paris Salon for his bronze monument to Chief Mahaska of the Iowa Indians, commissioned for the town square of Oskaloosa IA. He received numerous public commissions, and was regarded as one of the most promising sculptors of his time. After his return to

Facing page World War I camouflaged French tank (n.d.). Photo courtesy US Library of Congress, Prints and Photographs. *Above* (left) Sherry Edmundson Fry (n.d.) *Portrait Bust of Jerome* [sic] *Brush*. Photograph by De Witt Ward.Courtesy American Sculpture Photograph Study Collection, Photograph Archives, Smithsonian American Art Museum (S0001041). (right) Sherry Edmundson Fry in his studio (n.d.) as photographed by Peter A. Juley & Son. Courtesy Photograph Archives, Smithsonian American Art Museum (J0025514).

the US, he had a studio in the Lincoln Arcade Building (the current site of Lincoln Center), and a house in Roxbury CT.

It was Fry and Barry Faulkner who initiated the New York Camouflage Society, in advance of the Army's formation of the American Camouflage Corps, which both artists then enlisted in. Faulkner (1957, 1973) has written at length about this and their battlefield experiences in France, and there is also a published essay by Fry (1917) on "An American Corps for Camouflage."

Both Fry and Faulkner had studied with Augustus Saint-Gaudens, and both knew Abbott H. Thayer and the George de Forest Brush family. They were also at the American Academy in Rome at the same time. Faulkner (1973:66) describes Fry as "a tall, reddish young man, energetic, impulsive, full of tall stories of his experiences to which I delighted to listen." He also claims that Fry supported himself in Chicago by peeling potatoes in a hotel restaurant, and that he invented a potato peeler that was faster than any other (Ibid.:79-80).

Faulkner tells a great story about a boa constricter that lived in the walls of Fry's New York studio. It had been given to him by a woman whom he had met on a steamer as he was returning from Italy. Soon after landing in New York, she came to Fry to say that her partner had been arrested, and that she had no money. Fry gave her some cash, in exchange for, among other things, a boa constricter. Continues Faulkner: "The boa was gentle and contented, living most of the time in the jungle of water pipes and electric wires in the walls of Fry's studio, and only coming out to be fed an occasional squab or rabbit" (Ibid.:80).

Barry Faulkner: "...our friends in the French Army told us that 'camouflage' was the only word in the French language which we Americans could pronounce correctly."

...

Nature ("Camouflage in Modern Warfare" 1940:950): "...camouflage research and application are at the present time largely dominated by artists, or in the hands of Civil Servants and Army officers, and in either case controlled by people lacking the necessary scientific training and with no knowledge of the fundamental biological and psychological principles involved, the failure to achieve effectiveness which one sees on every side is rather to be expected than wondered about."

...

Time magazine ("Natural Camouflage" 1941): "Extremists like G.H. and Abbott H. Thayer (whose *Concealing Coloration in the Animal Kingdom* is often consulted by US Army camouflage experts) have claimed that all animals are camouflaged, 'the most gorgeous costumes being, in their own way, climaxes of obliterative coloring.'"

•

Louis Agassiz Fuertes
(February 7, 1874-August 22, 1927)
American ornithologist, illustrator

Fuertes' father was a professor of civil engineering at Cornell University; his mother was a pianist. When he was born, he was named after the famous Harvard naturalist, Louis Agassiz, who had died not long before. He made his first painting of a bird at age fourteen. In 1894, as a young architecture student at Cornell, he showed his illustrations to Elliot Coues, a prominent ornithologist and bird illustrator. After that meeting, according to his daughter (Boynton 1956:9), "he never thought of adopting any profession than that of bird portraiture." Three years earlier, he had joined the American Ornithologists Union, and it was at the 1896 meeting of that organization, at Harvard University, that Fuertes first met Abbott H. Thayer (at a demonstration of countershading), whom he came to regard as a father of sorts and referred to as "Uncle Abbott."

The young Fuertes often visited the Thayers in Dublin NH and sometimes went on field trips with them. While in Dublin, he would also visit the family of George de Forest Brush, as recalled by Brush's daughter in her vivid biography of her father (Bowditch 1970:189-190). He delighted the children because, as she remembers, he was "a most lovable personality" and "full of high spirits and fun." She recalls how he amused them by hanging from the rafters in the house, imitating an orangutan, or he would pretend to be Paul Du Chaillu, a famous French naturalist and anthropologist, lecturing in a comic French accent, describing his jungle encounter with a huge gorilla, complete with chest pounding and snorting.

Above Bird illustrations by Louis Agassiz Fuertes. On the left is a Great Horned Owl (*Bubo virgnianus*) for the US Department of Agriculture and Biological Survey (n.d.), and on the right is a Magpie (*Pica hudsonia*), which appeared in *The Second Book of Birds* (1901). Public domain.

As Abbott Thayer aged, he grew more insecure and became hypersensitive to any questioning (real or imagined) of his beliefs about camouflage. At the same time, his and Fuertes' relationship grew more complicated, largely because Fuertes was by then making his living as a bird illustrator for book publishers. His clients insisted on illustrations that continued the tradition of John James Audubon and Ernest Thompson Seton, in which clarity was of the utmost importance, with an obvious visual distinction between the subject and its background. In Thayer's view, animals should be portrayed as embedded in their natural setting, in which case they may not be easy to see. He pleaded with Fuertes to acknowledge in his own bird paintings the preeminence of protective coloration, and when Fuertes could not do that, Thayer began to regard it as a subversion of his own teachings. Letters pertaining to this and related issues can be found in Boynton (1956).

Thayer died in 1921. Six years later, as Fuertes and his wife were returning home from a trip, they were struck by a train at a crossing and he died at age 53.

•

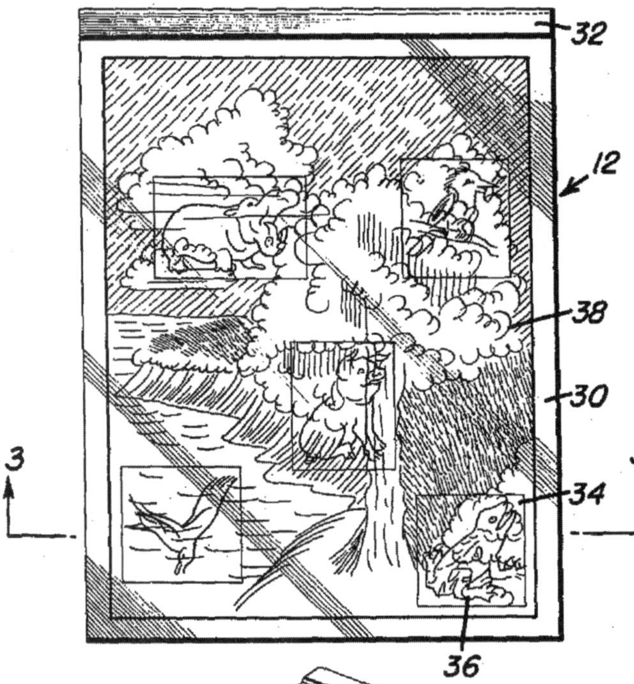
32
12
38
30
3
3
34
36

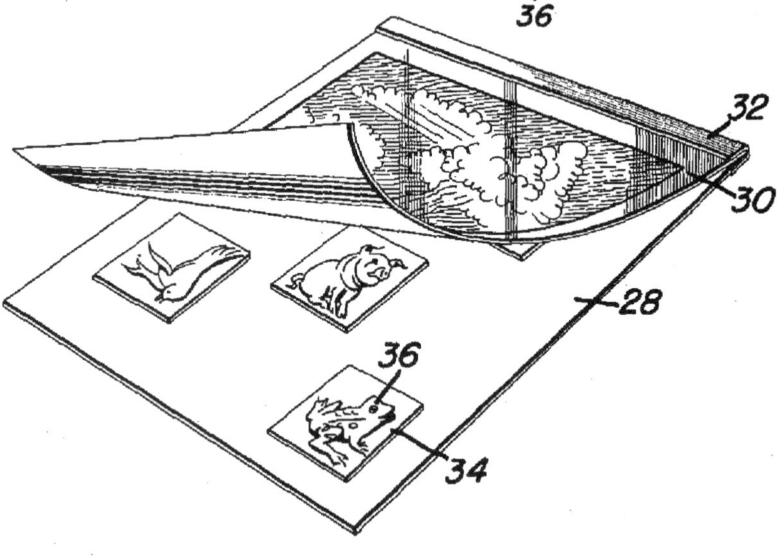
32
30
28
36
34

G

Josué Gaboriaud
(1883-1955)
French painter, illustrator

He is listed in Coutin (1988) as a World War I French camoufleur. He was a student of Maurice Denis, and a friend of Cubist painter Albert Gleizes.

•

Edwin C. Galligan
(dates undetermined)
British advertising artist

Galligan is listed in Reit (1978:138) as having served with Geoffrey Barkas in the Middle East Forces of the Royal Engineers in North Africa in WWII. He is also mentioned in Barkas (1952:121) as having done "excellent work at Tobruk," and in Sykes (1990:55) as having been "an exhibition designer in peacetime."

•

(Leslie) James Gardner
(1907-1995)
British designer

Nicknamed "G," he studied at Westminster School of Art, where American-born designer Edward McKnight Kauffer was one of his teachers. He worked

Facing page Drawings submitted for US Patent No. 2670961 titled "Puzzle," invented by Charles E. Winters (1954). See also embedded figure.

Farrer (1918:68): "As far as the guns go, the war [WWI] is developing into a colossal fancy dress ball, with immunity for the prize...It must be the most endless joy to go on planning these disguises. One would lie awake at night wondering how to make one's gun look like a dog kennel, or a dog kennel conceal a gun."

Right World War II-era advertisement for Chesterfield cigarettes.

Facing page Camouflaged French ocean liner, *La Lorraine*, during World War I.

as an apprentice at the London branch of Cartier, as well as at Carlton Studios, a prominent graphic design firm. Among his best-known works are posters designed for Shell, under the art direction of Jack Beddington, the brother of camoufleur Frederick Beddington. In later life, his major accomplishments were in the area of exhibition design. During World War II, he was associated with the Camouflage Development and Training Center at Farnham Castle, England.

MacCarthy (1972:146): "Not for nothing was James Gardner Chief Development Officer for Army Camouflage during the war: in 'Britain Can Make It' he camouflaged austerity. He made a post-war fantasy. The public was convinced."

...

Woodham (1995:313-314): "Selected for camouflage work alongside other notable contemporaries such as the couturier Victor Stiebel and the set designer Oliver Messel, he became 'Chief Development Officer' under Colonel [Richard McLean] Buckley. A perfect vehicle for Gardner's wide-ranging talents, such work involved the design and construction of prototypes for inflatable 'dummy' tanks and landing craft as well as the doctoring of photographs and documents for the purpose of wartime 'misinformation.'"

Goossen (1973:35): "While [American artist Ellsworth] Kelly was staying in France at the villa La Combe during the summer of 1950, his eye was drawn to the way the shadow of a handrail broke on the irons treads of the stairway leading down from his balcony room. His attraction to this phenomenon probably goes back to his experience [during WWII] of the camouflage nets woven with strips of oznaberg in random patterns to disrupt the tell-tale shadows of guns and tanks." See umbrella camouflage.

Jacob Bronowski (quoted in Gardner 1993:13): "...the brain...operates by rules of logic that mirror the logic of Nature, so it is inherently mathematical...

I am afraid, G, that someone must have muddled the mathematical linkages in your brain at a very early age."

•

Robert (Franklin) Gates
(1906-1982)
American artist

Born in Detroit, he attended the Detroit School of Arts and Crafts, then the Art Students League in New York. He worked as a WPA artist, and served as a US Navy camoufleur during WWII. He was head of the Art Department at the American University in Washington DC. His wife was artist Margaret Casey Gates (1903-1989), whose papers are in the Archives of American Art, as are his own.

•

Everett L. Warner (n.d.), referring to WWI: "For a nation so clever and versatile as the French, their achievement in ship camouflage was a trifle disappointing. Inferior to [the] Brit[ish]."

Pierre-Louis Gatier
(January 12, 1878-October 15, 1944)
French artist, naval camoufleur

Born in Toulon, Gatier was a watercolor painter who designed ship camouflage during World War I in the Camouflage Section of the French Navy at Rochefort. Following the war, he made studies of contrasting colors (resulting in works that have some resemblance to Josef Albers' *Homage to the Square* paintings), and irregular, broken shapes. From c1931-1934, he served as a camoufleur for the French Navy, in the course of which he worked on plans to camouflage the naval port at his hometown of Toulon. There is considerable information about him in Blechman (2004:29, 32-33; 168), including reproductions of his paintings of camouflaged ships. He was awarded the Legion of Honor.

•

Farrer (1918:67-68): "In certain respects the war [WWI] is nothing but a glorious, gigantic game of hide and seek—*camouflage* is nothing else. It is not only the art of making things invisible, but also of making them look like something else...What glory it must be to splash your tents and lorries all over with wild waggles of orange and emerald and ochre and umber, in a drunken chaos, until you have produced a perfect Futurist masterpiece which one would think would pierce the very vaults of heaven with its yells."

Above World War I camouflaged German field artillery. *Facing page* A German airplane from the same period, with a lozenge-patterned camouflage scheme.

John Gaitha Browning (Toliver, 1994:39): "We're at camouflage work again...It isn't difficult work, but so few of our officers or men seem to have any idea of it at all. With a good group of workers, we got along well with the project today. At least one of the jobs is the best I have worked at so far, possibly because I stuck with every phase of it clear through. I hate to leave a camouflage job and come back to find that some monstrosity has been constructed that wouldn't fool a one-eyed man who was five miles away."

Adolphe-Louis Gaussen
(May 12, 1871-1954)
French painter

Born in Marseilles, Gaussen was a landscape painter who served as an artist for the French Navy. He is listed in Coutin (1988) as a World War I French camoufleur.

•

Jean Genet
(December 19, 1910-April 15, 1986)
French writer

In his book *Miracle of the Rose* (1946:12), which is often described as a semi-fictional memoir of Genet's detainment in a penal colony and prison, he mentions camouflage several times, having been "assigned to the workshop where camouflage was being made for the German army which was then occupying France."

•

Paul Gerchik
(1913-August 13, 1998)
American artist

Born in New York, he studied at the Art Students League. He was President of the Artists' Union during the Depression. He taught at Brooklyn Museum of Art, moved to Los Angeles prior to World War II, then joined the US Army. *LA Times* ("Paul Gerchik" 1998): "Named an official war artist, he drew political cartoons and posters to promote the war effort, taught camouflage to military personnel and pioneered art therapy for wounded soldiers in military hospitals." He later worked as an art director and production designer for various motion pictures, including *Spartacus*.

•

Charles Gerhardt

(March 19, 1863-June 5, 1957)
American military officer

Gerhardt was born in Baltimore MD, and educated at the US Military Academy. He was an active participant in both the Spanish-American War and World War I. He also served as a Professor of Military Science and Tactics at Cumberland University in Lebanon TN (1894-97). He was a brigadier general in the US Army, and the father of Major General Charles H. Gerhardt.

National Encyclopedia of American Biography ("Charles Gerhardt":54): "While at the [Cumberland] university he was secretary of the US Military Wheelmen, a civilian organization for the development of the bicycle for military purposes. He devised a uniform of limited visibility which was one of the first steps made in this country in the use of camouflage. The uniform used by the wheelmen in Lebanon was of a drab color and consisted of a long buttoned-up jacket, trousers, high leggings, and gauntlet gloves."

•

German camouflage

Bement (1917:459): "Probably the first practical use made of it [camouflage] was the placing of black bags in sandbag breastworks by the Germans to confuse the enemy into thinking them loop holes, the enemy fire thus being partially distracted from the real openings."

Faulker (1957): "Once across the Marne I had the chance to study the German camouflage in their captured gun positions. It was simpler and more effective than ours. The guns were usually set up under trees near a road where ammunition could be dumped. A wired-in duckboard walk led to the guns, which sat in an enclosure of wire, and no man was allowed to step outside these boundaries. The strong point of their method was camouflage discipline, a thing of which we in the American Army were only dimly conscious."

New York Times ("Allied War Salon" 1918): "A very interesting section of the exhibition [held in New York at the American Art Galleries, December 9-24] is given to paintings and drawings by the marine camoufleurs of the US Shipping Board, Second District. Nearly all the examples have to do with what is called the 'baffle' system of camouflage as distinguished from the concealing system. By this baffle painting the outlines of the ship are distorted and the submarine watchers are misled as to its size, character, and course."

Above Detail from a photograph (c1897) showing two hunters, with the one on the left wearing a ghillie suit. Courtesy of the US Library of Congress, Prints and Photographs.

Sun Tzu (*Art of War*, 500 BC): "When able to attack, we must seem unable; when using our forces, we must seem inactive; when we are near, we must make the enemy believe that we are far away; when far away, we musst make him believe we are near. Hold out the bait; entice the enemy. Feign disorder and crush him."

Hartcup (1980:14-15) states that, during World War I, the first to experiment seriously with aircraft camouflage were the Germans, and that their methods seemed to be influenced by the writings of Abbott H. Thayer.

•

A(lfred) H(orace) Gerrard
(1899-1998)
British sculptor

Born in Cheshire, England, he attended Northwich Technical School, then served as a bomber pilot in World War I, and a camoufleur during World War II. In the 1920s, he studied under Henry Tonks at the Slade School of Art, where he himself taught sculpture both prior to and after World War II.

Buckman (1998): "With less than two hours' instruction he flew solo in Morris-Farmers and FE2Bs, bamboo structures and wings covered with cellulose-painted linen fabric. Because night bomber pilots were scarce he was sent up solo without night-flying instructions. For six consecutive nights fellow fliers in his billet were killed. On one occasion Gerrard took up a Morris-Farmer and noticed people running about below holding a wheel. Half his undercarriage was hanging off, so he crash landed but suffered an injured spine. During the Second World War, while working in camouflage, he again crash-landed, the resulting jolt curing the earlier injury. He was discharged in 1942 with multiple injuries after a plane crash while photographing. When surgeons wanted to amputate his badly crushed right arm, in a semi-conscious state he swore loudly at them, and thus saved it to sculpt again. Between 1944-45 he was a war artist."

•

Gestalt theory
Holistic branch of psychology

Gestalt theory is an influential branch of modern psychology that orginated in Germany prior to World War I. Its founders (Max Wertheimer, Kurt Koffka, and Wolfgang Köhler) were primarily interested in perception and problem-solving. They and their students investigated figure-ground relationships, formulated principles of perceptual organization, and stressed the contextual nature of experience ("The whole is greater than the sum of the parts"). The word *Gestalt* (pronounced *gush-TALT*) is now commonly used to imply an overall structural pattern, a layout. It is all but synonymous with "design," and in Germany, design schools are often referred to as *Hochschules für Gestaltung*. The Gestaltists were always interested in camouflage, and, in 1932, American psychologist Angeline Myra Keen published an article (Keen 1932) in which she discussed the resemblance between Abbott H. Thayer's camouflage principles (which he called "laws of disguise") and the findings of the Gestalt psychologists (who postulated "laws of perceptual organization" or "unit-forming factors"). See also Rudolf Arnheim.

•

Ghillie suit
Camouflage outfit that resembles foliage

Ghillie suits are usually said to have originated in Scotland, to enable a hunter to blend in with surrounding foliage. The practice was adopted by snipers in the British military at the end of the 19th century. Manufactured ghillie outfits are readily available now and widely used by snipers and hunters, although it is common for individual soldiers to construct their own by attaching twine and burlap strips to an underlying fabric.

•

Ghost Army
US Twenty-third Headquarters Special Troops, WWII

Goossen (1973:12): "The 603rd Engineers Camouflage Batallion was, according to official records, 'composed mainly of artists from New York and Philadelphia with an average IQ of 119. At full strength it comprised 28 officers, 2 warrant officers, and 349 enlisted men, the last of whom included [American painter Ellsworth] Kelly.'"

In Naversen (1989:141), when George Diestel was asked for names of others who served with him in the Ghost Army, he mentioned the following (in no apparent order): David Gray, Harry Reader, Frederick E. Fox, Cecil Spangler, Charles Kagel, Ned Harris, John Garvey, John Mason, Bill Blass, Art Kane, Arthur Singer, Ellsworth Kelly, Arthur Shiltone, John Hapgood and Ed Haas.

•

Ted Gilien
(March 17, 1914-1967)
American painter

Gilien was born in Brooklyn, New York. He studied at the Art Students League and the National Academy of Design, then worked as a muralist for the WPA. He was a combat artist and camoufleur during World War II.

Betty L. Hoag, in conversation in Charles W. White (White 1965:9): "Your speaking of being a camouflage artist and fighting floods instead made me think of Ted Gilien. I interviewed him earlier this week and he was a camouflage artist sent to New Guinea and what he did was to move ammunition dumps."

•

Colin (Unwin) Gill
(May 16, 1892-November 16, 1940)
British painter

Born in Kent, he studied at the Slade School of Art and in Rome. During World War I, he served in France as a camoufleur with the Royal Engineers (1915-1918) then as an Official War Artist (1918-1919). In Blechman (2004:130-131), there is a color reproduction of a magnificent painting by Gill of camouflaged artillery (1919).

•

Henry McBride (from his 1918 notebooks, as quoted in Watson 2000:118): "The ship at the opposite dock was camouflaged. In spite of being in New York all winter it was the first I had seen. The moment we swung into midstream for the voyage we saw others. In the lower harbor there were many small boats all camouflaged. Only one seemed to lose itself in the background and that was a boat that happened to be colored like a landscape and could hardly be seen against Staten Island. In the five days we have been at sea, I have seen the eight boats of our convoy in all sorts of atmospheric conditions but never once have I seen them much disguised. It is possible that zigzag shadings on boats might confuse the eye that directs the torpedo, but to an amateur it would appear that the mass and outline of the ship were at all times perfectly visible. The one first and sure effect was that the boats were completely cheapened by the camouflage."

...

Gimpel (1986:29) in a diary entry dated May 30, 1918: "[A dazzle-painted ship was] like an enormous cubist painting with great sheets of ultramarine blue, black, and green, sometimes parallel but more often with the sharp corners cleaving into one another, and although you don't quite make it out, you can divine a reason, a plan, a guiding principle, a scheme."

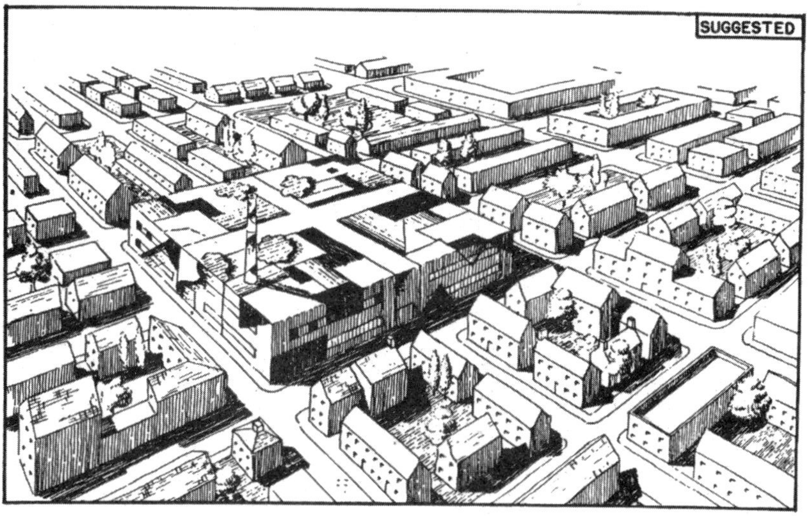

In both World Wars, much of industrial camouflage consisted of altering factories, architectural landmarks or other important structures, to prevent them from being recognized from the air. Shown here are World War II US government drawings that demonstrate an inept use of disruptive patterning (top), and the more effective use of coincident disruption (bottom), in which disruption and blending are used in combination. These and other aspects of industrial camouflage are discussed in detail in Breckenridge (1942) and Wittman (1942). Drawings from US War Department, *Civilian Defense: Protective Concealment* (1942:9).

Charles-Félix Gir

(November 1, 1883-1941)
French painter, illustrator

Gir was born in Tours, France. He is listed in Coutin (1988) as a World War I French camoufleur. Married to actress Jeanne Fusier-Gir, he was primarily known for his Art Deco style paintings of dancing figures in theatrical settings.

Beston (1919:123-124): "...destiny put in my path just the man I wanted to see, the captain of a British submarine. 'What do you think of camouflage?' I asked.

'Well,' he answered, after a pause, 'I can't remember that it ever hindered us from seeing a ship. Visibility at sea strikes me as being more a matter of mass than of color. The optical illusion tricks are too priceless silly. Must amuse the Huns [Germans]. You see if the eye does play him false, Fritz detects the error with his gauges.'"

•

Lancelot (Myles) Glasson

(1894-1959)
British painter

Hartcup (1980): "Glasson had served under Wyatt in France [during World War I], where he had lost a leg, and after the war had become a successful painter in the academic tradition. Convinced that camouflage was important, he became a source of inspiration to the same group of artists he gathered around him before the end of the year [1938]," including G.B. Solomon, Cosmo Clark, Richard Fairbairn, and Richard Carline. He was the grandson of British printmaker Myles Birket Foster.

Above The cover of the package for Jon Gnagy's famous *Learn to Draw Outfit*, which enjoyed great popularity in the 1950s and 60s.

•

Jon Gnagy

(1907-1981)
American artist, television personality

Kurt Vonnegut (in Allen 1988:171-172) [recalling World War II]: "When the Germans finally showed themselves, we saw they were wearing white camouflage suits [in the winter]. We didn't have anything like that. We were olive drab. No matter what season it was, we were olive drab."

Gnagy was born into a family of strict Hungarian-Swiss Mennonites near Pretty Prairie KS. From an early age, he learned rural crafts, then taught himself to draw and paint. At age seventeen, he began work as an art director for a public relations firm in Tulsa OK, then moved to Wichita KS, where he studied aerodynamics and illustrated aircraft ads. During the Depression, he took evening classes at the Kansas City Art Institute, then moved to New York with his wife and daughter, in search of art-related work. In the years that followed, he worked intermittently as an art director and advertising artist in New York and Philadelphia. During World War II, he instructed groups on exhibit preparation for the National Camouflage Exhibit at the Franklin Institute, under the direction of the Fourth US Army District. See Wallace (1942).

Using his newly acquired teaching skills, Gnagy offered lecture demonstrations to art schools and teachers colleges on art-related subjects like color harmony and the psychology of color vision. In May 1946, he began his television appearances through NBC, in which he taught the audience "how to draw" during ten-minute segments of a variety show called *Radio City Matinee*. Several months later, he was given his own program titled *You Are An Artist*, the success of which then led to the production of his immensely popular Jon Gnagy *Learn To Draw* instructional kits.

•

(Ralph) Harper Goff
(March 16, 1911-March 3, 1993)
American artist, musician, and actor

Born in Fort Collins CO, Goff was a muralist for the WPA in the 1930s. He studied art at Chouinard Art Institute in Los Angeles, then worked as a magazine illustrator in New York. In London in 1951, he happened to meet Walt Disney in a model-making shop, and the same year joined the Disney production team in Los Angeles. He made early renderings for Mickey Mouse Park, later renamed Disneyland, including Main Street USA and the Jungle Cruise. He also played a vital role in the production of such film classics as *Willy Wonka and the Chocolate Factory*, *Fantastic Voyage*, *20,000 Leagues Under the Sea*, *Sergeant York*, *Captain Blood* and *Pete Kelly's Blues*. He also worked as an actor, and as the banjo player in a seven-piece Dixieland band. In 1993, he was elected posthumously to the Disney Hall of Fame.

In Blechman (2004:34), Goff is described as having served during World War II as a paint and camouflage advisor to the US camouflage section at Fort Belvoir. He is quoted as saying that "We developed a series of basic colors, government issue you know, useful for camouflage."

In Naversen (1989:150), he is quoted as follows: "While working for Bishop and Company, I was making paint and working on a do-it-yourself painter's kit. While working on the brochures, I was contacted by the Army. We had done some advertising for the War Department. They needed a paint that wouldn't last very long for some camouflage work. It started there."

•

Erno Goldfinger
(November 1, 1902-November 15, 1987)
Hungarian-born architect and designer

Born in Budapest, Hungary, he became interested in architecture after reading *Das Englische Haus* by German architect Hermann Muthesius. Following World War I, he moved to Paris, where he became acquainted with Le Corbusier, Auguste Perret and Mies van der Rohe. In the early 1930s, he married the British heiress Ursula Blackwell, and established an architectural practice in London.

In Warburton (2003), there is an amusing lengthy account of the combative relations between Goldfinger and other members of the Industrial Camouflage Unit, with which he was briefly associated (mostly as an irrate landlord), and which included William Stanley Hayter, Roland Penrose, John Buckland Wright, and Julian Trevelyan.

It is said that novelist Ian Fleming named Auric Goldfinger, the villain in his James Bond novels, after Erno Goldfinger.

•

Zacarias Gomez
(dates undetermined)
American artist?

As discussed in Williams (2002:68), during World War I Gomez submitted a proposal for US ship camouflage that featured irregular, horizontal zones of color. Its apparent purpose was the simulation of the bands of stratification of clouds and the surface of the water. After the model was tested by Loyd A. Jones and his associates at the Eastman Kodak Research Laboratories, the proposal was abandoned. For details, see Jones (1919).

•

Above The dust jacket of Henrietta Goodden's study of World War II British camouflage artists (Goodden 2007).

Charles Bittinger (1940:1398): "For some reason nature has camouflaged most living things which move. If one believes in evolution, the natural conclusion is that the survival of the species is due to some extent to this protective coloring. However, an apparent anomaly in this reasoning is the human race. Man in his present birthday aura is about the only animal nature has neglected to camouflage. But man has been endowed with intelligence, and he has survived by producing the equivalent of camouflage for defense, and for offense has developed instruments to deal with his enemies at a safe distance."

Above Model of a WWI ship camouflage scheme, as proposed by Zacarias Gomez, from Jones (1919).

Henrietta Goodden
(1948-)
British textile designer, historian

Born in London and raised in East Anglia, she is the daughter of Robert Goodden, who was a major contributor to World War II British camouflage. She taught for many years at the School of Fashion and Textiles at the Royal College of Art, where her father had also been on the faculty. In 2007, she published *Camouflage and Art: Design for Deception in World War II* (Goodden 2007), in which she documented the extraordinary contributions by artists associated with the RCA to the development of wartime camouflage.

•

Robert (Yorke) Goodden
(May 7, 1909-March 24, 2002)
British architect and designer

Goodden was the father of Henrietta Goodden (see Goodden 2007). An architect and silversmith, he served in the British camouflage unit during World War II, in the company of other artists and architects, including Robin Darwin, Hugh Casson and David Pye. As the war ended, he was appointed to a team, along with fellow camoufleurs James Gardner and Basil Spence, that comprised the Council of Industrial Design. From 1948 until his retirement, he taught silversmithing and jewelry at the Royal College of Art (Margrie 2002).

•

Paul Goodman
(September 9, 1911-August 2, 1972)
American sociologist, writer

Born in New York, Goodman was the brother of architect Percival Goodman, with whom he wrote *Communitas*. His own most influential book was *Growing Up Absurd: Problems of Youth in the Organized System.*

Among his various writings are cryptic references to an "architect-turned-camoufleur," a character partly derived from the experiences of his architect brother, who formed a civilian camouflage company during World War II. He used this idea in an essay on "The Meaning of Abstraction in Literature" and in *The State of Nature* (Goodman, 1946).

Percival Goodman

(January 13, 1904-October 11, 1989)
American architect, urban theorist

Everett L. Warner (1919:102): "It [camouflage] has long been practiced in the realm of feminine wearing apparel, and some of our most successful camoufleurs have been among the ladies. When a woman appears on the street dressed in vertical stripes, she is certainly not trying to be invisible, but it is a safe guess that she is trying to give the impression of a figure more slender than she actually possesses, and she is usually quite successful. In the later British 'zebra' period [of WWI] this identifical device was employed on ships—vertical lines to raise the apparent height of certain parts and horizontal lines elsewhere to give the opposite effect."

Above World War II disruptive camouflage, as applied to the *USS Lamson* (1944). Courtesy US Naval Historical Archives (Navy photo 3092-44).

Beston (1919:120): "The subject which stirs up all the discussion is camouflage. To ask a crowded table: 'What do you think of camouflage,' is to hurl a very apple of discord down among your hosts."

The older brother of Paul Goodman, he was primarily known for his design of more than 50 synagogues. He taught architecture at Columbia University for more than 25 years.

In Elman (2001:33), he is described as having volunteered for the US Air Force in late 1941. "After failing the physical, he formed a camouflage company to do his bit against Hitler..." According to the same source (Ibid.:165): "So he turned to camouflage, a necessary part of the war effort in which he could help. He formed the Camouflage Engineering Company with [British printmaker] S[tanley] W[illiam] (Bill) Hayter...David Reeves, a mathematician and furniture builder whom they [Goodman and Hayter] had known in Paris; Hector Hoppin, Jungian analyst with excellent government connections (perhaps also a Paris connection); Milton Gendel, an art historian; Jack Straus, the former client who was a stockbroker and thus believed to be helpful on business matters; plus Kate Abbot, an artist, who was Percy's girlfriend at the time...The company bid successfully and completed a number of projects for the US Military, including camouflaging Fort Belvoir in Virginia. Percy and Bill Hayter worked out theories and methods of camouflage based on their knowledge of visual effects, and the group then went on location to supervise the actual work." The business failed financially when they advertently underbid a large project by an exorbitant amount. Subsequently, Hayter co-founded in London the Industrial Camouflage Research Unit with Roland Penrose, Julian Trevelyan, John Buckland Wright and Erno Goldfinger. See Stanley William Hayter.

•

Jan Gordon
(March 11, 1888-February 2, 1944)
British artist, author, musician and lecturer

Born Godfrey Jervis Gordon, he was the author of *Modern French Painters*. He was educated at Marlborough College , then, having decided to become a mining engineer, at the Truro School of Mines. He also studied briefly with Frank Brangwyn at the Kensington School of Art. With his wife (née Cora Turner), Gordon toured various countries, observing, writing, drawing and collecting folk music, referring to themselves as "vagabond artists." They first came to New York in 1927, touring parts of the US in the same spirit, and then writing about it (published January 1928).

New York Times ("Jan Gordon Dies" 1944): "In the first World War he was chief designer of naval camouflage and also saw service with the Red Cross in Serbia during the typhus epidemic."

Jan and Cora Gordon website: "[During World War I] Jan was classed as unfit for military service but volunteered for service with the Red Cross in Serbia... [Later] Jan, who by then had been 'directed' into war work, spent 18 months making airplane engine parts at Rolls Royce, Derby...He then went on to work on dazzle camouflage schemes for ships at the Royal Academy, for which he was commissioned a lieutenant in the Royal Navy Volunteer Reserve."

•

Arshile Gorky
(1904?-July 21, 1948)
Armenian-born American painter

Gorky was born near Van, Turkey. He fled the country in 1915, to avoid being killed, and escaped to Russia. He immigrated to the US in 1920, changed his name from Vostanik Manoog Adoyan to Arshile Gorky, and claimed to be related to Russian novelist Maxim Gorky. His daughter Maro married Matthew Spender, the son of British writer Stephen Spender.

Ethel Schwabacher (1957): "[During World War II] Gorky applied to the draft board to serve in camouflage, but was rejected as past the draft age.

He then organized a course in camouflage at the Grand Central School of Art through the cooperation of Mr. [Edmund W.] Greacen, its director. For this course, he prepared with his usual eagerness, voraciously consuming all literature on the subject, a literature that covered 'data on protective coloring in zoology, optical illusions in the physics of light, and visual reactions to movement in Gestalt psychology'...

Gorky threw all his energies into this task, delivered several excellent lectures, constructed models and designs to demonstrate the principles of camouflage, and stimulated his students to make original contributions. But somehow the course bogged down. The venture was, unfortunately, not a success and had to be abandoned."

Julian Levy (Seitz 1972:8): "There was much thought given to the problem of a class in camouflage he organized as a contribution to the war effort. The unconscious is, so to speak, the domain of camouflaged objects, and Gorky was to discover that if the realistic object can be camouflaged, so can the unreal, or 'Surreal' object be decoded and decamouflaged."

In an autobiographical chronology (Reinhardt 1975:6), American painter Ad Reinardt states: "1943 Refuses to help Arshile Gorky start a camouflage school."

In an interview with New York gallery director Betty Parsons (Parsons 1981), conducted by Gerald Silk, the following brief exchange occurs:

"Ms. Parsons: I used to work with [Arshile] Gorky. He taught me a hell of a lot.
Mr. Silk: At the camouflage school was that?
Ms. Parsons: Yes. I went there. Also, we used to have a little drawing class together."

•

Clarence W. Barron (1919:126): "That great modern hotel, the Adelphi at Liverpool, is empty except for a few stragglers in the tea room, but every few days 30,000 [US] troops debark, camp on Lord Derby's estate, Knotty Ash, and soon move south to accommodate the arrivals from the next convoy of more than a dozen ships that in their zebra stripes and camouflage color gyrations will give you optical delusions and brain problems even when you view them from the land and they lie still in the sunlight at their anchorages."

...

David Grubbs (2005): "'Camoufleur' is a term of derision coined by the abstract artist Ad Reinhardt. It was his name for an artist who gives a programmatic title to an abstract work. Barnett Newman, having entitled a series of paintings *Stations of the Cross* was a camoufleur."

(Leon Allen) "Goose" Goslin
(October 16, 1900-May 15, 1971)
American Major League baseball player

In 1932, while playing left field for the St. Louis Browns, Goose Goslin had a terrible year. In an attempt to raise his batting average, he introduced what journalists called a "camouflaged bat." Designed by Willis Johnson, the team's secretary, it was simply a standard bat with black and white lengthwise stripes.According to news reports, Johnson was planning to design comparable bats

Above A news photo (c1932) showing Goose Goslin with his "camouflaged" bat.

"decorated with cross-rings, blocks and triangles," for use by other team members—until use of the bats was outlawed. See "Camouflaged Bat" (1932).

•

Tom Gourdie
(1913-2005)
Scottish artist

Educated at Edinburgh College of Art. During World War II, he was a prolific painter of everyday life in the RAF, and an authority on handwriting. "He became involved in camouflage work and the production of three-dimensional maps."

•

Oliver Grahame-Hall
See Claude Muncaster

•

Laurence W. Grant
(1886-1930)
American painter

Born in Cleveland OH, he studied with Jean-Paul Laurens at the Académie Julian, and with William Merrit Chase at the Art Students League. He is mentioned as a World War I camoufleur by Faulkner (1957) in his radio talks.

•

George Gray
(1907-?)
American mural painter

Gray was born in Harrisburg PA. In Reit (1978) he is described as having worked on camouflage at Richmond Air Base in Virginia, with Jo Mielziner during World War II.

According to Coleman (2008), he was "a combat artist for the US Coast Guard during World War II and also did a lot of private, commissioned art projects before and after World War II. He was a staff artist for the *New York National Guardsman* and such publications as the *US Infantry Journal*, creating front cover paintings..." In 2008, his mural of a World War I battle scene, originally made for the Hendrick Hudson Hotel in Troy NY, was installed at the New York State Military Museum in Saratoga Springs NY.

Joseph Gray

(dates undetermined)
British illustrator

In Lewis (1994:55), he is described as "an elderly and rather bibulous illustrator...whose greatest claim to fame was that later on he nearly burnt down Farnham Castle, which had become the Camouflage Center." He is also mentioned in Goodden (2007:64) as having developed a "steel wool camouflage material (a different product from the domestic version)."

•

George Bernard Shaw (*Daily Chronicle*, March 5, 1917): "I was in khaki by way of camouflage."

...

Everett Warner (n.d., Part 2, p. 3): "John Gregory, a sculptor, never lost sight of the relation of design to form and though not our most prolific designer was easily the most thoughtful student of the question, and helped greatly in solving many knotty problems."

Ray Greenleaf

(1877-February 14, 1950)
American artist and color consultant

He studied at the Art Students League in New York, then worked as an art director for Ward and Gow Advertising Company. He was a member of the Society of Illustrators. *New York Times* ("Ray Greenleaf" 1950): "In the first World War he aided the national effort as a camouflage artist." He is listed in Gallatin (1919:35) as a member of the executive committee of the Division of Pictorial Publicity.

•

Above Photograph of John Gregory (n.d.) by Peter A. Juley & Son. Courtesy Photograph Archives, Smithsonian American Art Museum (J0001671).

John Gregory

(May 17, 1879-February 21, 1958)
British-born American sculptor

Born in London, Gregory was the son of a shipmaster. He came to the US in 1893 to join his mother, who was living in Washington DC. He studied in New York at the Art Students League (1900-03), and, in Europe, at the École des Beaux-Arts (1904-06) and the American Academy in Rome (1912-14). At various times, he worked as an assistant to J. Massey Rhind, Hermon A. MacNeil, Gutzon Borglum and Herbert Adams. He taught sculpture at the School of Architecture at Columbia

Above This painting of WWI camouflaged artillery was credited to S. Ugo (his signature is in the lower right corner), when it was published in *The Sphere* (August 3, 1918:86-87), but in Laffin (1991: facing 123) it is attributed to a different artist, Fortunino Matania.

University, and the Beaux-Arts Institute of Design in New York. In 1915, he established his own sculpture studio in 1915, and produced award-winning, commissioned work throughout his life. *National Encyclopedia* biography: "During the First World War he was assistant inspector of naval construction camouflage for the US Navy Department, Washington DC, designing dazzle camouflage." He designed bronze medals for the Cunard Building interior in New York (working with Barry Faulkner and Ezra Winter), and Shakespearean figures for the Folger Library in Washington DC.

•

Ralph E. Griswold
(dates undetermined)
American architect

A World War I camoufleur, he is listed in Embury (1917:138) as "a landscape architect in New York," who was "a Cornell graduate and has been practicing in New York."

•

William Griswold
(dates undetermined)
American painter

Gestalt psychologist Kurt Koffka (1935:77): "As a matter of fact, it is possible to interfere with the real units in such a way that they will no longer look like units, an effect which we try to produce when we want to conceal certain well-known objects. If a gun is covered with paint in such a way that one part of it will 'fuse' with the bole of a tree, another with the leaves, a third with the ground, then the beholder will no longer see a unit, the gun, but a multiplicity of much less important objects. Camouflage was an art well developed during the war [WWI], when even big ships were destroyed as real units in the behavioral world of the scouting enemy."

...

Anon (*New York World*, September 1917): "Little was thought of camouflage at the onset of the present big conflict [WWI]. There were officers' white-kid gloves—fatal targets for German snipers—and waving plumes; the burnished cuirass and pennoned lance. Then the two contending lines dug themselves in and locked horns. Concealment became all important—concealment from the aero with the eagle eye; from the artillery observation-station, binocular-eyed; from the practiced glance of the sharpshooter and the keen vision of the patrols. Artists in the ranks busied themselves, a new branch of the art military was born—camouflage."

He served with the Ghost Army during World War II. He is mentioned in Goossen (1973:13) as having studied architectural history. He spoke French, and was one of the many American GIs who tried to meet Picasso after the liberation of Paris. "But he did not take [his friend Ellsworth] Kelly along because he 'would not look good in a drawing room.' (Picasso, in his cold studio in rue des Grands Augustin, had no 'drawing room.')." Later, Goossen (1973:103) also notes: "Bill Griswold actually posed for Picasso, who painted him as a French sailor."

•

Louis Guingot
(January 2, 1864-December 16, 1948)
WWI French camoufleur

Born in Remiremont, France, he studied in Paris at the École des Beaux-Arts and the École des Arts Décoratifs. He was influenced by Victor Galland, director of the Manufacture des Gobelins. Guingot married in 1892 and settled in Nancy. He is buried in the cemetery at Bouxieres-aux-Dames.

Cordier (1980:174): "During this time, Guingot participated—along with Daum, Majorelle, Gallé, Prouvé, Vallin, Corbin and Friant—in an important artistic movement that profoundly marked his age and ours, called l'École de Nancy."

Cordier (Ibid.:174): "1914: Louis Guingot was now fifty years old. The 'revenge' was approaching. He knew the all too visible blue and red uniform of our soldiers, and he foresaw the impending massacre. The process of permanent coloring having been resolved, only one step remained for the creation of camouflage.

Eugene Corbin, businessman and patron of the arts, saw in this humanitarian invention an appreciable financial opportunity. He persuaded the army to adopt the idea (while Guingot would continue to hold the patent) and sold them rolls of canvas. The camoufleurs would have a chameleon as their insignia. Two workshops were established: the first in Nancy, the second in Toul.

This story could be one of fortune and glory, but Guingot's heart was broken. The war had separated him forever from his wife and children, and his invention had created jealousies. Certain 'friends' sometimes took credit for an idea that was not theirs, that had never been theirs."

Blechman (2004:128) also talks about the collaborative efforts of Guingot, his brother Henri, and Jean-Baptiste Eugene Corbin in the development of hooded capes and other camouflage attire. In Blechman are two recent photographs of Corbin's original cagoules (pp. 28 and 127). Comparable outfits are depicted in a painting on the cover of *Scientific American* (September 29, 1917). See cagoules.

Left A commemorative plaque in Nancy, France, honoring the painter Louis Guingot as an originator of modern camouflage, as photographed by Cal Kowal (1991). Courtesy the photographer.

Lucien-Victor Guirand de Scevola
(November 14, 1871-March 29, 1950)
French painter

According to Kahn (1984:166), an early memo lists him as one of the first French camoufleurs in World War I, with five others. He is often cited as the originator of modern camouflage (instead of Guingot), with the result that he was put in charge of all French camouflage operations. He was known for wearing white gloves as part of his officer's dress uniform, and in Goodden (2007:10) there is a wonderful photograph of him (sporting his gloves) conversing with a French actress and a British Army officer on their visit to the Section de Camouflage in 1915. He was awarded the Legion of Honor.

According to Kahn (1984:14), Guirand de Scevola was initially a telephone operator for an artillery division. He transmitted orders from headquarters to the front line unit. He soon noticed that these artillery positions were being spotted through aerial observation, and it was then that he proposed the use of hooded, camouflaged outfits, called cagoules.

Right Abbott H. Thayer was an early advocate of high contrast, disruptively patterned sniper suits in place of khaki field service uniforms. Shown here are two Thayer photographs illustrating that proposal, which he published in an article on military camouflage (Thayer 1918). His son used the same illustrations five years later in another article (Thayer 1923). It turns out that the figure on the right is the elder Thayer himself (confirmed by superimposing this photo onto one of him without camouflage), very likely wearing William James' Norfolk hunting jacket, which the philosopher's sons (Thayer's students) had given him when their father died. To break up its shape, he had attached to it a "crazy quilt" of discarded pieces of fabric, including one of his wife's stockings.

....

Gerald H. Thayer (1923:163): "In certain cases soldiers... [used] the deceptive help of *appendages*, which figure so largely in the 'camouflage' of nature. They *stuck things onto themselves*, projecting beyond their natural outlines, and falsifying the contour of the human figure. For, just as trick painting 'breaks' the figure inward, appendages 'break' it outward."

Richard (Gerald Talbot) Guyatt
(May 8, 1914-October 17, 2007)
British designer

Son of a British consul, he was born and raised in northern Spain. He was educated at Charterhouse, then served as an apprentice at the studio of (British stage designer and camoufleur) Oliver Messel. At age 19, he began his career as a graphic designer, producing book illustrations and posters, including several memorable works for Shell. Beginning in 1948, he taught design at the Royal College of Art for 34 years.

Frayling (2007): "During the second world war, he was recruited by the camouflage directorate to help hide Scotland's factories from the Luftwaffe, a sort of advertising in reverse. Much of his time was spent in aerial reconnaissance, crouched in the hull of a Sunderland flying boat, sketching possible targets in pastel with a view to blending them into their surrounding landscapes. Several of the postwar professors of the Royal College of Art became friends while working in camouflage." For more information, see Goodden (2007).

Above Adapted from a series of lithographs in a French publication titled *La Guerre Documentée* (Paris c1920), in this case illustrating (top row) the hooded cagoules worn by the French artillery teams and disruptively patterned sniper suits; and (bottom row) a deceptive front line ploy, using life-sized painted cut-outs of charging soldiers, as a way of drawing the enemy's fire. For reasons that are unclear, this became known as "the Chinese attack."

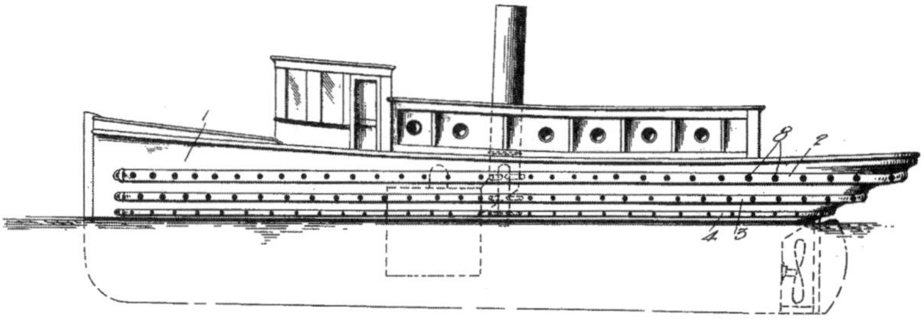
1
8
2
4
5

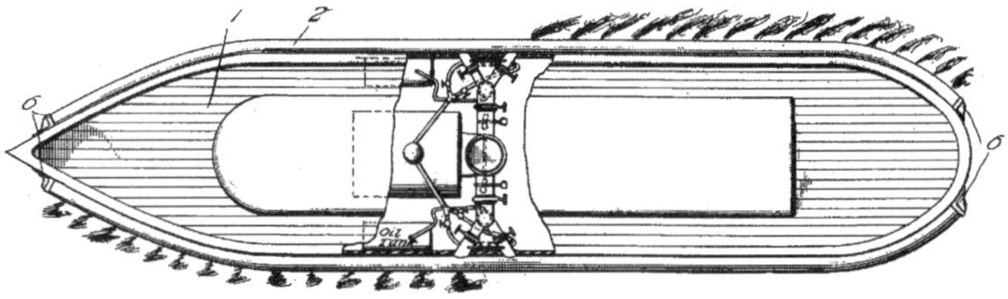
1
2
6
6
Oil Tank

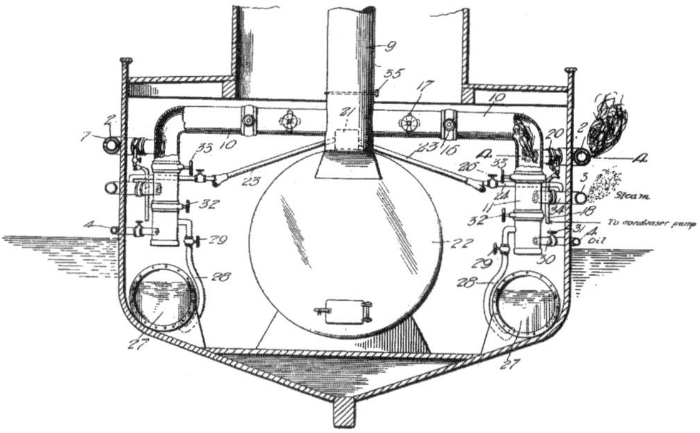
9
35
17
10
21
2
7
10
33
23
23
16
20
2
A
A
26
33
24
3
Steam
32
11
32
18
4
29
To condenser pump
31
A
Oil
22
30
29
28
28
27
27

H

Ed Haas
(dates undetermined)
American television writer

After serving in the Ghost Army during World War II, he was a developer (with Norm Liebmann) and script consultant for *The Munsters* television show in the mid-1960s.

•

Walter Hale
(August 4, 1869-1917)
American painter, journalist

Born in Chicago, Hale studied at the Minneapolis School of Fine Art, then worked as an illustrator and actor in New Orleans. It was reported in the *New York Times* ("Made Films at Verdun" 1916) that he was sketching on the French front, and making a film at Verdun. Bement (1917:460): "...to quote Mr. Walter Hale, artist and newspaperman, 'Nothing, that is of value, goes up to the front without its bath of camouflage, except the men and the horses, and they do not need it, being covered with mud.'" He died during World War I in France.

•

Facing page Drawings submitted for US Patent No. 1,301,087 titled "System of Camouflage," invented by C.H. Barkdull (1919).

Edith Gregor Halpert

(1900-1970)
American art dealer

Born in Odessa, Russia (née Edith Fivosiovitch), she immigrated to the US at age six, eventually enrolling at the National Academy of Design. She worked as a consultant for department stores, and married businessman Samuel Halpert, with whom she co-founded the influential Downtown Gallery of Contemporary Art in New York in 1926. In Pollock (2006), there is an account of her involvement in camouflage, including her support of a camouflage exhibition that opened at Macy's department store on February 5, 1943. An earlier camouflage exhibition, titled *Camouflage for Civilian Defense* had been held August 12 through September 13, 1942, at the Museum of Modern Art. Halpert arranged for Alfred H. Barr, Jr., James Thall Soby, herself and others from MOMA to visit the US Air Force camouflage research studio at Mitchell Field on Long Island.

Crowell (1921:502): "A popular idea among [WWI naval camouflage] inventors was to fit finely perforated pipes along the exterior of a hull and force water into these pipes at a good pressure so as to form a moving cloud of mist that would continuously obscure the ship." Compare with Barkdull patent on previous page; also see Douglas Volk.

•

Richard Hamilton

(February 24, 1922-)
British painter and collagist

Hamilton was born in London, and studied art at the Royal Academy. During World War II, he worked as a technical draughtsman, and was also trained in camouflage. In studies of 20th century art, it is often said that he was one of the first practitioners of Pop Art, largely because of a famous collage titled *Just What Is It That Makes Today's Homes So Different, So Appealing?* (1956).

Alister Hardy (1965:125-126): "During the first World War I became a camouflage officer, and, as a very young man, some 47 years ago, was applying the principles of animal coloration to the arts of war."

...

New York Times ("Training in Camouflage" 1942): "It is a sign of the times that the Museum of Modern Art has opened a camouflage exhibition displaying various methods of concealment, while Yale University announces that its department of architecture will start a course in industrial camouflage."

•

John Hapgood

(?-1995)
American artist

Hapgood was born in Boston, and moved to New York in 1938. He served with the Ghost Army in World War II. According to the website for an anticipated film on the subject (www.ghostarmy.org), he worked as an illustrator for *The New Yorker*, was a friend of Katharine Hepburn, and lived in the same building as Bernard Baruch. He had a house on Block Island that he named "Hap Hazard."

•

Felix Harbord
(dates undetermined)
British interior designer

Sykes (1990:40-41): "Another 'artist camoufleur,' Felix Harbord—a top interior designer in peacetime—was posted to Tobruk during the seige and found time there to paint a complete *trompe l'oeil* period interior in one of the few reasonably intact houses."

•

George M. Harding
(October 2, 1882-1959)
American illustrator, mural painter and architect

Born in Philadelphia, he was a student of Howard Pyle at Pennsylvania Academy of Fine Arts. He served on the Poster Committee for the US Navy Recruiting Service, and was an Official War Artist during World War I. He was also an artist in the US Marine Corps in World War II. See Krass (2007).

Faulkner (1957) mentions Harding and Louis Rosenberg in his radio talks in connection with the American Camouflage Corps during World War I, referring to them as "paint manufacturers."

•

Alister (Clavering) Hardy
(February 10, 1896-May 22, 1985)
British zoologist

Leytham (n.d.): "In 1914, aged 18, he [Hardy] gained a place at Exeter College, Oxford, but decided to volunteer for the army. He became an officer in the Northern Cyclist Battalion, whose duty was to patrol the coast of Lincolnshire, look out for espionage, and prepare defenses. He became a camouflage expert, due to his artistic talents (his father had been an architect)."

By Hardy's own account (Hardy 1965), he was chosen as a camoufleur by a fortuitous error. He was mistaken for another Hardy, who was a practicing artist. During World War I, he worked in the camouflage unit headed by Solomon J. Solomon. Later, he had a long, distinguished career as a marine biologist, with a particular interest in biological camouflage. He was knighted for his work.

Margaret Vandenburg (2001), speaking as a fictional Gertrude Stein: "More than any snapshot of trenches or painting of widows sobbing in vineyards, camouflage captures the essence of Europe during the Great War."

...

Fritz Heider [Gestalt psychologist] (in Henle 1973:71): "...[Cubism] consisted partly in destroying the natural units of familiar objects by opposing one unit forming factor to another. One specific part of a picture may make a good unit with a table according to one factor, but according to another factor it belongs to the wall."

Above Juan Gris, *Harlequin with Guitar* (1919). Oil on canvas. Note the use of coincident disruption (the break-up of the figure, combined with figure-ground blending) in this Cubist painting. Public domain photo.

Hardy was also interested in spirituality and extrasensory perception. In Hardy (1953:106-107), he tells about an incident during World War I by which he was first convinced of "the reality of telepathy": "I was now attached to the Royal Engineers as a camouflage officer and was attending a special course at the school set up in Kensington Gardens under Solomon J. Solomon, R.A. Mrs. Wedgwood [a woman who claimed to be psychic] came to stay in London whilst I was there and I went out to dinner with her...That afternoon at school we were doing experiments in dazzle effects. I had taken a large sheet of white cardboard and then painted it all over with a most vivid pink distemper. I was then going to cut it up into all sorts of shapes to use in our experiments, but I found it took much longer to dry than I had expected so that I had it in front of me and kept looking at it to see if it was ready for some considerable time before I actually cut it up. Again I am quite certain that no one could have told Mrs. Wedgwood what I had been doing, for no one at the camouflage school knew her or knew that I was going out to dinner with her. I had not sat down at the dinner table with her for more than a moment or two when she suddenly said, 'Oh, what have you have doing? I see a large pink square on the table in front of you.'"

Perhaps Hardy's finest book is *The Living Stream: A Restatement of Evolution Theory and its Relationship to the Spirit of Man* (Hardy 1965).

•

A.R. Harker
(dates undetermined)
British theatre designer

He served as a camoufleur in World War I. Chesney (1943:99) says the following: "And I must not forget Harker (young then), a scenic artist, who 'ticked' me off in the early days for painting the black lines between the colors of the canvas."

Harlequin
A traditional trickster or clown in the Commedia dell'Arte, typically dressed in multi-colored, diamond-patterned leotards

Richardson (1996:349): "On February 7, 1915, the artist [Pablo Picasso] sent the poet [Guillaume Apollinaire] some tips on camouflage: 'I'm going to give you a very good tip for the artillery. Even when painted gray, artillery and cannons are visible to airplanes because they retain their shape. Instead they should be painted very bright colors, bits of red, yellow, green, blue, white like a harlequin."

Penrose (1973:205): "A convoy of heavy guns on their way to the front passed them, and they [Pablo Picasso and Gertrude Stein] noticed to their astonishment that the guns had been painted with zigzag patterns to disrupt their outlines. 'We invented that,' exclaimed Picasso, surprised to see that his discoveries in the breaking up of forms should have been pressed so rapidly into military service. Realizing the possibilities of using Cubist technique to make recognition of an object less easy in warfare, he said later during the war to Cocteau: 'If they want to make an army invisible at a distance they have only to dress their men as harlequins.' Harlequin, Cubism and military camouflage had joined hands. The point they had in common was the disruption of their exterior form in a desire to change their too easily recognized identity."

•

Julian Harris
(August 22, 1906-January 25, 1987)
American architect, sculptor

Born in Carrollton GA, Harris earned a BS degree in Architecture in 1928 at the Georgia School of Technology. He then accepted a position in an architectural firm in Philadelphia, where he studied sculpture at the Pennsylvania Academy of Fine Arts in Philadelphia (1930-33). He taught in the School of Architecture at Georgia Tech from 1936 until his retirement in 1972. During World War II, he served as a major and Chief of the Special Planning Section at the Headquarters of the Army Air Force India-Burma Theatre (perhaps in the same unit in which Myer Abel served). According to Reit (1978:78), he also worked on camouflage with William Pahlmann at Jefferson Barracks in St. Louis.

•

Ned Harris
(dates undetermined)
American photographer, designer

Harris studied at Pratt Institute in New York. He served during World War II in the Ghost Army. Later, he was the co-owner of Wallack and Harris, a New York design firm. He is featured on the website for a film about the Ghost Army (www.ghostarmy.org).

•

Guy Hartcup
(dates undetermined)
British historian

After earning a degree in history at Cambridge University, he served in the Royal Air Force. As a military historian, especially in the areas of air warfare and combined operations, he has written a number of excellent books. Of particular interest is his important early study of Modern-era military camouflage (Hartcup 1980).

•

S.N. Hartell
(dates undetermined)
American architect

He is listed in Embury (1917:138) as a World War I American camoufleur, and is said to have "conducted his own office in the Architect's Building."

•

Vincent A(ndrew) Hartgen
(1914-November 27, 2003)
American artist and curator

Educated at the University of Pennsylvania, he was a long-time art professor and curator of the art collection at the University of Maine, where he joined the faculty in 1946. During World War II, he served as a camoufleur in the US Army.

Above The dust jacket of Guy Hartcup's early, excellent book on *Camouflage: A History of Concealment and Deception in War* (1980).

Crowell (1921:503): "Another novel idea [for WWI ship camouflage]—and one that possessed more practical merit—was submitted almost simultaneously by A.C. Perry Jr., and by George C. Evans, the latter a carpenter in the [US] Navy Bureau of Construction and Repair. Nearly everyone has observed the advertising signs which change their appearance when a person passes, presenting one picture or wording as one approaches, another when one stands squarely in front, and a third from the opposite direction....The proposal was to give a ship three shifting suits of camouflage by this method. The camouflage experts of the [US] Navy pronounced the scheme one that would treble the effectiveness of dazzle painting; but they rejected it on the ground that it was impracticable—too expensive, and too difficult for ordinary painters to manage."

(George) Kenneth Hartwell
(June 6, 1891-December 12, 1949)
American painter and lithographer

Hartwell was born in Fitchberg MA. He studied at the Art Students League under Kenneth Hayes Miller, George Bellows and Edward Hopper. He was an illustrator for *The Century*, *Scribners* and *Theatre Arts*. He also exhibited at Art Institute of Chicago, Society of Independent Artists, and the Carnegie Institute of Technology.

New York Times: ("George K. Hartwell" 1949:31): "In the first World War he served in the camouflage section of the Engineers Corps."

•

Herbert Haseltine
(April 10, 1877-January 8, 1962)
American sculptor

Born in Rome, he was the son of the landscape painter William Stanley Haseltine. He graduated from Harvard in 1899, then studied art in Munich, Rome and at the Académie Julian in Paris. He was known primarily for his statues of horses and other animals, notably one of the race horse Man O' War (1947), installed at the stallion's grave at Faraway Farm near Lexington KY. Among his other works are a WWI memorial called *The Empty Saddle* in London, an equestrian statue of George Washington in front of the National Cathedral in Washington DC, and various animal figures in the Field Museum in Chicago. *New York Times* ("Herbert Haseltine" 1962): "During World War I he served as a captain of Engineers in the American Expeditionary Force and helped organize the camouflage section."

He is mentioned in Love (1999:318); and in Blechman (2002:133) he and camoufleur Francis Jaques are said to have been sent to Europe in 1917 by the US government to report on what the French and British had accomplished with regard to camouflage. "They went on to form a fledgling section with a workshop in an old dancehall in Paris."

•

Ernest Haskell
(1876-1925)
American painter and printmaker

Born in Woodstock CT, Haskell studied in Paris from 1897-1899, then settled in New York, where he designed posters and painted portraits. He produced four hundred prints in 25 years. His landscape paintings, along with works

Rasmussen (1964:94): "For most people the Cubist camouflage [of WWI] was a demonstration of visual effects they had never seen before. But by the time the war was over everybody was familiar with them and new experiments with Cubist forms were made in architecture as well as other arts. One of these was the German film *Dr. Caligari's Cabinet*, made in 1919, in which the action takes place inside the brain of a lunatic where all forms are disintegrated into crooked triangles and other weird shapes. Buildings too were constructed with bizarre lines and shapes."

...

Stockbridge (1920): "The use of camouflage by both sides [in WWI] was general before the United States became a belligerent. American ingenuity, however, made one of the most important contributions to the art—it can hardly be called a science, although many of its scientific principles were ultimately developed—of camouflage. This Yankee contribution was in the camouflaging of ships."

Above Photograph of Childe Hassam (n.d.) by Peter A. Juley & Son, in the Photograph Archives, Smithsonian American Art Museum (J0105969). As reproduced in Cirker (1967), Dover Pictorial Archives.

by John Marin and Max Weber, were exhibited by Alfred Stieglitz. Fourteen of his Art Nouveau and Arts and Crafts posters (mostly *Scribner's* and newspaper work) are in the New York Public Library Digital Gallery.

Haskell was a friend of illustrator John Held, Jr., and of playwright Marc Connelly, who writes in his introduction to *The Most of John Held, Jr.*: "When we entered World War I in April, 1917 [Ernest] Haskell was commissioned to help develop the new science of camouflage. I remember the astonishment of John, Hal [Burrows] and myself one day that summer when Haskell, on leave, told us the remarkable achievements of the Camouflage Corps.

'We can paint battleships in such a way that at sea they seem invisible a mile away, or even less. We're also devising camouflage of artillery concealment and even uniforms. Come over to my flat and I'll show you a raincoat a Heinie [slang for German] can't recognize two hundred yards away.'

Haskell left his living room a moment and reappeared wearing a doughboy helmet and raincoat, both painted with a Dadaistic confusion of reds, greens, yellows, blues, and browns. He stood about 6'2" and now gave the appearance of a nightmarish Pierrot. While the rest of us stared at his weird appearance, John stretched out his hands like a man groping in the dark.

'My God!' John cried, 'Where's Ernie?'"

•

(Frederick) Childe Hassam
(October 17, 1859-August 27, 1935)
American painter

Childe (pronounced "child") Hassam was an important American Impressionist, and a member of a group of artists known as The Ten. He was an admirer of Abbott H. Thayer and was well acquainted with the latter's theories of natural camouflage. He was also an early supporter of the efforts of younger artists to serve as World War I camoufleurs.

As reported in the *New York Times* ("Childe Hassam Arrested," 1918), he was arrested and turned over to federal authorities on April 16, 1918, in Riverside Park when a New York policeman observed him making a sketch of a camouflaged US transport ship, anchored in the harbor. He was quickly released, then later used the drawing to produce a lithograph titled *Camouflage*, which he signed "arrested for this Childe Hassam April 18th, 1918."

•

Robert R. Hays
(1918-)
American painter and printmaker

Born in Pennsylvania, Hays earned his undergraduate degree at the Carnegie Institute of Technology (now Carnegie Mellon University), where one of his teachers was US Navy camoufleur Everett L. Warner. He went on to graduate school at the University of Pittsburgh, University of Southern California, and MIT. During World War II, he enlisted in the US Navy, and was trained in anti-submarine warfare, but, at Warner's request, was assigned instead to camouflage in the Bureau of Ships.

Hays (1999): "Everett [Warner] was released from Carnegie Institute of Technology in the summer of 1942...to join his old friend Commander Charles Bittinger, who was in charge of the Camouflage Section...I received a letter from Everett, mid-summer, seeking my assistance to join them."

•

Stanley William Hayter
(December 27, 1901-May 4, 1988)
British printmaker and painter

Born in London to a family of painters, he chose instead to study chemistry and geology, with the result that he initially worked for a Persian Gulf oil company, until he contracted malaria. He returned to London to recover, after which he moved to Paris in 1926 to study at the Académie Julian, and in 1933, to establish his own studio, called Atelier 17. It was during the same period that he became interested in Surrealism. In response to the Spanish Civil War, he was actively supportive of the Republican government and opposed to Fascism.

During World War II, Hayter co-founded the Industrial Camouflage Research Unit in London with Roland Penrose, John Buckland Wright, Julian Trevelyan, and (somewhat later) Dennis Clarke Hall. Trevelyan (1957:112): "It so happened that Bill Hayter and his new wife Helen Phillips had come to England from Paris in the first days of the war, and he now proposed to start, together with Roland and the engraver Buckland Wright, a group to produce camouflage designs for factories, and they asked me to join them. Already the rash of squiggly green patterns was breaking out in every part of the country, and Bill felt himself qualified to take some part in this, since he had seen something of it in the Spanish Civil War, and already his creative mind was leaping ahead with theories as to how it could be improved." See also Percival Goodman.

Rasmussen (1964:93): "The heavy hulls [of WWI camouflaged ships] became light and airy in their new harlequin dress. Incidentally, it is remarkable to see how strongly this painting—seemingly laid on quite at random—was determined by the artistic idiom of the day."

•

William Heaton Cooper
(October 6, 1903-July 26, 1995)
British landscape painter

Heaton Cooper studied at the Royal Academy School. He was a landscape painter and an authority on rock climbing. He served as a camouflage officer in World War II.

Heaton Cooper (1984:64): "In July 1940 I was called up by Air Ministry for training as a camouflage officer. This took place in Bush House, Kingsway, London, where several other artists and architects were taking the same course, including Hugh Casson, now President of the Royal Academy...

After only a month's training I was sent for a week to the North West area [of England] to see the work in action. The camouflage officer, Ronald Dann, had been there since before the war, and was very helpful in taking me around and explaining everything. One morning a message came from Air Ministry, London, saying that I was to take over the area, and that Ronald Dann was to stay on as my assistant! The position was so ridiculous, but we had to obey orders, so we agreed that I should do the designing and flying and receive any knocks that might come, while Ron managed the work of the 400 men who were spread over all RAF installations between Cheshire and Cumberland, including the Isle of Man."

•

Stockbridge (1920): "[During WWI] Yankee ingenuity found application in the devising of methods of putting on the camouflage with the utmost speed and efficiency. In the gun-carriage plant of the American Car and Foundry Company [shown above], for instance, I watched the girls in khaki overalls spraying the camouflage on the finished caissons and wheels one afternoon. The paints of different colors, in huge tanks, were deposited by means of a gigantic reproduction of the air brush, used by artists everywhere."

William (Penhallow) Henderson
(1877-1943)
American painter, architect and furniture designer

Born in Medford MA, Henderson grew up in Texas and Kansas. He moved back to New England in 1891, where he studied at the Massachusetts Normal Art School, and with Edmund C. Tarbell at the Boston Museum of Fine Arts. He then spent two years in Europe (1902-03), in the course of which he became friends with John Singer Sargent. Henderson taught at the Academy of Fine Arts in Chicago (1904-1910). He was married to Alice Corbin, the editor of *Poetry* magazine. It was he who created the murals, at the request of Frank Lloyd Wright, for Midway Gardens in Chicago. He moved to Santa Fe NM in 1916, and two years later (according to his papers in the Archives of American Art) "was employed by the US Shipping Board Emergency Fleet Corporation in San Francisco to paint camouflage onto the hulls of ships in World War I." In the mid-1930s, he worked for the WPA designing murals for the Santa Fe Federal Court Building. He was the father-in-law of wartime cartoonist Bill Mauldin. His papers at the Archives of American Art include "Camouflage Service Records, 1917-1918 (2 folders)."

•

Auguste Herbin
(April 29, 1882-1960)
French painter

Born in Quiévy, France. He studied at the Lille Academy de Bellas of Art, but rebelled against tradition and was drawn instead to Impressionism. He was the same age as Georges Braque and one year younger than Pablo Picasso. After settling in Paris in 1901, he was drawn to the Cubist experiments of those two artists and, later, to the work of Juan Gris and others.

According to ArtFact.com: "In October 1914, Herbin was drafted into the auxiliary force of the French army. He was asked to decorate the military chapel in Mailly-en-Champagne and then to design camouflage materials for the air force, which left him to continue his own [Cubist] work."

•

Claude Herndon
(dates undetermined)
American textile designer

He is listed in Reit (1978:79) as a camoufleur who worked with William Pahlmann during WWII at Jefferson Barracks, St Louis MO.

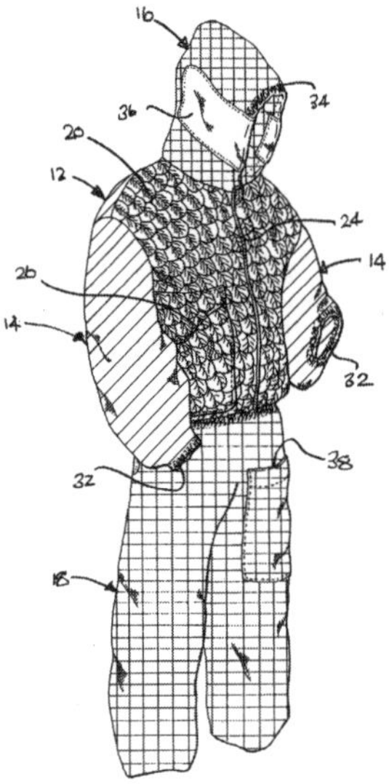

Above Drawing submitted for US Patent No. 5,920,923 titled "Waterfowl Hunting Apparel," invented by Roger and Bonnie Koehntop (1999). The outfit features surface coloring similar to that of certain bird species "in order to entice the waterfowl to approach the concealed hunter." See Mooney (1997).

There are three photographs of a model of Lewis Herzog's ship camouflage proposal in Williams (2001:38). When tested by a research team, led by physicist Loyd A. Jones, at Eastman Kodak in Rochester NY, its effectiveness was questioned. As a result, as Williams concludes: "It is not known whether the system was only tried experimentally or whether it was in fact ever actually applied... As with other small patterned schemes, paintwork maintenance would have been arduous and time-consuming."

Everett A. Herter

(?-1918)
American architect

Everett Herter was the son of a prominent artist and teacher, Albert Herter, and the brother of American diplomat Christian Herter, who would later become Governor of Massachusetts and Secretary of State. His grandfather was a well-known New York designer also named Christian Herter, who together with his brother Gustave formed the Herter Brothers interior design firm in the 19th century.

Everett Herter was the first member of the American Camouflage Corps to be killed in action in France. According to the *New York Times* ("New York Artist Killed in France," 1918:4), he was also "the first man to be accepted for the camouflage section when that branch of the engineering arm was authorized by the War Department." He is listed in Embury (1917).

•

Lewis (Edward) Herzog

(October 15, 1868-1934)
American artist

Born in Scarsdale NY, he was the son of German-born landscape painter Hermann Herzog, with whom he exhibited near the end of the latter's life. He studied art in London, Rome, Berlin, Dusseldorf, Munich and Venice. During World War I, the Herzog system of ship camouflage was one of five adopted by the US Shipping Board. It is described by Sumrall (1971:58) as consisting of broad bands of colored arcs and circles (blue, green and violet, on a gray background), with abrupt changes in direction and the avoidance of straight lines. The purpose, according to Sumrall, was "to produce a shimmering effect, like heat waves."

•

Dan Hisel
(dates undetermined)
American architect

Hisel studied architecture at the University of Kentucky and Yale University. He has won various architectural awards, including a Progressive Architecture Award Citation (2003), a Boston Society of Architects Award for unbuilt architecture (2003), and the Young Architects Forum award (2005). He owns Dan Hisel Design in Boston. He has long been interested in architecture in relation to camouflage, and especially in the concealment of the Douglas Aircraft Facility by Hollywood set designers during World War II.

Hisel's best-known architectural camouflage is the Cadyville Sauna, located in a forest setting in the Saranac River gorge in upstate New York. It has been described by Pietro Valle (on the Artland website) as "a little stereometric structure that leans on a natural cliff, incorporating it within the interior. The outer shell is completely clad in mirror panels and the path leading to it is positioned diagonally so that no person approaching the building is reflected. The sauna is camouflaged within the trees and the visitor discovers it by colliding with his/her own double at the last moment with a strong effect of displacement."

•

Laurance W. Hitt
(dates undetermined)
American architect

He studied architecture at Cornell University. He is mentioned in Embury (1917:138) as "an architect who had been driving an ambulance abroad and had been decorated by the French Government for gallantry..."

Rickard (1942:190) notes that Hitt came in as an "additional officer" (along with Carrol Berry and Victor White) to the camp of the American Camouflage Corps, which was located at Camp American University on the grounds of the American University in Washington DC.

•

Harvey C[hilds] Hodgens
(?-July 30, 1954)
American architect, theatre designer

A partner in the architectural firm of Hodgens and Hill, he designed the Allen Theatre in Philadelphia, and the Cathaum Theatre in State College PA. *New York Times* ("Harvey C. Hodgens" 1954:17): "Mr Hodgens specialized in theatre design. During World War II, he worked for the Navy on camouflage for ships."

Charles Constantin Joseph Hoffbauer

(June 28, 1875-1957)
French painter

Facing page Photo postcard of an unidentified dazzle painted US ship during WWI. Author's collection.

He was the son of a well-known architect, Théodor-Joseph-Hubert Hoffbauer, and a student of Gustave Moreau. He is listed in Coutin (1988) as a World War I French camoufleur.

•

Greville Rickard (1942:198): "*Razzle Dazzle*—Here is a phrase of confusion, generally more applicable for nighttime than daytime camouflage...As a means of dazzling and confusing the eyes of the bomber in order to take them away from whatever objective he is after, it might consist of some form of scintillating movement, say a simulation of the glistening of water, a flashing of reflectors, or any kind of wild movement calculated to attract the eyes by its strangeness or intensity."

...

Francis B[urrall] Hoffman

(March 6, 1882-November 27,1980)
American architect

Hoffman was born in New Orleans. He graduated from Harvard University in 1903, then studied art at the École des Beaux-Arts. He was one of the designers of a two-story wing that was added to the Gracie Mansion (residence of the Mayor of New York) in the 1960s. He was the Captain of the Camouflage Corps during World War I, and is spuriously mentioned in Faulkner (1973:105) as "Captain Hoffman Burrill."

•

Harry Horner [a sampling of the lyrics from a WWII camouflage musical, titled *You Bet Your Life*, performed at March Field in California](in Gilder 1944):

"It's so confusing,
But so amusing,
The ruses
One uses
Are nature's own scheme...
We're all camouflages—
Things Are Not What They Seem...
No,
Things are never quite what
they seem!"

...

James Denton Hogan

(dates undetermined)
American art professor

Hogan was on the art faculty at the University of Illinois. *College Art Journal* ("Wartime Arctivities" 1946:363): "Captain James Denton Hogan served as base camouflage officer at Camp Pinedale, Fresno, California."

•

W.S. Gilbert (in *HMS Pinafore*):
"Things are seldom what they seem,
Skim milk masquerades as cream."

Walter (Frederick Clarke) Holden

(1882-1953)
British architect

He studied at Cambridge University. According to the website for the Public Sculpture of Warwickshire: "During the First World War, he

joined the Royal Engineers, gaining the Military Cross for work he undertook as one of the first camouflage officers."

•

Henry Holzer
(December 31, 1907-July 8, 2007)
British painter

Holzer was born in Tottenham, England. He was educated at the Central School of Arts and Crafts and the Regent Street Ploytechnic in London, then taught at Hornsey College of Art. Collins (2007): "War service [during WWII] saw a posting to peaceful northern India, where he spent most of time painting the spectacular scenery. He ended the war as a second lieutenant in the Royal Artillery producing camouflage work and, in the weeks after VE Day, lithographs of anti-doodlebug defenses on the Suffolk coast while stationed at Walberswick."

•

Harry Horner
(July 24, 1910-December 5, 1994)
Austrian-born American theatre designer

Horner grew up in Vienna,where he was trained both as an actor and an architect. He immigrated to the US (with Max Reinhardt, the Austrian-American theatre and film director) in the mid-1930s, but because his English was inadequate, he gave up his acting career and turned instead to architecture. Combining the two interests, he became a set designer in New York. He was a member of the Camouflage Society of Professional Stage Designers, and joined the US Army during World War II. While in the service, he was responsible for various aspects of two stage productions, *Winged Victory*, and an instructional musical production about camouflage titled *You Bet Your Life*. See interview and other material in Naversen (1989), as well as in Horner (1943) and Gilder (1944).

•

Right Photograph of the *USS Henderson*, with dazzle camouflage designed by Stephen J. Hoxie (1919). Courtesy US Naval Historical Center (NH 53506).

John Lewis (1951:43): "He [Hughes-Stanton] served in the Middle East and went to Greece. At the end of that ill-fated expedition he was captured. Whilst in a prison cage he was shot through the jaw by a trigger-happy German sentry. By some miracle he was not killed, and after a prolonged progress through various military hospitals, he was eventually repatriated. Whilst in prison hospitals he did some moving and interesting drawings of patients."

...

Farrer (1918:67-68): "Down from the sky peers the microscopic midget of a Boche [German] plane: he sees a tree—but it may be a gun: he sees a gun—but it may be only a tree. And so the game of hide and seek goes on, in a steady acceleration of ingenuity on both sides, till at last the only logical outcome will be to have no camouflage at all. You will simply put out your big guns fair and square in the open, because nobody will ever believe, by that time, that anything really is what it looks like."

Stephen J. Hoxie
(dates undetermined)
Unknown

In the Everett L. Warner collection of WWI-era lantern slides, Hoxie is listed as having designed camouflage for the *US Transport Henderson*, but he is unidentified otherwise.

•

Burnham F. Hoyt
(1887-April 6, 1960)
American architect

Born in Denver CO, he studied in New York at the Beaux Arts Institute. He served as a camoufleur in the same unit as Barry Faulkner in France. Following the war, he moved back to Denver and formed an architectural partnership with his brother, Merrill Hoyt. He is listed in Embury (1917) as "B.G. Goodhue's head draftsman" (p. 138), and is also mentioned in *Literary Digest* ("Company A" 1917) as having been a camoufleur. He taught at New York University in 1929-1933, and designed the interior woodwork of Riverside Church in New York.

•

Benny Hubbard
(1910-March 27,1964)
American floral designer

Hubbard joined the florist company of Irene Hayes Wadley and Smythe in the 1940s. He created floral designs for advertising agencies, and lectured widely on the subject. *New York Times* ("Benny Hubbard is Dead" 1964:29): "During World War II he served in the Army Corps of Engineers as a camouflage expert."

•

G(eorge) T(emperley) C. Hubble
(c1903-1960)
Australian illustrator, advertising artist

Born in Petersham, Sydney, New South Wales, Hubble worked as an advertising salesman, comic writer and artist, producing comics signed as "Hub." Among his chief interests was science fiction. Kerr, "G.T.C. Hubble" (n.d.): "He was a camouflage artist during World War II."

•

Blair Hughes-Stanton
(February 22, 1902-1981)
British painter, wood engraver

Hughes-Stanton was a student of Leon Underwood. During World War II, he served under Geoffrey Barkas in the camouflage unit for the Middle East Forces of the Royal Engineers. With fellow artists John Hutton and P.E. Phillips, he was instrumental in planning the large-scale deception at El Alamein in the desert of North Africa. He is mentioned in Barkas (1952) and Hartcup (1980).

•

E[dward] O[lson] Hulburt
(dates undetermined)
US Navy officer

Amato (n.d.:99-100), regarding Hulburt of the US Naval Research Lab: "In these years, camouflage remained largely in the realm of optics and that meant Hulburt would be involved. He hired his next door neighbor, Washington DC artist Charles Bittinger, to help out in this regard. With Bittinger's assistance, the division determined the optimum colors and patterns for all kinds of naval vessels. Hulburt compiled these results in the 'Handbook of Camouflage,' which became the Navy's bible of camouflage during the war."

Right and facing page These WWI-era photographs were published in a book titled *Yankee Ingenuity in the War* (Stockbridge 1920). They illustrate two phases in the process of designing ship camouflage. The person in the photo on this page appears to be US Shipping Board camoufleur William Andrew Mackay.

John D. Whiting (1928:78-79): "Here, in this queer old loft [William Andrew Mackay's New York headquarters], paint-bedaubed artists were engaging in devising those outrageous patterns of dazzle camouflage which were to puzzle the world. A fleet of tiny steamer models lined the shelves, and on these designs were being painted. They were tried out on a miniature sea, against a sliding canvas sky, observed through a real periscope arranged some twenty feet from the vessels. Hugh was seated at the periscope and a model placed upon the greenish burlap sea. Marvelously real it looked against the drifting clouds, which gave to it a vivid sense of motion. The illusion was perfect."

...

Lawrence Perry (1919:138-139): "A mural painter of New York, William Andrew Mackay, who had long experimented in the chemistry of color (he is now a member of the staff of navy camoufleurs), had applied a process of low visibility to naval vessels long before war broke out in Europe...

...the first trials of the system were conducted at Newport [RI], in 1913, in conjuction with Lieutenant Kenneth Whiting, of the submarine flotilla. After a period, experiments were continued at the Brooklyn Navy Yard."

Edward Huntsman-Trout

(July 31, 1889-1974)
Canadian-born American architect

Huntsman-Trout was born in Ontario, Canada, where, following his mother's death in 1892, he was taken in by relatives in Toronto. In 1908, his adopted family moved to Los Angeles. He attended the University of California at Berkeley, majoring in botany, and then entered Harvard, where he earned a degree in landscape architecture. He eventually returned to California, where he became a prominent landscape architect. He was in France during World War I, but may not have been a camoufleur. "Edward Huntsman-Trout" (n.d.): "During World War II, Huntsman-Trout's career was briefly placed on hold. Although he was too old to join the military, he supported the war effort by working for the Douglas Corporation as a camoufleur."

•

Hugh Hurst
(dates undetermined)
British writer

Hurst was the author of "Dazzle-Painting in War Time," in *International Studio*, September 1919, pp. 93-99.

•

John (Campbell) Hutton
(August 8, 1906-?)
New Zealand artist, glass engraver

Born in Clyde, New Zealand, he moved to England in 1936 to study art. He engraved the glass panels for the Shakespeare Center at Stratford-on-Avon. According to his biography in the Online Directory of New Zealand Expatriates: "During the Second World War he served as a camouflage officer in the British forces in Egypt, Palestine, Cyprus, and the European campaign, and was mentioned in dispatches three times." He is mentioned in Sykes (1990) and in Barkas (1952:26), who describes him as a "painter, tall, slim, dark, and reminding me somehow of Dante."

•

2
11
10
11
14
22
12
13
2

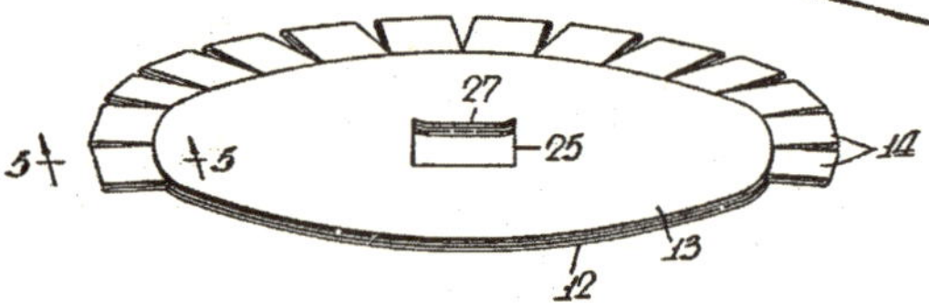
27
25
5
5
14
13
12

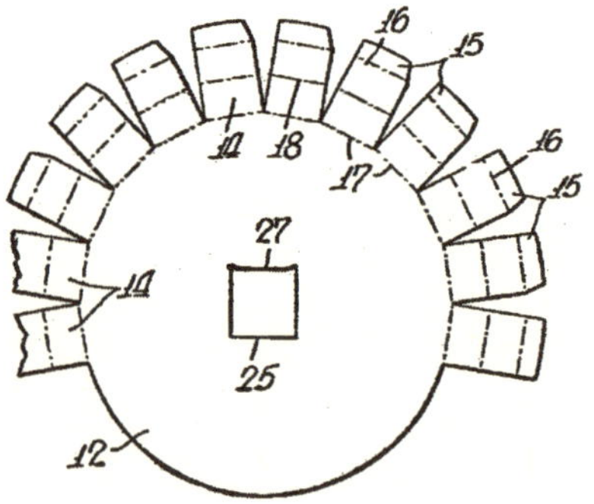
16
15
14
18
17
16
15
27
14
25
12

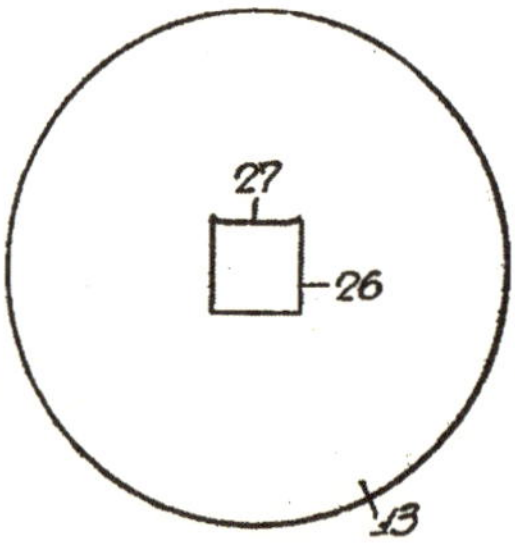
27
26
13

IJK

Robert Bruce Inverarity
(1909-1999)
American artist, administrator, anthropologist

Born in Seattle WA, Inverarity spent much of his childhood in Canada, where he developed lifelong interests in puppetry and Northwest Coast Indian art and culture. He studied briefly with Mark Tobey, then taught at the Cornish School of Art. Later, he taught puppetry at the University of Washington, and in 1938, published a pioneering book titled *Manual of Puppetry*. He also served as a regional director for the WPA (1936-39). Following World War II, he completed a BA degree in art and anthropology at the University of Washington, then continued his education at Fremont University (Los Angeles), earning an MA in art and a PhD. In 1949-54, he was the first director of the Museum of International Folk Art (Santa Fe NM), then served as director of the Adirondack Museum (Blue Mountain Lake NY) in 1954-65. He also worked at the University of California Press as a book designer and illustrator.

In a 1975 interview, Inverarity "talks about designing camouflage and terrain models [during WWII], describes his camouflage research center, and explains how he provided camouflage for his district. He also remembers teaching camouflage techniques and recalls his period as a naval, studio artist" (Archives of American Art 1995:82).

•

Facing page Drawings submitted for US Patent No. 2,393,776 titled "Cardboard Camouflage Structure," invented by J.V. Horr (1946).

Erling F. Iverson

(dates undetermined)
Camouflage Instructor at Pratt Institute

During World War II, Iverson developed a course on air raid defense at Pratt Institute in New York, as described by him in Iverson (1940). His efforts are also acknowledged in "War and the Art Schools" (1940:28), which states: "Probably the most extensive courses in this new field are those offered at Pratt Institute. Grouped under the heading of Air Raid Precautions courses, Pratt's curriculum includes work in the design and construction of shelters, the protection of old and new buildings, blackout provisions, ventilation and camouflage. Coordinating instructor in charge of these ominous-sounding courses is Erling F. Iverson, a Pratt alumnus who has spent the past two years studying air raid precautions in both England and Germany."

•

Geoffrey Barkas (1952:92): "There is nothing new about camouflage. It is as old as war itself. The dream of every commander has always been to achieve complete surprise over his enemy, just as the thought of being himself the victim of surprise has always been his nightmare. Camouflage is simply one factor of surprise. It means deceiving the *eyes* of the enemy."

Above WWI camouflaged tank.

James (Ranalph) Jackson

(July 3, 1882-September 9, 1975)
New Zealand artist

Jackson was born near Palmerston North, New Zealand. He moved to Australia to study drawing, and then moved on to London, where he worked with Frank Brangwyn. He also studied in Paris, where he was influenced by Impressionism. He eventually returned Australia to live, where during World War II, according to Pearce (1996), "he joined the camouflage section of the Department of Defense in 1942 until dismissed when the department discovered his real age—a fact he had always concealed."

•

Henry Berry (1978:206-207): "[During WWI] Someone got the bright idea that our [US Army] division should have some camouflage men, so they started putting together a special squad of creative types who had some connection with painting. I'd been studying at Pratt Institute before I enlisted, so I volunteered. I must confess that the whole thing was somewhat of a lark, but what a deal it turned out to be.

They rounded up twenty of us, under the command of a captain who'd been a sign painter in New York City and told us to go to it. None of us, including the captain, knew a goddamn thing about camouflage, but it got us out of all the drilling and what have you."

Michel Jacobs

(1877-1958)
Canadian-born American artist

He studied at the National Academy of Design in New York, at the École des Beaux-Arts and Académie Julian in Paris, as well as in Berlin.

He lived in Washington DC in 1914-1920, during which he served as a camouflage expert for the US Army. Following World War I, he returned to New York, where he founded the Metropolitan School of Art. He also lived briefly in Los Angeles.

•

Alain Jacquet
(February 22, 1939-)
French artist

Jacquet briefly studied architecture at the École des Beaux-Arts in Paris, then turned instead to painting. In the 1960s, he began to create disruptively patterned images in which other images are embedded, often consisting of "take-offs" on familiar works by other well-known artists, such as Édouard Manet, Jasper Johns, and Roy Lichtenstein. For more information, see Blechman (2004:284-287) and Mendez Baiges (2007).

•

Louis Jamme
(1913-?)
American illustrator and cartoonist

Born in Chicago, Jamme was a magazine illustrator and cartoonist prior to World War II. Crane (1944:228): "He is attached to the camouflage platoon of the 909th Engineer Air Force Headquarters Company stationed at Mitchell Field, New York. During the last year he has completed publicity work, posters, booklets and camouflage demonstration work for the First Air Force."

•

Francis (Lee) Jaques
(September 28, 1887-July 24, 1969)
American naturalist and illustrator

Jaques was born in Genesco IL, and attended the Milwaukee School of Engineering. In 1917-19, he served in the US Army, during which he was assigned to camouflage. He is mentioned in Love (1999:318) and Blechman (2004:133).

Following the war, Jaques worked as an illustrator at the American Museum of Natural History in New York, from 1924 to 1942. He also produced dioramas for the Bell Museum of Natural History in Minneapolis. He and his wife, Florence Page Jaques, collaborated on books about nature, which she wrote and he illustrated. In 1946, their work was awarded the John Burroughs Medal for the best literary work on nature. He was also a muralist, and was responsible for the dome in the bird hall in the American Museum of Natural History, as well as dioramas there. Following his death, his wife produced a book about him, titled *Artist of the Wilderness World* (Jaques 1973). Also see Luce (1982) and Sterling (1997). There is a Jaques Art Center in the former Carnegie Library in Aitkin MN, for which there is a website at (www.jaquesart.com). Also see Herbert Haseltine.

•

John Jarvie
(dates undetermined)
American designer

During World War II, he served with the Ghost Army. He had been trained in art at Cooper Union, and later worked as an art director. He is featured on the website for an anticipated film about the Ghost Army at (www.ghostarmy.org).

•

Gustave Louis Jaulmes
(1873-1959)
French painter, interior designer

Born in Lausanne, Switzerland, he studied with Jean-Paul Laurens at the Académie Julian. He is listed in Coutin (1988) as a World War I French camoufleur. Ducasse (1962) cites his family name as Jeaulmes.

•

Arthur B(ruce) Jensen
(dates undetermined)
American inventor

During World War I, Jensen served as a lieutenant in the infantry, United States Reserves, and senior instructor in the camouflage section of the 86th Infantry Division (Blackhawk Division) at Camp Grant, near Rockford IL. At the end of the war, he published a book titled *Jensen System of Modeling:*

Everett L. Warner, quoted in White (1951:135): "In their final report at the end of the First World War the French pointed out that there was no such thing as a low visibility shade of paint, since the color used to make a ship less visible at one time was the very means of making her more conspicuous at another. This is why protective coloration as practiced on land has seldom proved suitable for ships at sea."

...

Skerrett (1919:177-178): "The Submarine Defense Association, after considerable research, decided that a ship that is hard to see is hard to hit, and, therefore, it set about seeking a protective coloring which would give the lowest practicable degree of visibility beyond the average long-distance gun range of the German U-boats, i.e., 5000 yards, and short of that distance would present a confusing appearance that would make it extremely difficult for the periscopic observer to determine the vessel's range, speed, and course." Also see Lindon W. Bates.

Above As was common in the 19th century, Abbott H. Thayer "tested" his hypotheses not through rigorous scientific experiments, but through demonstrations. Shown here are two views of one of a number of demonstrations that he, his son, and his students devised for exhibitions at the Art Institute of Chicago (c1919) and other locations. On the right is a Thayer painting of the typical background of a warbler in which the bird is all but invisible; however, it has a hinged door-like panel , with the silhouette of the warbler cut out of it, while a second bird is painted on the outside front of the panel. When the panel is closed, one sees the final result on the left.

Employing the Superskill Modeling Device (Jensen 1919). While the book indicates that the patent for his modeling device has been "applied for," it was apparently never approved.

In the book's foreword, Jensen writes (1919:3-4): "*Pläste papier* modeling has a glorious past. The US Army used this method in the World War for making dummies to attract enemy fire and so learn their location, and in many other ways. In several instances, a dead horse lying on No-Man's Land near the enemy trench was replaced by a good duplicate made in *pläste papier*, and the observer would crawl into the paper horse before dawn each day and observe the enemies action all day at close range. Great guns were imitated in this material and placed in the vicinity of real guns that were carefully concealed through camouflage. The Boche [German] observers sent out to discover the location of the real batteries that were doing the havoc to the enemy would discover the paper guns only, and so the enemy artillery would train their fire on the dummy batteries and allow the real guns to continue their fire without being molested. The real guns were concealed by having their outlines painted out by means of broken forms of various colors painted all over the guns, and also covering the guns with foliage. A great statue of Blackhawk, 25 feet high, was modeled in this material and waterproofed, just before the Blackhawk Division left Camp Grant IL, for France in July 1918. This work was done by the Blackhawk Division's Camouflage Department, of which the writer was a member. It still stands, after these many months, defying the elements; the winds and rains and snows could not destroy it; and it still looks as permanent

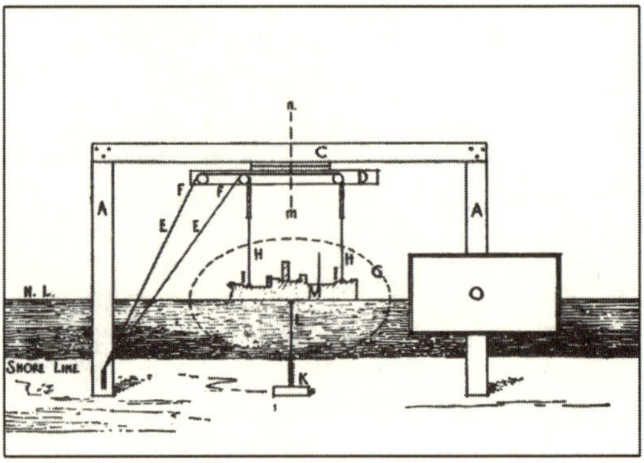

Right Diagram of a device, invented by Loyd A. Jones, for testing the visibility of camouflage-painted ship silhouettes. This was actually installed outdoors on the shore of Lake Ontario. The cut-out shape of the ship was suspended from the crossbeam with piano wire and positioned at a height at which it appeared to rest on the water. Each model was viewed through a telescope-like instrument called a *visibility meter* (developed by Jones) and measurements were made of its reflected light. As reproduced in Skerrett (1919:176).

Matthew Luckiesh (1922:222-223): "In the early days of unrestricted submarine warfare many schemes for modifying the appearance of vessels were submitted...The question of low visibility for vessels could be thoroughly studied on a laboratory scale, because the seascape and natural lighting conditions could be reproduced very closely. Even the general weather conditions could be simulated, although, of course, the experiments could be prosecuted outdoors with small models, as indeed they were. Mr. L.A. Jones carried out an investigation on the shore of Lake Ontario..."

as though carved out of solid rock. It stands as a monument to those brave Blackhawks who gave up their lives overseas; to their memory this book is solemnly dedicated."

•

Eric E(rnest) Jolliffe
(1907-November 2001)
British-born Australian cartoonist and painter

Born in Portsmouth, England, Jolliffe grew up in Australia. According to his entry in the Dictionary of Australian Artists Online (Kerr n.d.), "Jolliffe was a camouflage artist with the RAAF during the War [WWII], traveling to Arnhem Land and the Kimberley, where he got to know a number of indigenous people. He specialized in bush humor, both black and white, and was admired by both races... He first met Aborigines from Arnhem Land and the Kimberley while working in a camouflage office with the RAAF in the Northern Territory during the war." He was accused of racism in 1980 by the Federal Anti-Discrimination Board because of his cartoon portrayals of Aborigines. He collected and republished his cartoons in more than 130 anthologies.

•

David Jones
(1910-2000)
British car designer

Born in Birmingham, England, Jones studied privately and at the Birmingham Art School. He was awarded a scholarship to attend the Royal College of Art, where he worked with Henry Moore. In 1932, he became a designer at Vauxhall, the British subsidiary of General Motors, during which he served as an apprentice at General Motors with American designer Harley Earl in Detroit. With the retirement of Eric Kennington from Vauxhall, Jones became director of styling, a position he returned to after World War II.

Dixon (2000): "[During World War II] He was commissioned in the Royal Engineers with the rank of captain in the camouflage section, and divided his time between camouflaging airfields and working on troop carrier design at Vauxhall."

See also Ledyard Towle. Harley Earl. McClelland Barclay.

•

Loyd A[ncile] Jones
(April 12, 1884-May 15, 1954)
American physicist

Jones was born in York NE. He studied at the University of Nebraska, then moved to Washington DC, where he worked as a laboratory assistant in the Bureau of Standards in 1910-12. He joined the Eastman Kodak Company in Rochester NY as an assistant physicist, and in 1916, he was appointed head of that company's physics department, a position he continued in until his retirement. He is credited with eighteen patents and dozens of articles, related to such subjects as photometry, physical optics, illuminating engineering, colorimetry, photography and motion pictures.

Van Buskirk (1919): "On March 25, 1918, the US Navy (its Bureau of Construction and Repair) was placed in charge of all camouflage designs, while the civilian-based Emergency Fleet Corporation (US Shipping Board) was responsible for the actual application of the camouflage designs to ships' surfaces. I had the honor of being placed in general charge of the Navy's organization and was fortunate indeed in being given the assistance of Lt. [Everett L.] Warner and Lt. [Loyd] Jones, who were placed in charge of the subsection of design and the subsection of research respectively."

Ackerman (1930:310): "For some time scientists engaged by the [US] Submarine Defense Association, under the direction of Lindon W. Bates, had been conducting experiments to determine the best methods of camouflaging all types of ships. At the beginning of the war [World War I] this subject had been studied by artists, but, as it was primarily a question of optics and physics, [George] Eastman had lent Loyd A. Jones, of the research forces, to that association. One of his early contributions was the invention of a visibility meter, which measured scientifically the visibility of any object on the ocean." See US Patent No. 1,437,809.

Ackerman (ibid.:311-312): "Jones and his staff constructed at the [Kodak] Park a complete 'experimental ocean,' reproducing, by means of a large tank, a movable sky, miniature ships, and artificial lights, a working model, where the various types of deceptive colorations of ships could be studied through a submarine periscope. These observations and studies evolved eventually a scientific scheme of camouflage which was placed at the service of the United States and Allied authorities."

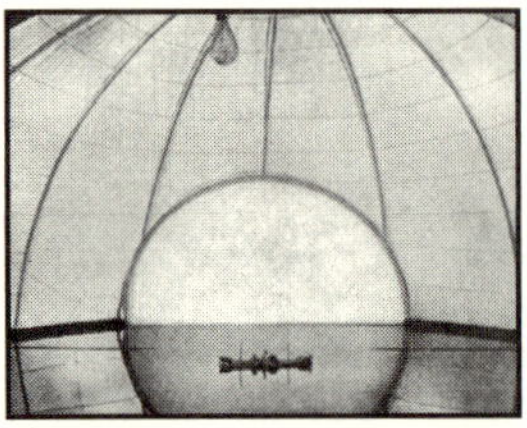

Above Two views of the indoor "experimental ocean" or camouflage testing theatre that was devised by Loyd A. Jones and constructed during World War I at Eastman Kodak in Rochester NY. The domed observation tank (top) included an artificial sun and other components that simulated outdoor lighting conditions when observed through a periscope. For details, see *Scientific American* ("A Theatre for Studying Camouflaged Ship Models" 1919).

•

Roy C. Jones
(dates undetermined)
American architect and educator

He is listed in Embury (1917) as a World War I American camoufleur. He later became a professor and head of the School of Architecture at the University of Minnesota. In 1932, he was the co-author (with F.H. Bosworth, Jr.) of *A Study of Architectural Schools* (New York: Charles Scribner's Sons).

•

Howard Joyner
(1901-1996)
American art professor

Joyner studied in Paris at the École des Beaux-Arts, and in the US at the University of California, Harvard University and the University of Iowa. In 1937, he was hired as Head of Art Department at North Texas Agricultural College (now the University of Texas at Arlington). He founded the Arlington Museum of Art.

UTA Magazine Online, Fall 2004 ("Yesteryear: And the rest is...art history"): "World War II decimated class enrollments in the art program. But Joyner had a novel idea that would not only boost the war effort but increase campus participation.

He sought and received classified material from the War Department and started teaching camouflage painting to the Marines and Navy young men of V-12 Unit, which was established at the college.

'This camouflage painting was very important to the War Department, and it brought more and more students into the art department,' [faculty member Jack] Plummer said.

The move is widely credited with saving the department."

•

Béla Julesz
(February 19, 1928-December 31, 2003)
Hungarian-born American scientist

Born in Budapest, Julesz immigrated to the US in 1956. He worked at Bell Laboratories, where he specialized in vision research until his retirement in 1989. He then became director of the Laboratory of Vision Research at Rutgers University. He is particularly known for his development of the random-dot stereogram, which he credited to his experience with camouflage detection using stereoscopic aerial photographs, during his military service in Hungary.His book on the subject, titled *Foundations of Cyclopean Perception*, is regarded as a classic in scientific literature.

Julesz (1995:18-19): "As a former radar engineer, I knew that...in order to break camouflage in aerial reconnaissance, one would view aerial images (taken from two different positions, called *parallax*) through a stereoscope, and the camouflaged target would jump out in vivid depth. Of course, in real life there is no ideal camouflage, and after stereoscopic viewing one can detect with a single eye a few faint cues that might discriminate a target from its surroundings. So I used one of the first big computers, an IBM 704 that had just arrived at Bell Labs, to create ideally camouflaged stereoscopic images..."

K

Sheffield Kagy
(1907-1989)
American printmaker

Born in Cleveland OH, Kagy studied at the Cleveland School of Art and at the Corcoran School of Art in Washington DC. As a WPA artist in 1934-36, he was commissioned to create several prints. After moving to Washington DC in 1936 to teach at an art school, he worked for the US Treasury Department

Above American artist Everett L. Warner was in charge of the design subsection (in Washington DC) of US naval camouflage program during World War I, while Loyd A. Jones oversaw the scientific subsection (at Eastman Kodak in Rochester NY). During World War II, Warner was once again responsible for naval camouflage, during which his team included (above, left to right) Bennet Buck, Sheffield Kagy, William Walters, Warner (with ship model), Arthur Conrad, and Robert Hays. The seventh person (left background) is unidentified. US Navy photograph.

(1937-40), during which he painted murals for two post offices. After serving as head of the art department at Chevy Chase Junior College in 1940-43, he worked with Everett L. Warner on the design of ship camouflage until 1945. He returned to teaching, then served as an exhibition designer for the US State Department in 1959-73. See his Wikipedia online biography, and Behrens (2002).

•

Elizabeth Louise Kahn
(1948-April 20, 2004)
American art historian

Kahn completed her undergraduate work at the University of Wisconsin-Madison, then earned further degrees at the University of Pittsburgh and UCLA. She joined the art history faculty at St. Lawrence University in Appleton WI in 1978, where she held the L.M. Flint Chair in the Fine Arts. Her extensive study of the social impact of World War

I French camouflage, titled *The Neglected Majority: "Les Camoufleurs," Art History, and World War I* (Kahn 1884) was a pivotal contribution to the early research of art and camouflage.

•

Art Kane
(April 9, 1925-February 3, 1995)
American photographer

Born Arthur Kanofsky, Kane was well known for his portrait photographs of jazz and rock musicians. During World War II, he volunteered for camouflage and was assigned to the Ghost Army. He is mentioned in Park (1985).

•

Ezra Jack Keats
(March 11, 1916-May 6, 1983)
American author and illustrator

Born in Brooklyn, his birth name was Jacob Ezra Katz. He studied at the Art Students League with Jean Charlot. In 1937, he was commissioned to paint murals for the WPA. He was also a comic book illustrator at Fawcett Publications, where he drew the backgrounds for *Captain Marvel* comic books. In 1943, he joined the US Army, where "taking advantage of his skill as an artist, the army trained him to design camouflage patterns" ("Ezra Jack Keats," n.d.). He was one of the first children's book authors and illustrators to assign a central role to minority children. At the time of his death, he had illustrated 85 children's books, and had both written and illustrated an additional 24.

•

Erna (Laura) Keighley
(June 25, 1891-July 16, 1955)
British-born Australian civil rights advocate

Born in Manningham, Bradford, England, Keighley was president of the women's auxiliary of the National Defense League of Australia during World War II. According to Oppenheimer (1996), "Her followers drilled, made camouflage netting, studied military-related subjects such as mechanics, map-reading, signaling and convoy-driving, and formed a comfort group which she headed."

•

György Kepes (1944:45): "A snake camouflaged by nature is no longer a snake. It is an aggregation of small units of color-shape."

Above Milwaukee-based conceptual artist Harvey Opgenorth "camouflages" himself in art museums by dressing in clothing patterns that partially blend in with a particular artwork in the museum's collection. In this photograph (1998), he is standing in front of an Ellsworth Kelly painting at the Metropolitan Museum of Art. Courtesy the artist.

JoAnn Deak (2002:99): "[Tweens (preadolescent girls)] cluster together and need to act alike, talk alike, and look alike for the protective camouflage it provides. It reminds me of a Discovery Channel program about running zebra: The stripes made them all blend together so that any predatory animal would have a hard time singling out any one zebra."

H(arold) Roy Kelley
(1913-1965)
American architect

A graduate of Cornell University, he was associated with Donald Douglas, Sr., of Douglas Aircraft and apparently worked for him (c1940) in collaboration with Frank Collbohm, in devising ways to disguise the Douglas Aircraft factory. See Rasmussen (2002) and Kelley (n.d.).

•

Donald H. Kelly
(1923-1997)
American scientist

Kelly studied optics and visual science at the University of Rochester. During World War II, he was drafted into the US Army, then commissioned as a Navy officer at the Patuxent River Naval Air Station, near Chesapeake Bay, within the Bureau of Aeronautics (BuAer). As explained in detail in Kelly (1990), he was assigned to the Aircraft Camouflage Section of the Tactical Test Unit, where he tested aircraft camouflage schemes.

•

Ellsworth Kelly
(May 31, 1923-)
American painter

Kelly is widely considered to be a leading American painter. Born in Newburgh NY, he studied at Pratt Institute from 1941 until he was drafted into the US Army in 1943. During World War II, he served in the Ghost Army. After the war, he studied art in Boston and then at the École des Beaux-Arts in Paris. See Goossen (1973).

•

György Kepes
(October 4, 1906-December 29, 2001)
Hungarian-born painter, designer, art theorist

In art and design education, few people have been more influential than Kepes, who was born in Selyp, Hungary, and studied at the Royal Academy of Fine Arts in Budapest. In 1930, he moved to Berlin, where he began a long association with Hungarian photographer and designer Laszlo Moholy-Nagy, who had taught at the Dessau Bauhaus. In 1937, both Kepes and Moholy immigrated to the US, when the latter became the director of the New Bauhaus (later called the Institute of Design) in Chicago, while Kepes was in charge of the courses on Light and Color. Beginning in 1942, he also taught a camouflage course, and, at the request of the city's mayor, researched the feasibility of disguising the aerial landmarks of its Lake Michigan shoreline.

Wechsler (1978:10): "...[While teaching] in Chicago, during the war, he [Kepes] explored the possibilities of transforming the large scale image of the city [to protect it from enemy bombing]. To dislocate the night landmarks, he proposed to float on cables a network of lights that would hover over Lake Michigan so that the pattern would take on an apparent reality confusing to potential raiders. The camouflage research gave Kepes an extraordinary opportunity to envision the possibilities of an environmental art."

György Kepes, 1942-1943 Chicago Institute of Design course description, as cited in Kostelanetz (1970:179): "The Principles of Camouflage: Research in natural camouflage; surface covering; mimicry; visual illusions; basic photography; investigation of camouflage techniques. Conducted by George [sic] Kepes."

Kepes (1978): "For some of your questions I could give only negative answers. I never met Col Homer Saint-Gaudens...I was for a few weeks at Fort Belvoir with the Army Engineers but we had not studied George de Forest Brush or [Abbott H.] Thayer's work."

In 1944, he published an influential book on design theory titled *Language of Vision*. Three years later, he joined the faculty at MIT, where he became founder/director of the Center for Advanced Visual Studies.

•

John Graham Kerr

(September 18, 1869-April 21, 1957)
British embryologist

C.B. Williams (quoted in Cott, 1940:332): "In the district round Amani NE Tanganyika a small moth resembling a bird dropping was not uncommon. On one occasion I observed what I thought to be one on a leaf, but after a close examination from a distance of only a few inches I discovered (to my own satisfaction) that it was after all only a bird dropping. Just as I turned away the said bird dropping flew off!"

A British naturalist and a Member of Parliament from Scotland, Kerr was educated at Cambridge and Edinburgh Universities. During two expeditions to South America, he became particularly interested in animal coloration and, in turn, its application to military purposes. He was the teacher of (camoufleur) Hugh B. Cott. As early as the 1890s, Kerr advocated ship camouflage and openly expressed support for some of Abbott H. Thayer's beliefs about the subject, particularly countershading. On September 24, 1914, shortly after the start of World War I, he wrote to Winston Churchill, recommending that British warships be countershaded. He later claimed to have anticipated dazzle painting in advance of World War I, by a method he referred to as "parti-coloring." See *New York Times* ("John Graham Kerr," 1957).

Nature ("Camouflage in Modern Warfare" 1940:949): "The general application of these [natural camouflage] principles for war purposes was first suggested to a Government Department by Prof. (now Sir John) Graham Kerr, who in September 1914 communicated to the First Lord of the Admirality methods of rendering ships at sea less easily recognized—in particular, by means of countershading, and disruptive coloration or 'dazzle' as it was called later."

Above *William Kienbusch, May 1956*. Martha Holmes, photographer, 1 photographic print: b&w; 26 x 21 cm. Courtesy of the William Kienbusch papers, 1920-1079, Archives of American Art, Smithsonian Institution.

•

William Kienbusch

(1914-1980)
American painter

Pierre-Louis Gautier, quoted in (Blechman, 2004:278): "In 1917 two British painters, Norman Wikinson and Cecil King, worked out a method to slow down the destruction of Allied naval ships. What they produced was more or less fantasy—arbitrary and fortuitous graphic ingenuity."

Kienbusch grew up in New York, and studied at Princeton University, while also taking summer courses in painting from (camoufleur) Eliot O'Hara. He then studied at the Art Students League, where he was influenced by Henry Varnum Poor, Stuart Davis and (camoufleur) Abraham Rattner. During

World War II, he volunteered for military service, and was placed in the US Corps of Engineers, first in support of the Air Force, then in support of the Army.

Kienbusch (Selvig 1968): "...I got in, and in the first year I was in this lousy camouflage outfit outside of Richmond, Virginia, in the Air Force.

...But I was fortunate in the Army. As I said, first of all I was in about a year doing construction work in camouflage for the Air Force. For example, camouflaging Air Force installations, gun emplacements, buildings, all these things.

...I think the major emotion I had [while in the Army] was one of terrible boredom. This was less true when I taught camouflage, when I actually was teaching it.

...the only painter I knew well [in the Army] was [camoufleur] John Walley who is now a professor at the University of Illinois. I knew him for a good two years in the Army. However, I had one buddy, a young man from somewhere in Brooklyn–I don't really know what Sammy Katzen did–we just got along well. We were both in the same field, camouflage."

•

B(enjamin Sayre) Cory Kilvert
(1879-1946)
Canadian-born illustrator

Born in Hamilton, Ontario, he moved to New York, where he studied with Robert Henri at the Art Students League. His primary achievements were as a book and magazine illustrator, marine artist, and map illustrator.

Reed (2001:106): "During World War I, Kilvert served with the Canadian Expeditionary Force, based in London, devising techniques for camouflage."

•

Cecil (George Charles) King
(August 6, 1981-December 9, 1942)
British marine artist

King was educated at the Westminster Art School, and later studied under Jean-Paul Laurens in Paris. He was appointed Lieutenant RNVR in March 1917, to serve under Norman Wilkinson in the British Dazzle Painting Section.

He is described by Williams (1989:63) as Wilkinson's assistant. "These [painted ship models] were then observed, using a periscope device, in a viewing theatre specially constructed by Norman Wilkinson and Cecil King."

Lisle (1941:97): "Cecil King, who was associated during the World War in British camouflage research with Professor Abel [?], the inventor of dazzle painting, has made some interesting comments during the past few years, from which these notes have been made."

Above This photo of an apparent field office for US camoufleurs was released to American news services near the end of World War I. It shows a quaintly decorated hut, with a sign that reads "Chateau Camouflage."

"Sea Camouflage" (1919:20): "The originator of 'dazzle-painting' was Lieut.-Commander [Norman] Wilkinson, the well-known marine painter, who has been assisted in his work by a fellow artist, Lieut. Cecil King."

•

Paul B[ernard] King
(February 9, 1867-November 25, 1947)
American painter

Born in Buffalo NY, he studied art in Buffalo and at the Art Students League in New York, then later in Paris and Holland as well. While living temporarily in Philadelphia, he was acting president of the School of Design for Women. He settled in New York permanently in 1921. *New York Times* ("Paul B. King," 1947): "In the first World War, Mr. King served in the camouflage branch under the US Shipping Board." His camouflage service is also noted in the description of his papers in the Archives of American Art (King, n.d.).

Gertrude Stein (1938:76): "When I was in America I for the first time traveled pretty much all the time in an airplane and when I looked at the earth I saw all the lines of Cubism made at a time when not any painter had ever gone up in an airplane. I saw there on the earth the mingling lines of Picasso, coming and going, developing and destroying themselves.
I saw the simple solutions of Braque. I saw the wandering lines of Masson, yes I saw and once more I knew that a creator is contemporary, he understands what is contemporary when the contemporaries do not yet know it..."

•

Edward (Frederick) Kohler
(May 27, 1890-June 28, 1964)
Australian sculptor

Born near Lowood, Queensland, he was the son of Prussian immigrants. He served with the Australian army in World War I. during which he fought at Gallipoli. He studied art at the Académie des Beaux-Arts in Lille, then worked as a sculptor in France. After moving back to Western Australia in 1932, he became known for commemorative sculpture and religious works.

Taylor (2000:38): "During World War II he was employed as a camouflage officer with the Department of Home Security."

•

Gabriel Kohn
(June 12, 1910-May 21, 1975)
American sculptor

Kohn was born in Philadelphia, studied in New York at Cooper Union and the Beaux-Arts Institute of Design, then worked in Los Angeles for Paramount Pictures. After World War II, he studied in Paris with Ossip Zadkine. He also taught briefly at Cranbrook Academy. According to his biographical entry at GroveArt.com, "From 1942 to 1945 he was a camouflage designer in the Army Air Corps."

•

Konstantin Alexeyevich Korovin
(1861-1939)
Russian painter and designer

He studied at the Moscow School of Painting and, briefly, at the St. Petersburg Academy of Arts. He was the head artistic designer of the Russian pavilions for the International World's Fair in Paris in 1900. According to Art-Roots.com, "During the First World War Korovin worked as a camouflage consultant at the headquarters of one of the Russian armies. Despite his poor state of health (an old nervous illness and heart disease) he was often at the front line."

•

(Reginald) Guy Kortright
(1877-1934)
Canadian-born British painter

Born in Barrie, Ontario, he was a navy lieutenant in World War I, during which he made paintings of dazzle-painted ships in dock. He was associated with L. Campbell Taylor. For examples of his paintings of camouflaged ships, see Hurst (1919).

•

J(ack) L(ouis) Koskie
(June 3, 1914-?)
British-born Australian artist, graphic designer

Born in Yorkshire, England, Koskie moved to Australia in 1939, where he served as chief designer for the Commonwealth Office of Education and the Tasmanian Government Printer. He was the author of *The Ships that Shaped Australia* (1987). According to the Dictionary of Australian Artists Online, "During World War II he served with the camouflage engineers."

•

A. Newnham (quoted in Cott, 1940: 330): "I was stretching across to collect a beetle and in withdrawing my hand nearly touched what I took to be the disgusting excrement of a crow. Then to my astonishment I saw it was a caterpillar [*Notodontid*] half-hanging, half-lying limply down a leaf...[What] struck me was the skill with which the coloring rendered the varying surfaces, the dried portion at the top, then the main portion, moist, viscid, soft, and the glistening globule at the end. A skilled artist, working with all the materials at his command, could not have done it better."

Above During World War II, apparently some US Army training centers were still using *papier maché* rocks for foxhole covers (1942). Photo courtesy US Library of Congress, Prints and Photographs.

A[lexander] J[usserand] Kostellow
(1898-August 31, 1954)
Industrial designer

Kostellow was born in Ispahan, Iran. He studied at the University of Berlin (1912-1916), then immigrated to the US. There, he studied at the Art Students League, National Academy of Design and the Kansas City Art Institute. He served as head of Industrial Design at Carnegie Institute of Technology in Pittsburgh, then joined the industrial design faculty at Pratt Institute in New York in 1938, where, six years later, he became head of the department. *New York Times* ("A.J. Kostellow" 1954): "From 1940-1943, he was coordinator of military camouflage work for the government at Pratt." Lewis (1994) recalls a conversation with Kostellow in which he learned more about design than from any other source. He mistakenly says that Kostellow had been a student at the Bauhaus.

•

Myron Kozman
(1916-2002)
IIT professor, designer, painter

Kozman earned his degree from the Chicago School of Design (originally called the New Bauhaus) in 1941, where he studied with Laszlo Moholy-Nagy. Earlier, he had also participated in the

Above Undoubtedly one of the reasons why military officers were resistant to camouflage is that during their inspections too often it made them look stupid. In October 1917, there was an incident shown here during an inspection of camouflage work by US President Woodrow Wilson and Commanding General John J. Pershing, in which neither one could find a sniper who was hiding within ten feet of them. The sniper, as it turned out, was hiding in a foxhole, shown here in the foreground, with a *papier maché* fake rock as a lid. Standing in the center in the background (with hat and beard) is American painter John Singer Sargent, who was painting the president's portrait that day. For the full story see Behrens (2002:59-60). In the same war, camoufleur Charles Burchfield talked about the pleasure of watching a major "who spurred his horse over one of the Camouflage Corps's artificial hills with disastrous results" (Bauer 1982:92). US government photograph, 1917.

WPA program, for which he was commissioned to produce several abstract paintings. During World War II, he served in Europe for three years as a camoufleur. Following the war, he joined the faculty at the Chicago School of Design (known by then as the Institute of Design), where he taught until c1953. He married artist Lois Field (born 1923) in 1948, and served as head of design at the Milwaukee Institute of Art and Design in the 1960s.

Boxer (1993:7): "Myron Kozman is an accomplished educator, architect, designer and lifelong advocate and practitioner of Bauhaus teaching principles. After graduation and a stint in the US Army, where he served in a camouflage battalion employing skills first developed in a camouflage course initiated at the School of Design in Chicago by György Kepes, he remained on the faculty of the Institute of Design until 1953."

Kozman (in Boxer 1993:15): "Having completed my education in Grand Rapids, I went to Chicago to earn a living as a commercial artist. In browsing through

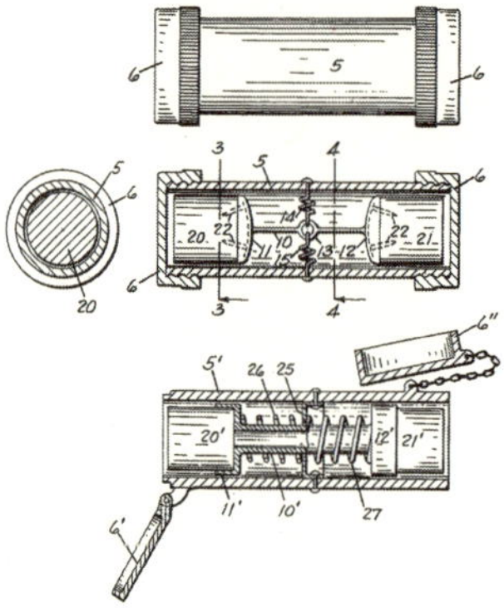

Above (right) WWII instruction sheet for a portable camouflage face painting for individual American GIs. (left) Some of the drawings submitted for US Patent No. 2,334,527 titled "Camouflage Pencil," invented by C.S. Althouse (1943), by which is provided a convenient carrier for two sticks of camouflage face paint.

Facing page Illustration from *Concealing Coloration in the Animal Kingdom* (Thayer 1909), produced by repainting the background of a photograph of a captive Jaguar.

a *Time* magazine, I observed on Monday, October 18, 1937, the New Bauhaus was opening in Marshall Field's south side mansion.

Knowing only a minimum about the German Bauhaus, curiosity found me attending the opening. A gentleman with a thick Hungarian accent led me on a tour of the workshop areas and asked if I were interested in attending the school. I told him that I was just observing. Being a brash young man, I explained that I was already an artist, planned on getting a job and had no time, no money nor the inclination to go back to school.

Moholy[-Nagy] (my Hungarian tour guide) must have spotted a spark of interest in a potential student. He suggested that I bring in a portfolio of representative work. I again responded negatively, reiterating that I was already an artist and needed no more schooling. I thanked him for the tour. He said that if I brought in some examples of my work and if it was promising, there was the possibility of a scholarship. To remove myself from his persuasive power, I agreed to prepare a portfolio. I promised to return and said good-bye. 'Where are you going?' he admonished. I reaffirmed my agreement to return and tried to leave. Moholy explained, 'If you leave now, young man, you will have missed the day's activities. So you will go to class.' I obeyed, ultimately received a four year scholarship,

continued to become one of the five first graduates, stayed an additional twelve years on the faculty (including three years in Europe in an Army camouflage battalion) and have been teaching and promoting the Bauhaus philosophy ever since."

•

George Kruger-Gray
(1881-1943)
British painter, designer

Born in Kensington, Kruger-Gray attended the Royal College of Art. According to the *New York Times* ("Kruger-Gray" 1943) he served in "the Royal Engineers Camouflage School" during World War I.

•

Steven Kyle
(1913-October 19, 1979)
American designer

Kyle was born in New York, where he studied at the Art Students League. He was the founder of Americraft, a supplier of decorative accessories. His wife was actress and playwright Betty Comden. *New York Times* ("Steven Kyle" 1979): "During World War II, during which he was in service for four years, he was attached to a camouflage unit."

•

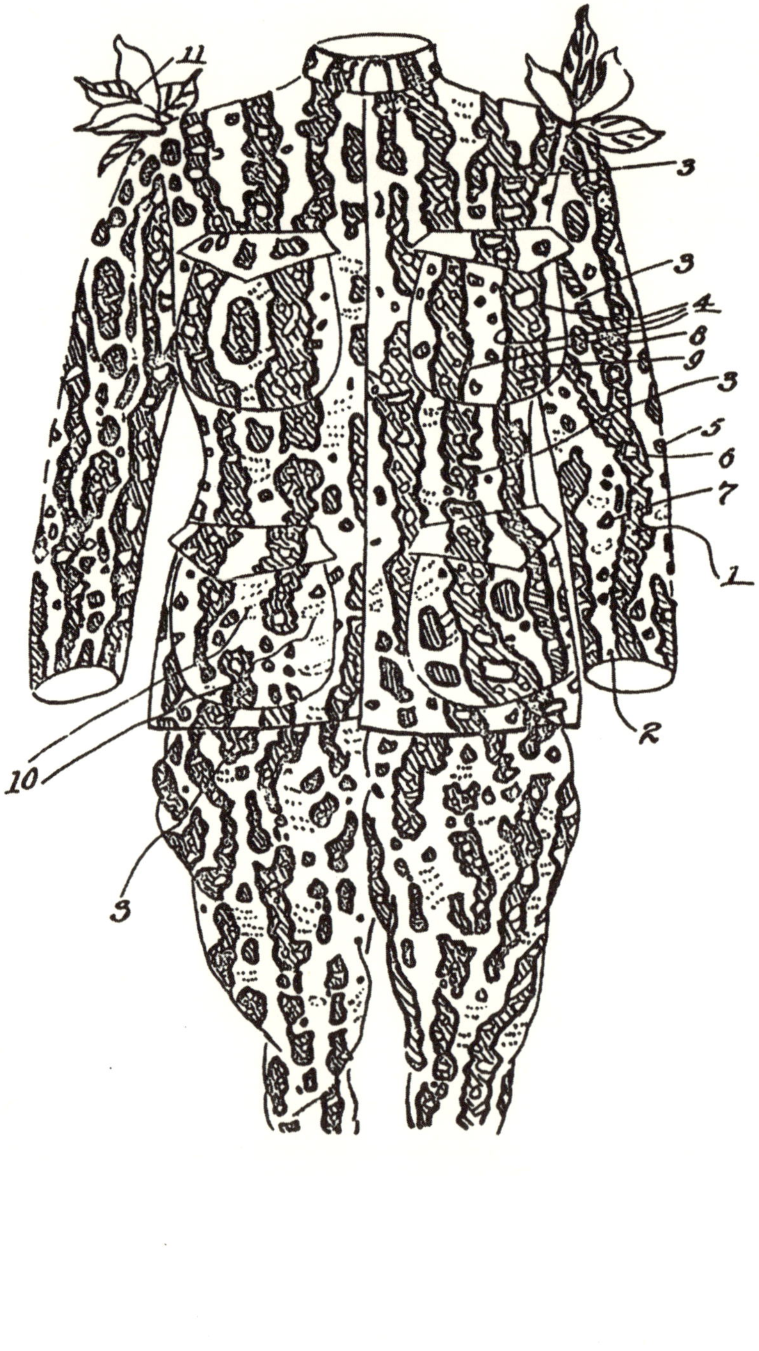
11
3
3
4
8
9
3
5
6
7
1
2
10
3

L

Jean Labatut
(1899-1986)
American architect, historian

Labatut was an architectural critic and historian who taught at Princeton University as a Distinguished Professor of Architecture from 1928 to 1967. His papers are housed at Princeton, where the online finding aid refers to materials he used in teaching a course on camouflage. He and his students conducted camouflage experiments in and around Palmer Stadium on the Princeton campus. During World War II, he was described in the *College Art Journal* ("Wartime Activities," 1946:364) as "Professor of Design, Consultant and Lecturer for the Army in Camouflage."

•

Roger de La Fresnaye
(July 11, 1885-November 27, 1925)
French painter, illustrator, sculptor

La Fresnaye was born in La Mans, France, and studied art at the Académie Julian, where he became friends with three other French artists who would later serve as camoufleurs: Jean-Louis Boussingault, André Dunoyer de Segonzac and Luc-Albert Moreau. An admirer of the work of Paul Cézanne, he met Georges Braque in 1912 and became interested in Cubism, and especially the artwork of Robert Delaunay. As listed in Coutin (1988), he served as a World War I camoufleur from 1914 to 1917, but was discharged when he contracted tuberculosis. See also Sonia Delaunay.

Facing page Drawing submitted for US Patent No. 1,139,642 titled "Military Uniform or the Like," invented by A.S. Cox (1915).

Richard (Francis) Lahey
(June 23, 1893-1979)
American artist

Lahey's papers, in the collection of the Archives of American Art, include some camouflage-related documents, including a diagram for applying dazzle camouflage to a World War I ship (which may or may not have been his design), and a news clipping about American ship camouflage in which one of the two men in the photograph is Everett L. Warner. The caption reads: "A corner of the model room of the camouflage section at the Navy Department. Scale models of every type of vessel, decorated with the weirdest of designs, are kept here for test and experimental purposes. It is easy to imagine what unbounded joy would be Young America's were he once let loose in this toy shipyard."

Johnson (1999:84): "[During WWII] Camouflage, a hitherto neglected art, was suddenly offered in landscape architecture schools. Harold Atkinson wryly noted, 'The necessary reversing of the design process to distract, disperse, and conceal rather than to attract, organize, and enhance brings up numerous problems.' Because it was uncharted territory, camouflage became a true collaborative opportunity."

•

Paul-Maximilien Landowski
(June 1, 1875-April 1961)
French sculptor

Facing page Photo of captured WWI German artillery. The camouflage consists of multi-colored splotches, reminiscent of Pointillist painting. At a sufficient distance, it is reduced to a nondescript gray.

Landowski was born in Paris. He is listed in Coutin (1988) as a World War I French camoufleur, for which he was awarded the Legion of Honor. He served as the director of the Académie de France in Rome, and of the École des Beaux Arts in Paris.

•

Albert Landry
(dates undetermined)
American art administrator

He was founder and director of the Albert Landry Gallery in New York until 1963, when appointed director of the J.L. Hudson Gallery. During World War II, he served in the Ghost Army. See Goossen (1973:103).

•

William-Julien-Émile-Edouard Laparra
(November 25, 1873-September 5, 1920)
French painter

Born in Bourdeaux, he was a student of Jules Joseph Lefebvre and William-Adolphe Bougereau at the École des Beaux Arts. He is listed in Coutin (1988) as a World War I French camoufleur.

•

Pierre Laprade
(July 25, 1875-December 23, 1931)
French painter, printmaker

Born in Narbonne, he was a painter, book illustrator, and a set and costume designer. He was an associate of Henri Matisse, André Derain and (camoufleur) Jean Puy, with whom he started Fauvism, and founded the Salon d'Automne. He is listed in Coutin (1988) as a World War I French camoufleur.

•

Loma (Kyle) Latour
(May 16, 1902-December 19, 1964)
Australian sculptor

Her birth name was Turnbull. She was the wife of Raymond Lindsay, son of Australian artist Norman Lindsay. During World War II, according to Roberts

(2000), "For a time Latour made camouflage at Bankstown before returning to inner Sydney."

•

Facing page During World War II, as in any war, it was advantageous as well as patriotic to find ways of using ones civilian skills to support the war effort. In this US government poster, which advertises war bonds and shows "the contribution of art to the war effort," sixteen art-related means of participating are suggested, including posters, picture-graphs, maps, mechanical drawing, lettering, cartoons, explosion drawings, emblems, figures, symbols, illustrations, pen and ink, caricatures, watercolor, display—and camouflage. War Savings Program, US Treasury Department.

Paul-Albert Laurens
(1870-1934)
French painter

Laurens was the son of Jean-Paul Laurens, who was a French painter and and influential teacher at the École des Beaux-Arts. He was also the brother of Jean-Pierre Laurens, who, like himself, became a painter and teacher at the Académie Julian. Paul-Albert was born in Lyons, and studied with Léon Bonnat. He is listed in Coutin (1988) as a World War I French camoufleur. He is also mentioned in Ducasse (1962).

•

John Lavalle
(1896-November 13, 1971)
American portrait painter, landscape artist

Above Cover of Neil Leach, *Camouflage* (Cambridge MA: MIT Press, 2006). According to Leach, an architectural theorist, "We human beings are governed by a chameleon-like urge to blend in with our surroundings—to 'camouflage' ourselves within our environment." See also Roger Caillois.

He was the great-grandson of General Juan de Lavalle, one of the founders of the Republic of Argentina. Born in Nahant MA, he was educated at St Paul's School, and graduated from Harvard in 1918. Following service in World War I as a bomber pilot for the Royal Air Force, he studied art at the Académie Julian in Paris and the School of the Boston Museum of Fine Arts. *New York Times* ("John Lavalle," 1971): "In World War II he was a camouflage artist for the 12th Air Force before the Sicily invasion and painted many portraits and combat scenes of its subsequent campaigns." He was known mostly for his commissioned portraits of university presidents.

•

Gaston Lavy
(dates undetermined)
French writer and amateur artist

Lavy served in World War I from 1914 to 1918, during which he was affiliated with the French camouflage

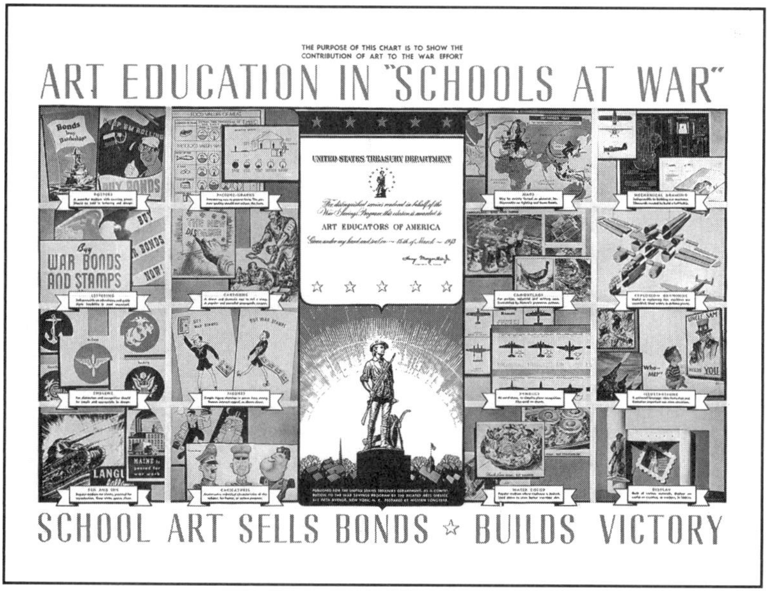

service. He is listed as a camoufleur in Coutin (1988). Following the war, he published a self-illustrated wartime journal, titled *My Great War*.

•

Robert Lawson
(October 4, 1892-May 27, 1957)
American illustrator

Born in New York, Lawson studied art at Parsons College. During World War I, he served in France with the American Camouflage Corps. Having launched his career as a children's book illustrator in 1922, he became widely known for his achievements as an illustrator (and later, as both a writer and an illustrator). He was the first person to receive both the Caldecott Medal (1940) and the Newberry Medal (1945). Among his most popular books were *The Story of Ferdinand* by Munro Leaf (1936), and *Ben and Me: An Astonishing Life of Benjamin Franklin by His Good Mouse Amos* (1939), which he both wrote and illustrated.

Faulkner (1957): "It was his [Lawson's] sense of fantasy and humor which made our [American Camouflage Corps] musical shows successful."

•

Frederick Roy Lear
(December 2, 1882-June 21, 1950)
American architectural designer

Born in Corning NY, Lear graduated in 1905 from the School of Architecture at Syracuse University, following which he joined the faculty. He remained at Syracuse until his retirement in 1947. He designed the memorial for North Pole explorer Vice Admiral Robert E. Peary in Arlington Cemetery. He was also a color consultant for the Onondaga Litholite Company. According to the online Syracuse Archives, "His love of color stemmed from his love of butterflies. He was an avid collector and sometimes drew inspiration from nature's patterns and colors. During WWII, Mr. Lear lectured aviation cadets on camouflage and taught concealment techniques at Syracuse University."

•

Above World War I rail-mounted artillery with a camouflaged barrel.

Hugh B. Cott (1940:438): "In these days of unrest and uncertainty, it should not be necessary to stress the importance of this whole question of camouflage. In nature visual concealment and deception have proved one of the main means of obtaining those two essentials of life—security and sustenance. In everyday matters, opposite devices making for conspicuousness possess also a power—whether in the form of street signs or lipstick—which is universally recognized: and on advertisement and cosmetics we are prepared as a nation to spend enormous sums annually. But camouflage, a product of the Great War, is still in its infancy—a child suffering from arrested development. Its importance and possibilities have yet to be fully appreciated in official quarters."

(Joseph) Fernand (Henri) Léger
(February 4, 1881-August 17, 1955)
French painter

While Léger had no direct involvement in camouflage, he is often mentioned in connection with it because a number of his artist-friends were World War I camoufleurs, and because, when given the chance to transfer from the infantry to a less dangerous camouflage unit, he declined.

Cork (1944:86): "When the opportunity arose to be transferred to a mobile team of [French] army camouflage painters at Noyon, he [Léger] turned down the invitation. Léger would have found himself in the company of fellow artists there, among them Dunoyer de Segonzac, André Mare and Jacques Villon. In addition to camouflaging airfields and establishing dummy observation posts, they all found time to make their own images of the military life around them. Léger, however, had no wish to join a specialized circle of painters removed from the ordinary soliders of the French Army."

•

Michael Lekakis
(March 3, 1907-November 18, 1987)
American sculptor

The son of a Greek flower street vendor, Lekakis was born in New York. According to the Kouros Gallery (*Michael Lekakis*, 1987), "[in 1942] he was inducted into the Army Air Corps and, like many artists, was placed in a camouflage unit. His was in Greensboro, North Carolina, where he remained for over two years, first studying camouflage technique, then lecturing on it... He rose to the rank of sergeant, constructing forty-five full-scale installations and, as training aids, sixteen three-by-four foot models. He still remembers these exact numbers and the painstaking detail he put into the models—camouflaged animals, birds, reptiles, even insects, as well as plants. It is not too much to suggest that the specific military techniques of camouflage, based on natural phenomena, had implications that effected his mature sculpture, the work in which he would strip the exterior of natural forms in his search for concealed structures and meanings." While serving in the Army Air Corps, Lekakis became acquainted with artist Thomas Chimes, with whom he later went to school. He was discharged from the Army Air Corps in 1945. His papers are in the Archives of American Art, including documents about his service as a camouflage instructor.

•

(Jules Marie) Auguste Leroux
(April 14, 1871-March 26, 1954)
French painter

The brother of Georges-Paul Leroux, he is described in Levitch (2006:38-39) as a World War I camoufleur and the youngest of the artists who contributed to the painting of a gargantuan panorama titled *Panthéon de la Guerre.*

•

Georges-Paul Leroux
(1877-1957)
French painter

He is listed in Coutin (1988) as a World War I French camoufleur, and is also mentioned in Cork (1994:171), which includes a reproduction of his horrific painting of soldiers trapped at night in a battlefield shell hole, titled *Hell.* He was the brother of camoufleur Auguste Leroux.

•

Maurice Leroy
(dates undetermined)
French illustrator

Leroy was a prolific book artist, and was particularly known for his erotic illustrations. He is listed in Coutin (1988) as a World War I French camoufleur.

Gerald H. Thayer (1923:161-162): "*Hiding*, and eye-trickery, of many kinds, *are*, doubtless, as old as warfare; and it would scarcely be safe to say of any one of the many types of 'camouflage' tricks used in this latest great war than nothing of the kind had ever been tried before. Yet on the whole there has unquestionably been, in the recent war-camouflage, as well as in this use of the word itself, a very considerable element of novelty."

•

Richard Levin
(1910-2000)
British television designer

His initial interest was in photography and film. He studied design at the Slade School of Art and in Paris, then turned to industrial and exhibition design. According to an online obituary in the Oldclaymorian Society Newsletter: "During the war, he worked for the Air Ministry as a camouflage officer and for the Ministry of Information as an exhibition designer." A prominent television designer and the author of *Television by Design*, he was the head of design for BBC Television for eighteen years (1953-1971).

Above World War I-era American poster, designed by Lloyd Harrison (1917), showing a camouflaged tank. Courtesy the US Library of Congress, Prints and Photographs.

•

John (Noel Claude) Lewis
(December 11, 1912-December 22, 1996)
British typographer, illustrator

Born in Rhoose, Glamorgan, Wales, Lewis initially entered medical school, then studied art at Goldsmiths College School of Art, where Denton Welch was one of his classmates.

According to an obituary in *The Independent*, "In 1935 he set up as a freelance illustrator, and was doing fairly well when war [World War II] broke out. This he spent in camouflage, a neglected art between the wars that had to be revived. It took him to Canada (posted to the far north, he pointed out that ten feet of snow was no training ground for Europe and transferred to the greens and browns of British Columbia) and Italy, where he spent a brief but enjoyable time on 'secret devices.' It also

Rubin (1943:9): "The study of camouflage provides a vast store of interesting and timely material that may be used in art and other school courses."

brought him in contact with like-minded friends: Victor Stiebel, Oliver Messel, Gabriel White and his brother-in-law Ted Ardizzone, Freddy Mayor of the Mayor Gallery, Basil Spence, Blair Hughes-Stanton and Lynton Lamb."

It was Lamb who suggested to Lewis that he become a typographer and book designer. (Camoufleur) James Gardner provided a connection to the publishing firm of W.S. Cowell Ltd in Ipswich, headed by Geoffrey Smith. The first result of their collaboration was *A Handbook of Printing Types* (1947). Lewis was on the faculty at the Royal College of Art in 1951-1963, under the direction of (camoufleur) Robin Darwin. He published a number of other influential books, among them *Printed Ephemera* (1962), *Typography: Design and Practice* (1978), *The Twentieth Century Book* (1984), and (his autobiography) *Such Things Happen* (1994).

•

Sol LeWitt
(September 9, 1928-April 8, 2007)
American artist

In Garrels (2000), there is reference to a three-part painted wood relief, made by LeWitt in 1963 but which "has never been catalogued, published, or professionally photographed," and which the artist described as a "camouflage piece." Garrels first learned of it in a telephone conversation with LeWitt on August 23, 1999, after the latter had read an essay in which Garrels "addressed the coincidental relationship between patterns in certain of LeWitt's wall drawings and the patterns used in military dazzle camouflage painting." According to Garrels, LeWitt was "not familiar with dazzle painting per se."

•

André Lhote
(July 5, 1885-January 24, 1962)
French painter and sculptor

Lhote was born in Bordeaux, and studied art at the École des Beaux Arts in that city. While living in Paris, he belonged to the Section d'Or group of Cubist artists, where he was affiliated with (camoufleurs) Roger de La Fresnaye and Jacques Villon. He is listed in Coutin (1988) as a World War I French camoufleur.

•

Alex[andre] Lippmann

(1881-1960)
French illustrator

If this is the same Alexandre Lippman, he was the grandson of Alexandre Dumas *fils*, whose daughter was Lippman's mother. This may be true since he illustrated books by Dumas. He is listed in Coutin (1988) as a World War I French camoufleur.

Russell Lynes: "Camouflage is a game we all like to play, but our secrets are as surely revealed by what we want to seem to be as by what we want to conceal."

•

Above Canadian artist Arthur Lismer, as photographed by M.O. Hammond (1930). Public domain.

Arthur Lismer

(June 27, 1885-March 23, 1969)
British-born Canadian painter

Lismer was born in Sheffield, England. He moved to Toronto, Ontario, in 1911, and worked as a designer. He was also the Principal at the Victoria School of Art and Design in Halifax, Nova Scotia. In 1918, with the intention of getting permission to sketch in the vicinity of Royal Canadian Navy docks, he was instead appointed an Official War Artist and assigned to document naval activities at the Halifax Harbor. Among the subjects he painted were dazzle-painted ships. After World War I, having moved to Toronto, he became associated with the Canadian painters known as the Group of Seven (formed in 1920). See Dykuis (1999).

•

Hugh B. Cott (1940:436): "In his posters and street signs man has only repeated and recently developed method of catching the eye which have been in use through the ages by members of the animal and vegetable worlds...The Automobile Association has adopted a system of coloration whose copyright belongs by priority to wasps and salamanders."

Philip Little

(September 6, 1857-March 31, 1942)
American artist

Born in Swampscott MA, Little studied at MIT (1875-76) and the Boston Museum of Fine Arts School (1881-82). He also served as a major in the Massachusetts National Guard. *New York Times* ("Philip Little," 1942): "During the first World War he aided the Navy Department in designing camouflage patterns for North Atlantic vessels."

Warner (n.d.) states that two photographs (*facing page*) of camouflage schemes, designed by Philip Little, were forwarded to him by Boston Yard on

January 10, 1918. They show the application of Little's designs to the *USS Yacona* (*top*) and the *USS Aztec* (*bottom*). Judging from US Navy records, both photographs were probably taken on December 20, 1917. Photos courtesy US Naval Historical Center (NH 102576 and NH 543).

•

Robert (Augustus) Lockwood
(January 24, 1893-January 24, 1949)
American painter

Born in California, Lockwood served in the American Camouflage Corps in France during World War I. Following the war, he studied at the Art Institute of Chicago, then moved to California, where he worked as an architectural draftsman (for the Myron Hunt firm) and film production artist. According to an online AskArt biography, in 1931 he was living in San Francisco "where his model, who was in love with him, tried to take his life." He survived that, but then died in 1949 in an automobile accident on his birthday.

•

Charles Loloma
(January 7, 1921-1991)
Native Indian Jeweler

Loloma was born in the Southwestern US to Hopi parents. His father was a weaver and moccasin maker; his mother was a basket maker. After completing high school in 1941, he was drafted into the US Army. "Charles Loloma" (n.d.): "He became a camouflage expert and was stationed in Missouri... With the help of the GI Bill, Charles attended the School of the American Craftsman at Alfred University. There he studied design, mechanical drawing, ceramic chemistry and marketing." He was later associated with Arizona State College and the Institute of American Arts in Santa Fe NM.

•

Will (William Frederick) Longstaff
(December 25, 1879-July 1, 1953)
Australian artist

Born in Ballarat, Victoria, Longstaff studied art at Grenville College in Australia and in London at the Heatherley School of Fine Art.

According to Gray (1986), "In October 1915 he joined the Australian Imperial Force, embarking as captain for Egypt in November with the 1st Australian Remount Unit. Longstaff was invalided to England in October 1917 and next year was selected for camouflage training and attached to the 2nd Division Headquarters as divisional artist under the Australian Records Section war art scheme... His thoroughness contributed to successful concealment of allied activity before the offensive of 8 August and probably saved many lives."

•

Fred R. Lorenz
(dates undetermined)
American architect

Lorenz studied architecture at the University of Pennsylvania. Following the war, he practiced architecture in New York. He is listed in Embury (1917) as a World War I American camoufleur.

•

Facing page Abbott H. Thayer believed that colorful high contrast natural patterns functioned not as advertising (in sexual selection) but as camouflage. It was largely on that basis that he advised against the use of khaki field service uniforms. The caption for this illustration (Thayer 1909:152) reads as follows: "Here the spectator will discover, if he recedes far enough...that all three of the monochrome butterflies, even the dimmest, can be seen further, or in less illumination, than the normally and brightly patterned one. This latter *fades first*. This shows how contrasted juxtaposed color-notes efface each other, so that contrary to the old theories they are not so good as monochrome for revealing the wearer, even in the open, while, seen through the average tracery of outdoor vegetation, they almost *guarantee* disguise and concealment."

Jacques Lacan: "[Mimicry] ...is not a question of harmonizing with the background but, against a mottled background, of becoming mottled—exactly like the technique of camouflage practiced in human warfare."

...

Thayer (1923:160): "A world, a universe almost, of veritable wonder and enchantment, awaits each new student of these consummate masterpieces of optical illusion in the field of nature. As an astonished friend of mine once remarked, 'These things [examples of protective coloration] are enough to give one a new belief in God!"

Amy Lowell
(February 9, 1874-May 12, 1925)
American poet

Lowell published a poem about World War I ship camouflage in *The Dial* (1918:403), titled "Camouflaged Troop-Ship: Boston Harbor" (see following pages). It was reprinted a year later in a book of her poetry titled *Pictures of the Floating World* (Lowell, 1919). Her interest in camouflage may in part have sprung from her association with the Dublin NH artists colony. Surely, she was well aware of her neighbor Abbott H. Thayer's involvement in the subject.

•

Matthew Luckiesh
(September 14, 1883-November 2, 1967)
American physicist

Luckiesh (pronounced LEW-kish) was born in Maquoketa IA. He earned degrees in electrical engineering at Purdue University and Iowa State University. In 1910, he began working for General Electric, and served for many years as Director of the GE Lighting Research Laboratory. In a period of fifty years, he originated eleven US patents, wrote nearly thirty books, and published more than 850 articles. During World War I, he wrote extensively on military camouflage, especially naval camouflage, in *Scientific American*. In his best-known book of essays, titled *Visual Illusions: Their Causes, Characteristics and Applications*, first published in 1922 (Luckiesh, 1965), there is a detailed and informative chapter on camouflage.

•

Camouflaged Troop-Ship
Boston Harbor

Uprightness,
Masts, one behind another,
Syncopated beyond and between one another,
Clouding together,
Becoming confused.
A mist of gray, blurring stems
Platformed upon horizontal thicknesses.
Decks,
Bows and sterns escaping fore and aft,
A long line of flatness
Darker than the fog of masts,
More solid,
Monotonous gray.
Dull smokestacks
Plotting lusterless clouds.
An ebb-tide
Slowing sucking the ruse of a harbor
Seaward.

The ferry turns;
And there,
On the starboard quarter,
Thrust out from the vapor-wall of ships;
Color.
Against the perpendicular:

Obliqueness.
In front of the horizontal:
A crenelated edge. A vessel, grooved and conical,
Shell-shaped, flower-flowing,
Gothic, bizarre, and unrelated.
Black spirals over cream-color
Broken at a half-way point.
A slab of black amidships.
At the stern,
Lines:
Rising from the water,
Curled round and over,
Whorled, scattered,
Drawn upon one another.
Snakes starting from a still ocean,
Writhing over cream-color,
Crashed upon and cut down
By a flat impinging horizon.

The sea is gray and low,
But the vessel is high with up-thrusting lines:
Hair lines incessantly moving,
Broad bands of black turning evenly over emptiness,
Importing upon their circuits,
Teasing the eye with indefinite motion,
Coming from nothing,
Ending without cessation.

Drowned hair drifting against mother-of-pearl;
Kelp-aprons,

Shredded upon a yellow beach;
Black spray
Salted over cream-gray wave tops.

You hollow into rising water,
You double-turn under the dripped edge of clouds,
You move in a hundred directions,
And keep to a course the eye cannot see.
Your terrible lines
Are swift as the plunge of a kingfisher;
They vanish as one traces them,
They are constantly vanishing,
And yet you swing at anchor in the gray harbor
Waiting for your quota of troops.
Men will sail in you,
Netted in whirling paint,
Held like brittle eggs
In an osier basket.
They will sail,
Over black-skinned water,
Into a distance of cream-color and vague shadow-
shotted blue.

The ferry whistle blows for the landing.
Start the engine
That we may not block
The string of waiting carts.

—Amy Lowell (1918)

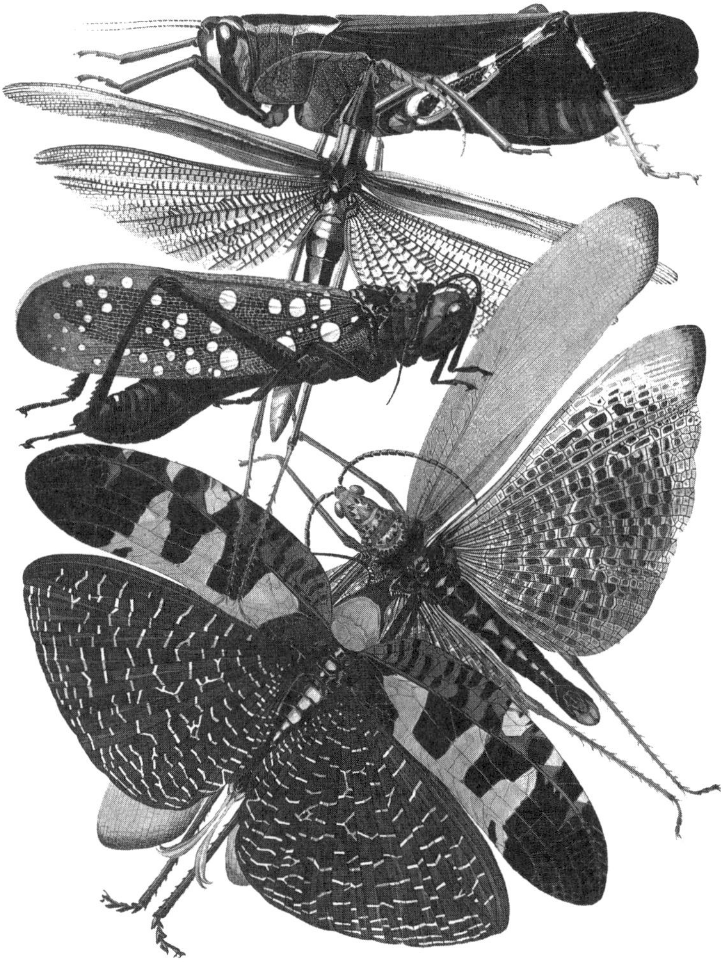

Above Disruptive patterns in various insects as portrayed by French designer E.A. Seguy (c1920), from *Seguy's Decorative Butterflies and Insects in Full Color* (New York: Dover Publications, 1977), in the Dover Pictorial Archives.

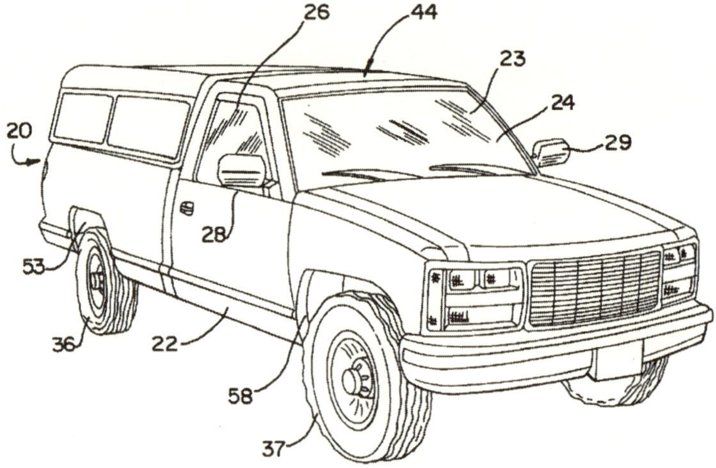
26
44
23
24
29
20
28
53
36
22
58
37

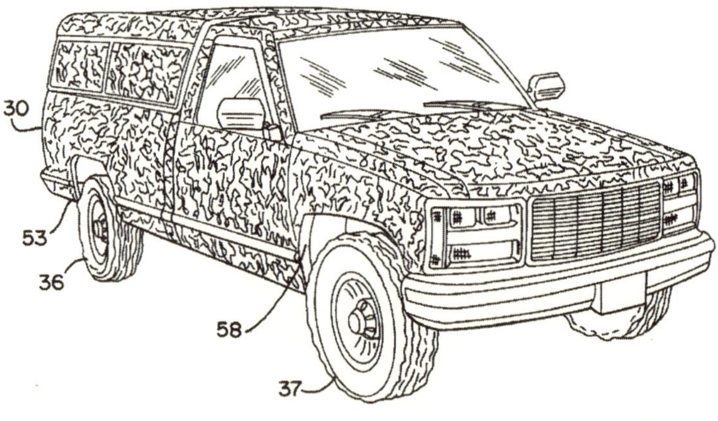
30
53
36
58
37

M

Jimmy (James Stuart) MacDonald
(March 28, 1878-November 12, 1952)
Australian art critic, gallery director

Born in Melbourne, MacDonald was educated in Australia at the National Gallery of Victoria School, in London at the Westminster School of Art, and in Paris at the Académie Julian. He then moved to New York, where he taught high school for about six years. According to Serle (1986): "[After his recovery from being seriously wounded at Gallipoli, he worked] as a pay sergeant in 1916-17, mainly in England. In June-August 1918 he worked as a camouflage artist with the 5th Division in France." He wrote several books, the first of which concerned the work of Australian painter Frederick McCubbin, the father of Australian artist and camoufleur Louis F. McCubbin. For many years, he was art critic for the *Melbourne Herald*. In 1928 he became director of the National Art Gallery of New South Wales, and, in 1936, director of the National Gallery of Victoria, a position that was not renewed four years later.

Continues Serle (1986): "In 1943 he was first witness on behalf of those who brought an action against the award of the Archibald prize to [Australian camoufleur] William Dobell for his portrait of [camoufleur] Joshua Smith: MacDonald described it as a 'pictorial defamation of character.'

...A convivial drinker at the Savage Club, he [MacDonald] was a man of charm, tactlessness and combativeness."

•

Facing page Drawings submitted for US Patent No. 6,017,0793 titled "Vehicle Camouflage," invented by Joseph Warner (2000).

Kenneth (Stevens) MacIntire

(1891-1979)
American painter and sculptor

MacIntire was born in Boston, where he studied at the Boston Museum of Fine Arts School, and the Architectural Club School of Design. During World War I, he worked under Everett L. Warner as a member of the staff of artists who designed US ship camouflage. In Warner (1919b:107), it is stated that "The work began in the model-making room, where about half a dozen skilled men under Ensign Kenneth MacIntire were kept constantly busy in the production of miniature wooden models, which were accurately made to a fixed scale from blue-prints of the vessels required. The reader may get a general idea of the size of these models from the dimensions of the [US] President's ship, the *USS George Washington.* It was one of the largest of them, and measured about twenty-two and a half inches in length."

Following the war, MacIntire lived in Washington DC for about ten years, then moved to Hollywood, where he was known for his portraits of film stars and other celebrities.

In the annual report of the Museums of the Brooklyn Institute of Arts and Sciences for 1919 (under Accessions) the following items are listed as having been loaned to the Department of Natural Science:

"*Bureau of Construction and Repair, Navy Department, Washington DC.* Ship models, diagrams, photographs and other naval camouflage records.

Cameron Clark. 7 airplane models, several of which are camouflaged.

Ensign Kenneth S. MacIntire. Painting of a camouflaged transport; by [Frederick] Waugh.

W.A. Mackay. 10 photographs. 6 drawings and diagrams. 2 periscopes. 1 turntable for ship models. 40 models. 300 government designs. 1 group showing submarine in action and other minor objects."

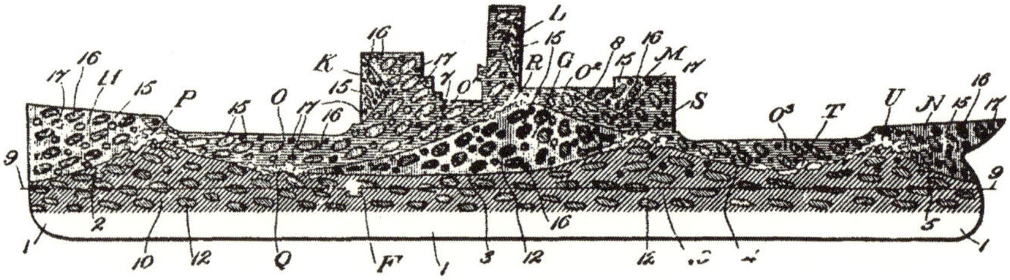

Above Drawing submitted for US Patent No. 1,305,296 titled "Process of Rendering Objects Less Visible Against Backgrounds," invented by William Andrew Mackay (1919). US Patent Office.

William Andrew Mackay

(1876-July 26, 1939)
American mural painter

Born in Philadelphia, Mackay was educated at City College in New York, the Académie Julian in Paris, and the American Academy in Rome. In 1935, he completed three large murals commemorating the achievements of Theodore Roosevelt for the

Above WWI-era US government photograph of the *USS DeKalb* painted in a Mackay disruptive low visibility camouflage scheme (1918). Photo courtesy US Naval Historical Center (NH 54662). *Below* Photograph of an aging William Andrew Mackay (n.d.) by Peter A. Juley & Son. Courtesy Photograph Archives, Smithsonian American Art Museum (J0004314).

Roosevelt Memorial Hall in the American Museum of Natural History. He painted other murals for the Library of Congress, the House of Representatives in St Paul MN, the 1939 New York World's Fair, and other locations.

New York Times ("William A. Mackay" 1939:11): "During the World War Mr. Mackay was Chief Camouflage Artist, Second District, United States Army."

Mackay invented one of five ship camouflage schemes (referred to then as "measures") approved by the US Shipping Board during World War I for use on merchant ships.

New York Times ("Shows New Ship Painting," 1917:20): "[In a Navy exhibition at the Electrical Show at the Grand Central Palace in New York, there are] two navy [ship] models, prepared by William Mackay, who has been making experiments in lessening visibility by color effects. One ship is painted flake gray, the conventional navy war paint. The other is painted with the new war paint. Back of the two ships is a scene representing the sea and sky. At thirty feet away the difference in the visibility of the two ships is noticeable, the gray one standing out against the scene behind, while the other fades into the background."

In Whiting (1928), there are repeated references to Mackay in relation to his duties as the head of the camouflage section for the US Shipping Board.

Because Whiting's writing style is novelistic, it is uncertain to what extent his book is factual. But Mackay's name is used in reference to US Shipping Board camouflage, as is the name of a (presumably actual) ship camoufleur named Phil Duroc. If Whiting's text (written largely in the form of conversations) is believable, there appears to have been some resentment between Mackay's US Shipping Board artists and those working in Washington DC under Everett L. Warner. For example, there is mention of Mackay "quietly evolving his own school for camoufleurs," without the knowledge of the Warner's section.

Niccolo Machiavelli: "Wise commanders never attempt to win by force what they can win by fraud."

Above Two views of the *USS Isabel* during World War I, painted in a camouflage scheme designed by William Andrew Mackay (c1917). Photos courtesy US Naval Historical Center (NH 00544 and NH 000545).

John D. Whiting (1928:80): "'The queerest lot of wights are working here, scraped from all the corners of Bohemia. But they work; [William Andrew] Mackay is a hustler.'

'Drives you, does he?'

'No, he innoculates us; he has more ideas than all your admirals together. And the boys are simply nuts about him.'"

Covert (2007:53): "In 1920, the United States Navy established a camouflage training school led by William Andrew Mackay. The school carried out camouflage experiments and created the *Handbook on Ship Camouflage*, published in 1937. The Handbook contained an evaluation of dazzle and directions for how to apply various camouflage patterns to ships."

•

Alister MacKenzie

(August 30, 1870-January 6, 1934)

British golf course architect

MacKenzie, one of the century's leading golf course designers, was born in Wakefield, Yorkshire, England. Initially trained as a medical doctor, he served as a surgeon with the Somerset Regiment in South Africa in the Boer War. In the process, he studied the effective use of camouflage by the Boers. MacKenzie (1919:42): "The brilliant successes of the Boers were due to great extent to their making the best use of natural cover and the construction of artificial cover indistinguishable from nature."

During World War I, MacKenzie returned to military service not as a surgeon but as a camouflage expert. Following the war, he turned his full attention to golf course architecture, with the result that he is remembered as the designer of some of the world's

Above The Mackay camouflage system applied to the *USS Winfield S. Cahill* during World War I. US Navy photograph.

finest golf courses, among them Augusta National Golf Club (Augusta GA), Cypress Point Club (Monterey Peninsula CA), Royal Melbourne Golf Club (Melbourne, Australia), Pasatiempo Golf Club (Santa Cruz CA), Crystal Downs Country Club (Ann Arbor MI), Lahinch Golf Course (Ireland) and Meadow Club (Fairfax CA). In 1920, he published *Golf Architecture*, in which he wrote that "the chief object of every golf course architect worth his salt is to imitate the beauties of nature so closely as to make his work indistinguishable from nature itself." See MacKenzie (1915) and (1919), and *New York Times* ("Alister MacKenzie" 1934).

•

(Leroy) Daniel MacMorris
(1893-1981)
American designer

MacMorris was born in Sedalia MO. He studied at the Kansas City Art Institute, at the Art Institute of Chicago with Alphons Mucha, and in Taos NM with Leon Gaspard. Following World War I, he lived and worked in Paris, then returned to the US, where he studied in New York with Joseph Pennell, Robert Henri and George Bridgman. His papers are at the University of Missouri in Kansas City (MacMorris, n.d.).

In Nibbe (n.d.), he is described as having worked for Eastman Kodak in Rochester prior to World War I, then serving as a photographer with the US Army aviation branch. During World War II, he was an "instructor at a camouflage camp in Utah" for three years.

In the late 1950s, MacMorris was in charge of the interior of Memory Hall of the Liberty Memorial Museum in Kansas City MO. In that capacity, he cut apart and rearranged selected portions of a huge World War I-era panorama, titled *Panthéon de la Guerre*. Originally painted in Paris by French artists, it had been the world's largest panorama, with five thousand full-length portraits of heroic wartime Allies.

According to Levitch (2006), MacMorris used less than seven percent of the original massive mural to produce a rearranged collage (sixty-nine feet in width) that is now on public view in Kansas City at the National World War I Museum (housed in the old Liberty Memorial). Few of the figures in the remaining mural can be identified, but among those who are recognizable are French camoufleurs Auguste Leroux (who also contributed to the panorama), Lucien-Victor Guirand de Scevola, Louis Abel-Truchet, Jean-Louis Forain, André Mare, and André Devambez.

•

Above World War I German soldiers examining a captured tank.

Adolf Portmann (1959:9): "Camouflage implies a seeing eye from which to hide."

...

Abbott H. Thayer (in a letter to Franklin D. Roosevelt, dated September 3, 1917): "Wm. A. Mackay's 'vibrating color' [ship camouflage plan] is only an accessory of a scheme and of trifling importance. If the checkerboard pattern is his, it simply makes a vessel a light gray."

...

John D. Whiting (1928:81): "The voice [of William Andrew Mackay] had a slight Irish brogue, put on, Hugh thought, for amusement. Peering curiously at him through the doorway, Hugh saw a fiery, dynamic little man with an odd twist to his mouth. He was glaring at a giant blackboard where the names and locations of all the vessels to be camouflaged were neatly marked in yellow chalk. It was strange enough, this wild-eyed little artist holding in his skillful hands the wires of a switchboard that embraced maritime New York."

William (Howard) Magoffin
(1859-January 30, 1941)
American manufacturer

Magoffin was the president of the Atlantic Dryer and Varnish Company, in Philadelphia, which he founded in 1880. According to the *New York Times* ("William H. Magoffin" 1942), "During the World War Mr. Magoffin's firm did much of the ship camouflage work in the Philadelphia area."

•

Robert Mahias
(October 12, 1890-August 24, 1962)
French painter, illustrator and designer

Mahias was born in Brussels, Belgium, of French ancestry. He studied at the École des Arts Décoratifs and the École des Beaux-Arts. He moved to Canada, where he taught at the École des Beaux-Arts in Montréal, then returned to France to teach. He is listed in Coutin (1988) as a World War I French camoufleur.

•

Charles Alexandre Malfray
(July 19, 1887-May 28, 1940)
French sculptor

Malfray was born in Orléans, France, and studied at the École des Beaux-Arts. He is listed in Coutin (1988) as a World War I French camoufleur. He sustained permanent injuries from being gassed at the Battle of Verdun. Following the war, he was responsible for several monuments erected to honor people he knew who had perished in the war.

•

Pavel Andreevitch Mansurov
(March 26, 1896-1983)
Russian painter and designer

Mansurov was a teacher at Inkhuk (Institute of Artistic Culture) in Moscow, where he briefly headed its experimental section. Much of his innovative work was devoted to straddling the boundary between art and science, often incorporating found materials from nature, such as twigs and butterflies.

According to Wescher (1971:110), in one of Mansurov's artworks "a butterfly with outspread wings was mounted on a crumpled piece of wrapping paper painted with white highlights that are echoed by similar white patches on the wooden plank making up the base of the relief..."

Mansurov himself said that in the course of his investigations into the phenomena of mimicry he became aware of how animals are protected by their colors, which blend with their surroundings. On the basis of this phenomenon he took to pasting white feathers into drawings on a white ground, and for one collage he deliberately chose a butterfly whose light color shadings matched the silvery tones of the birchbark he was using."

Wolfram (1975:63) describes the same experiments, noting that "during his [Mansurov's] investigation into this kind of 'visual mimicry' he became aware of the possibilities of camouflage in nature..."

Irvin Rock (1995:117): "... camouflage, both as it occurs in nature and as it is produced by humans, makes use of the [Gestalt] principles of organization to yield perceptual objects different from the actual objects present, either by obscuring the actual objects present or by creating perceptual objects not actually present."

Henri Marchal
(June 22, 1878-?)
French painter

This Henri Marchal should not be confused with a French archaeologist of the same name and time period, most of whose work was done in Southeast Asia. This Marchal appears to have been a painter of landscapes and still lifes, and is listed in Coutin (1988) as a World War I French camoufleur.

•

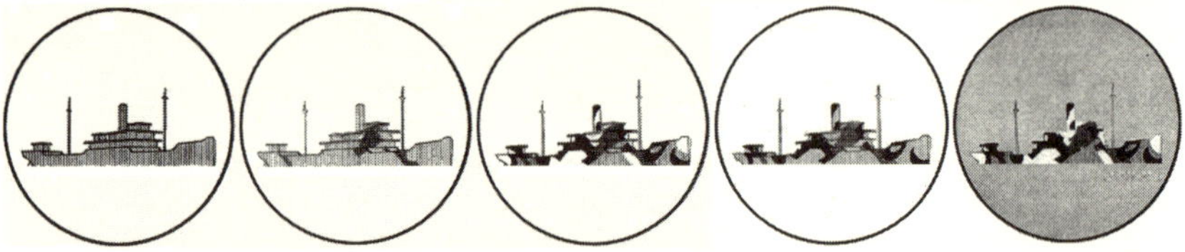

Above A sequence of illustrations from a WWI publication on dazzle ship camouflage, demonstrating successive stages of increasing visual distortion, as might be observed through a U-boat periscope. See Stockbridge (1920). *Below* The *USS Keresaspa* in camouflage, c1918. Courtesy US Naval Historical Center (NH 101959).

Pierre Mardrus
(dates undetermined)
American artist?

An American of this name, who may or may not have been an artist, is mentioned by Louis Bouché in his papers, on file in the Archives of American Art (described as the unfinished notes for an autobiography). Bouché (n.d.:3): "So I invited Pierre Mardrus who was in my camouflage outfit. They [Bouche's girlfriends] liked Pierre and he entered into the spirit beautifully. He is still a good friend."

•

Louis-Casimir Marcoussis (born Ludwig Markus)
(1883-1941)
Polish-born French painter

Marcoussis was born in Warsaw, Poland, and studied art in Krakow at the Academy of Art. He moved to Paris in 1903 to attend the Académie Julian, where he became acquainted with (camoufleur) Roger de La Fresnaye. He was a follower of the Cubists, both before and after World War I. He is listed in Coutin (1988) as a World War I French camoufleur. See Osborne (1981).

André Mare
(1885-1932)
French painter, designer

Mare was a member of the Puteaux Cubists, along with (camoufleur) Jacques Villon, Juan Gris and Fernand Léger (who had been a childhood friend). As listed in Coutin (1988) and elsewhere, he was a World War I French camoufleur.

According to Roshwald (2002:305), "Why did he [Mare] choose Cubism as his medium—an art form which was being denounced as 'Boche' [German] in Paris? Because he found it to be best suited to capturing the essence of this abstract and industrialized war. Designated as a camouflage artist, he was sent to Italy, where he was dazzled by what he described as a country 'whose people love life too much to do battle.' Upon his return to France, he abandoned Cubism in order to depict life in his art—that life which he knew would soon abandon him."

After the war, Mare turned to interior design, and, in partnership with Louis Sue, formed the Compagnie des Arts Francais (1919-1928), by which they designed interiors for such wealthy clients as Helena Rubinstein and Jean Patou, as well as the interior of the ocean liner *Normandie*.

•

Henri-Justin Marret
(February 15, 1878-July 25, 1964)
French painter

Marret was born in Paris. He is listed in Coutin (1988) as a World War I French camoufleur. He was one of the artists who contributed to the murals in the church of Notre-Dame-des-Missions-du-Cygne d'Enghien, near Paris.

•

Étienne de Martenne
(November 10, 1868-February 9, 1920)
French painter

Born in Lyons, he was greatly interested in the Impressionists. He is listed in Coutin (1988) as a World War I French camoufleur. He was a friend of Louis de Monard and Robert Noir.

•

Keith Martin

(January 22, 1910-?)
American arts administrator

Anton Ehrenzweig (1953:44): "An interesting borderline case between a combination game and artistic activity is the reading of a crime story...[in which] the author never allows the clues to narrow down so as to point definitely to a single culprint; he keeps them as ambiguous and disconnected as possible...I, not being a crime-story fan myself, am always surprised how little is necessary to achieve this last-minute logical satisfaction [of solving the crime]. A few odds and bits which have been carefully tucked away under a dazzling camouflage of insignificant details are triumphantly dragged to the surface and displayed as the logical fundament of the story."

...

Jo Mielziner (*New York Times*, 1976:30): "On the stage my job was to make people grasp a situation as quickly as possible. In camouflage, my job was to keep them from grasping it at all."

Right Hollywood motion picture personnel (c1918) demonstrating a camouflaged dummy cannon, which they "fired" with the use of special effects. Courtesy US National Archives (165-WW-71D-1).

Grenville Rickard (1942:190): "This nucleus [during WWI, in the American Camouflage Corps] of artists, sculptors, painters, advertising men, architects, and the like, was soon afterwards wholesomely leavened with the bringing in of technicians, men who could do things with their hands: carpenters, metal workers, scene shifters, electricians, scenic artists, cooks, et cetera, and a few weeks later by a fluent group from Hollywood... The personnel was unique and the whole made up a rather baroque institution."

Born in Perth Amboy NJ, Martin studied at Harvard University. He was director of the Kansas City Art Institute from 1939 to 1943, then headed the Art Department at Syracuse University from 1946 to 1948. As head of the Fine Arts Department at IBM from 1946 to 1953, he originated a traveling exhibition of working replicas of Leonardo da Vinci's inventions. Near the end of his career, he was the first director of the Roberson Center for Arts and Sciences in Binghamton NY.

According to *Time* magazine ("Camouflage School," 1940:28), during World War II, Martin initiated the "first class in industrial camouflage in the US" at the Kansas City Art Institute, while he was its director. Enrolled in the course were "some forty students: artists, sculptors, interior decorators, advertising designers, photographers, students of architecture and engineering. Lecturer: the Institute's director, husky, 30-year-old Keith Martin, onetime Harvard crewman and portrait painter. Purpose of the course: to explore ways & means of protecting vital industries, plants, stores from attack by air.

Beginning with a study of aerial photographs, Director Martin and his students will make scale models of buildings, take reconnaissance flights in TWA and Naval Reserve planes. From the No. 1 US camoufleur, Director Homer Schiff Saint-Gaudens of the Carnegie Institute's department of fine arts, Director Martin got a general line on the problem. Says Saint-Gaudens, a lieutenant colonel in the Engineers Reserve Corps: 'Camouflage is just plain Injun fighting.'"

•

Noel Martin
(April 19, 1921-)
American graphic designer

Martin studied at the Art Academy of Cincinnati in 1939 to 1941, and again in 1945 to 1947. His studies were interrupted when he served as a camoufleur in the Army Air Force. For many years, he was associated with the Cincinnati Art Museum, and served as its graphic designer beginning in 1947. He also taught at the adjoining Art Academy from 1951 to 1957, and later at the University of Cincinnati. He achieved national prominence in 1953, when his work was included (along with that of Leo Lionni, Herbert Matter and Ben Shahn) in an exhibition titled *Four American Designers* at the Museum of Modern Art. See Schoener (1989).

•

Raymond Martin
(dates undetermined)
French sculptor

He is listed in Coutin (1988) and Ducasse (1962) as a World War I French camoufleur. With (camoufleur) Marcel Damboise, he was a member of the Groupe des Neuf.

•

André Edouard Marty
(April 16, 1882-August 1974)
French printmaker, illustrator

Born in Paris, Marty studied at the École des Beaux-Arts. He is listed in Coutin (1988) as a World War I French camoufleur. He was a prolific illustrator for such prominent publications as *Vanity Fair*, *House and Garden*, *Harper's Bazaar*, and *Vogue*. He also designed posters, advertisements, books, costumes and theatrical sets. As a member of the jury for the Exposition Internationale des Arts Décoratif in 1925 (from which the name Art Deco would later be coined), he was a significant contributor to the emergence of that style.

•

Zerubavel (1993:1): "Separating entities from their surroundings is what allows us to perceive them in the first place. In order to discern any 'thing,' we must distinguish that which we attend [to] from that which we ignore. Such an an inevitable link between differentiation and perception is most apparent in colorblindness tests or camouflage..."

Above Cartoon advertisement for Pepsi-Cola, by American illustrator Otto Soglow, published in *Life* magazine on December 21, 1942. Author's collection.

Jack Masey
(dates undetermined)
American exhibit designer

Masey served in the Ghost Army during World War II. Following that, according to the website on that unit (www.ghostarmy.org), he attended Yale University and became an exhibit designer. As Director of Design for the US Information Agency from 1951 to 1979, it was he who designed the setting for the famous "kitchen debate" between Nikita Krushchev and Richard Nixon in 1959. He also designed the US Pavilion at the Montreal World's Fair in 1967. He shared his recollections of those years in Masey (2008).

•

Jasper Maskelyne
(1902-1973)
British stage magician

Maskelyne was the youngest of three generations of British stage magicians. During World War II, he joined the Royal Engineers, and became a leading contributor to a series of wartime deceptions. Under the direction of Geoffrey Barkas, these were carried out by a team of 14 camoufleurs, informally known as the Magic Gang. His own account of this is in his book *Magic: Top Secret* (Maskelyne, 1949), but its accuracy has been challenged. (Was it trickery?) He is listed in Reit (1978:138) as having worked with Barkas in WWII, as well as in Trevelyan (1957).

•

Frank (Henry) Mason
(1875-1965)
British marine artist

Mason was a self-taught painter, who worked in watercolor, printmaking, poster design, and the construction of ship models. Yardley (1996) discusses his service as a soldier in World Wars I and II, and as a camouflage painter.

Frank Bird Masters
(September 25, 1873-March 25, 1955)
American photographer, illustrator

Born in Watertown MA, Masters completed his undergraduate degree at MIT, then studied art with Howard Pyle. During World War I, he worked as a camoufleur with the US Shipping Board at Jacksonville and Tampa FL, Charleston SC, Savannah GA, and Washington DC.

•

Mateo Maté
(dates undetermined)
Spanish artist

In a series of artworks called *Uniformed Landscapes*, Maté makes use of the ambiguity of the word "uniform," in coincident allusions to military garments, repetitive overall sameness, conformity, and so on. As described by Maite Méndez Baiges (2007), he integrates military camouflage patterns into reproductions of 19th century landscape paintings for the purpose of returning camouflage to a non-military, artistic context, "where it justly belongs."

•

George Hans Eric Maunsbach
(1890-1969)
American painter

Also sometimes listed as Christian Erik Maunsbach or Erik Christian Maunsbach, he was a student of Anders Zorn, John Singer Sargent and William Merrit Chase. He is cited in Nibbe (n.d.) as camoufleur. See also *New York Times* ("Eric Maunsbach," 1969:47).

•

Stanley McCandless
(1897-August 4, 1967)
American stage lighting designer

McCandless earned his BA degree from the University of Wisconsin in 1920, and a Master of Architecture degree from Harvard School of Architecture in 1923. In the following year, he joined the New York firm of McKim, Mead and White. In 1925, he began to teach stage lighting at Yale University, a position he continued in until his retirement in 1965. He was the author of *Methods of Lighting the Stage*, and *Syllabus of Stage Lighting*. He was a lighting consultant for Radio City, National Gallery, Trans World Airlines Terminal at JFK International Airport, and United Nations Assembly Hall. He designed the lighting for theatres at Yale, Amherst, Sweetbriar and Williams College. *New York Times* ("Stanley McCandless," 1967:21): "During World War II he served with the Office of Civil Defense and the camouflage section of the Army Engineers."

•

Henry (Bainbridge) McCarter
(July 5, 1886-November 2, 1942)
American illustrator

McCarter was born in Philadelphia, and studied under Thomas Eakins at the Pennsylvania Academy of the Fine Arts. He also studied in Paris with Puvis de Chavannes and others.

His illustrations were often published in leading American magazines, including *The Century*, *Scribner's* and *Collier's*. He taught for many years at the Pennsylvania Academy. One of his students there was painter Elisabeth Spalding, who gave a lecture on McCarter in 1944. In the transcript of that, in the Archives of American Art, she talks about McCarter's involvement with camouflage during World War I (see Spalding 1944).

•

Herbert McClintock
(1906-c1985)
Australian painter, illustrator, singer

Born in Perth, Australia, McClintock worked initially as a sign painter and apprentice engraver. He then studied art at Melbourne. Like his friend (and later camoufleur) Roy Dalgarno, he joined the Communist Party, and throughout the 1930s and 40s, produced editorial cartoons for various trade union and political publications. While painting, he used the pseudonym Max Ebert, but he also performed as a singer called McClintock. According to Kerr ("Herbert McClintock," n.d.): "During World War II McClintock was rejected for active service on

Robert Muir Graves (2002:144): "Alister Mackenzie, the 'father' of wartime camouflage, never hesitated [in the design of golf courses] to obscure a problem for the less alert or less observant. On several occasions, he observed that camouflage and course design had aspects in common."

...

Annie Dillard (1974:17): "A fish flashes, then dissolves in the water before my eyes like so much salt. Deer apparently ascend bodily into heaven; the brightest oriole fades into leaves. These disappearances stun me into stillness and concentration; they say of nature that it conceals with a grand nonchalance, and they say of vision that it is a deliberate gift, the revelation of a dancer who for my eyes only flings away her seven veils. For nature does reveal as well as conceal: now-you-don't-see-it, now-you-do."

...

Irwin (1917): "As soon as this war [World War I] is over and Tommy [the British GI] resumes his civilian activities, the British will be getting out new editions of those dictionaries wherein, they vainly believe, is embalmed the standard English language of the world. And in the C section, probably without the comment of 'argot' or 'slang' or 'colloquial,' or any other mark of reputability, will appear 'camouflage.' Doubtless it will make its way, though more slowly, into those pure repositories of the tongue published in Boston or New York; for we are sending an army over to France just now, and the first new word they will learn at the Front will be *camouflage*."

medical grounds and conscripted into an iron factory. Later he was moved to camouflage. In July 1943 he began work with the Allied Works Council and Civil Construction Corps, painting images of labor beside [camoufleur] William Dobell."

•

Paul McCobb
(1918-March 10, 1969)
American furniture and display designer

Born in Boston, McCobb intended to be a painter, but at age 20 he took a job as a display and interior designer for a major Boston department store. Later, he served as a design consultant for Columbia Records, Singer Manufacturing Company, Bell and Howell, Alcoa, Goodyear Tire and Rubber Company, Philco Corporation, and Remington Rand. *New York Times* ("Paul McCobb," 1969:33): "During World War II, he served in the Army Camouflage Corps."

•

Louis F(rederick) McCubbin
(March 18, 1890-December 6, 1952)
Australian artist, gallery director

McCubbin was born in Melbourne, Australia, and was privately educated. According to Gray (1986), during World War II, he "was deputy director of camouflage, South Australia (1941-43), and was a member of the art committee of the Australian War Memorial."

•

Gil McElroy
(1956-)
Canadian writer and curator

McElroy is an independent curator, art critic, poet and visual artist. Among his publications are *Gravity & Grace: Selected Writing on Contemporary Canadian Art* (2001), and *Razzle Dazzle: The Uses of Abstraction* (McElroy 2008), the latter of which was the catalog for an art and camouflage exhibition.

•

John McIntosh Patrick

(1907-?)
Scottish landscape painter

James Elkins (1996:82-83): "Sometimes I dress to fit in, so that I can blend with the crowd and not attract attention. When I do that I am imagining the world as a picture, and I want to find a place in the picture where I can disappear. As I walk down the sidewalk or enter the conference room, I want to be nothing more than a blotch among other blotches. But on many other occasions I have exactly the opposite effect in mind. I dress in my best clothes and I make sure I look just right, because I want to make an impression: I need to stand out; I want to be noticed."

The son of an architect, he was born in Dundee, Scotland, and studied at Glasgow School of Art (1924-28). He was a member of the Royal Scottish Academy. According to his biography on the Easyart website, "In 1940, James McIntosh Patrick was called up into service with the Camouflage Corps, with whom he spent the war."

•

Jo (Josephine) Mead

(1918-October 23, 1999)
American interior designer, historic preservationist

Above Cover of Maite Méndez Baiges' study of camouflage in relation to experiments in contemporary art (Méndez Baignes 2007).

Born and raised in Lansing MI, Mead moved to Chicago at age 19. She attended the Institute of Design, where she studied with Laszlo Moholy-Nagy. There, according to the *Chicago Sun-Times* ("Jo Mead," 2000), "she was asked to mask US vessels in World War II." According to the same source, she "did such a good job camouflaging Navy destroyers that the enemy couldn't even tell in which direction the ships were going." After the war, she founded Jo Mead Designs, which produced large scale sculpture for homes, restaurants and hotels. In her later years, she lived in Galena IL. Many of the things she collected are now housed at the House on the Rock, near Spring Green WI.

•

Robert Medley

(1905-1994)
British painter

Medley studied at Byam Shaw School of Art and Slade School of Art. He also studied in Paris for two years. As an affiliate of the Group Theatre Project, he made important contributions to costume and set design. He was the Head of Theatre Design at the Slade from 1949 to 1958. According to the Tate

Archives Online, he served as a camouflage officer during World War II. He is also mentioned by Barkas (1952) as having served with him in North Africa.

•

Juan Menchaca
(1910-?)
American painter

Born in Fort Worth, Texas, Menchaca attended the Texas School of Fine Arts and the Denver Art Museum school. He also worked as a WPA artist, and served as the Chief Curator at the Denver Museum of Art.

There is an interview in the Archives of American Art (Menchaca, 1964) in which he states that during World War II, "I was interested in either getting in the service or to find a place where I could be of more value to the war effort and I made one application with the Remington plant that had just begun to operate, and the next thing I knew I was altogether in a different field. I was with Lowry Field and my job or my assignment was to become an instructor in camouflage. This was the time when the Army or any of the military operations could change from day to day, and after two months of training the camouflage unit was eliminated because it was felt that the camouflage was part of the engineers instead of the Air Corps."

•

Maite Méndez Baiges
(dates undetermined)
Spanish art theorist, historian

Méndez Baiges is a professor of contemporary art at the Universidad de Málaga in Spain. She has also taught at the Spanish Academy in Rome, the Independent University of Madrid, and the University of Michigan in Ann Arbor. Her writings on camouflage have addressed its impact on recent art, both Modern and Postmodern. In addition to various papers, she has also published *Camuflaje: Engano y ocultación en el arte contemporáneo* (Méndez Baiges 2007).

•

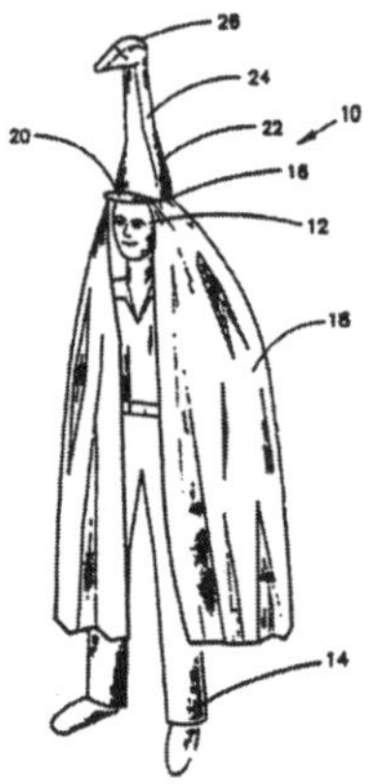

Above Drawing submitted for US Patent No. 5,197,216 titled "Combined Decoy and Camouflage Device," invented by Raymond E. Norris (1993).

Above Painting by WWI camoufleur (and Thayer student) Richard ("Wig") Meryman of coincident disruption in a male wood duck, reproduced in color in Thayer (1909). In its caption, Abbott H. Thayer states that the duck's "dark areas, with all their varied colors, here 'become a part' of the like-colored dark reflections in the water, and his white patterns exactly reproduce the bright sky reflections, so that he is so to speak 'dissolved' into the scene."

Ana Mendieta
(1948-1985)
Cuban artist

Mendieta left her Cuban homeland in 1961, separating her from her family. In her best-known performance works, the *Silueta Series* (c1977-1979), she covered herself with earth and other materials in outdoor natural settings, to create what she referred to as "earth/body sculptures [in which] I bond totally with the Earth." She took her own life in 1985. See Méndez Baiges (2007).

•

Richard [Sumner] Meryman
(1883-September 24, 1963)
American painter

Born in Boston, Meryman studied at the Boston Museum of Fine Arts School, and, independently, with Edmund Tarbell and Abbott H. Thayer. He taught at the Corcoran Gallery Art School, and succeeded Tarbell as its director. As Thayer's student, he contributed to the illustrations in G.H. Thayer's *Concealing Coloration in the Animal Kingdom* (1909). In 1936, he settled in Dublin NH. *New York Times* ("Richard S. Meryman," 1963): "He drove an ambulance in the French Expeditionary Force during World War I and later served as a member of the United States Army's camouflage corps." Meryman's son, also named Richard, published an excellent article on Abbott Thayer, with memories of his father, in *Smithsonian Magazine* (Meryman, 1999), and also appears in an interview in the film *Invisible: Abbott Thayer and the Art of Camouflage* (2008).

•

Oliver (Hilary Sambourne) Messel
(January 13, 1904-July 13, 1978)
British theatre designer

Born in London, Messel was from a prominent family. He was the uncle of photographer Anthony Armstrong-Jones, who became Lord Snowdon through his marriage to Princess Margaret. He studied art at the Slade School of Fine Art, then turned to theatrical and interior design. He is generally regarded as one of the century's finest stage designers.

During World War II, he served as a camouflage officer, in the course of which he specialized in the disguise of military bunkers (called "pill boxes"). Castle (1986:114): "He [Messel] disguised some pill-boxes as Gothic lodges and others as caravans, haystacks, ruins and wayside cafés..."

•

Jo Mielziner
(1901-March 15, 1976)
American theatrical designer

Jo Mielziner (pronounced MELL-zeener), who was born in Paris, was the son of Leo Mielziner, an American portrait painter. He was one of the most accomplished and influential scenic, costume and lighting designers of the century, winning five Tony awards, in the process of designing sets for *A Streetcar Named Desire, Death of a Salesman, The King and I, South Pacific, Guys and Dolls, Cat on a Hot Tin Roof*, and other major Broadway productions.

New York Times ("Jo Mielziner Dead," 1976:38): "In World War II, Mr. Mielziner, who by then had achieved fame and a comfortable living in the theater, volunteered for duty with the Army Air Force, in the belief that his special knowledge and training would be helpful in developing camouflage. His offer was early accepted, and he was given the rank of major."

•

Jules Migonney
(February 22, 1876-July 5, 1929)
French painter, printmaker

Sometimes listed as Jules Migonnet, he was born in Bourg-en-Bresse. He studied at the École des Beaux-Arts in Lyons, and in Paris with Eugéne Carriere and Léon Bonnat. He is listed in Coutin (1988) as a World War I French camoufleur. He was friend of the Fauvists, although his own work is considered to be Orientalist.

•

Robert (Andrews) Millikan

(1868-1953)
American scientist

Millikan attended high school in Maquoketa IA, and was later in contact with chemist and camouflage expert Matthew Luckiesh, from the same hometown. His papers in the archives at California Institute of Technology include a folder labeled "Camouflage" from 1942 (Box 20 Folder 7).

•

A. Wallis Mills

(1878-1940)
British illustrator

Mills was born in Sussex, and studied art at the South Kensington School of Art. A prolific illustrator, he was a frequent contributor to *Punch* and other magazines, beginning in the early 1900s. He is cited on the Royal Engineers Museum and Library website as a British camoufleur during World War I.

•

David B. Milne

(January 8, 1882-December 26, 1953)
Canadian artist and writer

Born in Burgoyne, Ontario, Canada, he moved to New York to study at the Art Students League in 1903. Five of his paintings were exhibited in the Armory Show. During World War I, he served as a war artist, and while he was not a camoufleur, he was greatly interested in the subject. According to Silcox (1996:82), "In several paintings...Milne's fascination with the events of the Great War in Europe show up in his references to camouflage, a convention of tricking the eye that Milne was often intrigued by." He returned to Canada to live in 1929.

•

Facing page Various examples of natural camouflage through mimetic resemblance (clockwise from top left): The head of a large South American plant hopper (*Fulgora laternaria*), commonly called a Lantern Bug, Peanut Bug—or Alligator Bug, because of its resemblance to an alligator. Photo from the *Illustrated London News* (April 5, 1924). A Giant Swallowtail butterfly (*Papilio cresphontes*) in its larval stage, at which time it convincingly looks like a moist bird dropping. Photo courtesy Mary Snyder Behrens. A Dead Leaf Mantis (*Deroplatys desiccata*), native to Madagascar, Africa, which looks amazingly like a decayed and rotting leaf. Public domain photo. A Chinese praying mantis (*Tenodera sinensis*), a grass mimic, positioned upsidedown on tree bark. Photo courtesy Mary Snyder Behrens. A Tawny Frogmouth (*Podargus strigoides*), an owl-like Australian bird with a striking resemblance to tree bark. Photo courtesy C. Coverdale, with permission as posted on Wikipedia.

Northrup Frye (1971:207): "[In the paintings of David Milne, one of the Canadian Group of Seven] The beholder is at once well inside the picture, where he finds that everything is on much the same pictorial level. In other words, he finds a certain amount of camouflage or dazzle-painting (notice how hard it is to see the human figure in the untitled landscape study...foreground and background seem to merge in an elaborate interlocking pattern...)"

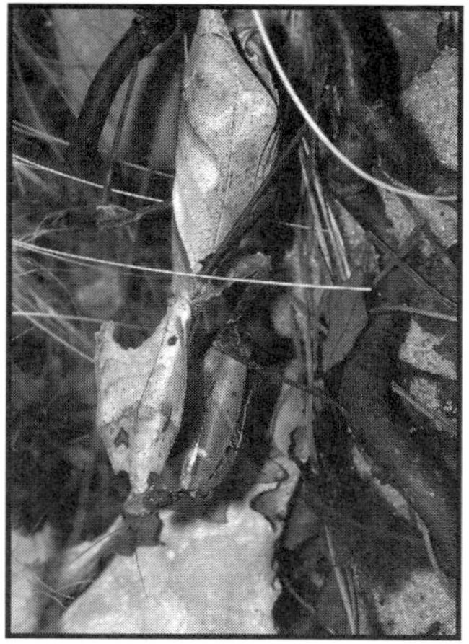

Mimicry

Resemblance of one kind of thing to another

In blending camouflage, a figure resembles a background. Mimicry is a related concept, since it too depends on similarity, but the resemblance is not between figure and ground, but between two different kinds of things. An insect may resemble a leaf, a stick, a piece of bark, or even a bird dropping. Or, a non-poisonous snake may gain a survival advantage from its resemblance to a poisonous snake.

Facing page Two examples of World War I uses of mimicry in ship camouflage. In the top photo is the *USS Antigone*, with the black silhouette of a destroyer on its side, so as to appear to be guarded. Below that is the troop ship *USS Von Steuben*, also with a destroyer painted on its side. Courtesy US Naval Historical Center (NH 57624 and NH 101626).

It is common to distinguish between Batesian mimicry and Mullerian mimicry. In the former, a mimic (which is neither unpalatable nor poisonous, and is fewer in number) benefits from its resemblance to an unpalatable, poisonous or otherwise harmful model. In the latter, both the mimic and the model are in some way harmful, and both may gain advantage from looking alike. Other kinds of mimicry have also been identified. Authoritative sources on the subject include Cott (1940) and Wickler (1968).

Crowell (1921:501-502): "On each of the light gray sides of the British warship *HMS Suffolk* was painted, in darker gray, the silhouette of a smaller vessel, by which device it was hoped to deceive the enemy as to the actual size of the ship. One of the former German boats, the navy transport *Von Steuben*, appeared in a camouflage coat the principal feature of which was the painted silhouette of a destroyer. The theory was that a submarine at a distance would believe that the transport was being escorted. This was a popular conception, and numerous ship owners adopted it."

...

There are countless examples of the wartime use of mimicry, but the number and variety of these increased substantially in World Wars I and II. The widest variety of fake objects were devised, such as imitation trees, rocks, horse carcasses, even enemy corpses, all of which were used to hide snipers and observers. During World War II, entire battlefield events were faked, most notably by units like the Ghost Army.

•

William C. Pahlmann, quoted in Reit (1978:80): "We kept emphasizing [in World War II camouflage training] that camouflage meant fooling the eye—not only covering things up, but *diverting* the enemy's attention."

Misdirection

Distraction or Shift of Attention

Hugh B. Cott (1938): "When a pickpocket intends to rob you of your watch or wallet, he, or his confederate, takes care to distract your attention from what he intends to do by creating a diversion. He draws your eyes from what is really happening to what seems to be happening."

•

Meldon Mitchell
(1897-February 11, 1967)
American advertising artist

Mitchell attended the Art Students League and Cooper Union in New York. He was a member of the Advertising Club of New York, and the Society of American Military Engineers. For 40 years, he headed Mitchell Studios in New York, providing package design for Colgate Palmolive, Continental Can, International Paper and other corporations. *New York Times* ("Meldon Mitchell" 1967): "Mr. Mitchell pioneered in camouflage techniques while serving in the Marine Corps in World War I and was a civilian advisor on camouflage to the Marines in World War II. 'The Mitchell Suit,' which he designed, was used by Marines in both world wars and in Korea. He also served as an advisor on industrial camouflage to several business concerns."

•

Laszlo Moholy-Nagy

(July 20, 1895-November 24, 1946)
Hungarian-born photographer, designer, author and educator

Moholy-Nagy was born Laszlo Weisz in what is now Mol, Serbia. He studied law in Budapest, then served in World War I. In 1919, he became interested in the works of the Russian Constructivists. In 1923, he joined the faculty of the Bauhaus in Dessau, Germany, where he directed the foundations program and taught the metals workshop. He left the Bauhaus in 1928 to work as a designer in Berlin. In 1937, two years after moving to London, he accepted an offer to become the first director of the New Bauhaus in Chicago, which later became the School of Design, then the Institute of Design. During World War II, he became involved with civilian defense camouflage, partly as a strategy to maintain the school's enrollment.

Sibyl Moholy-Nagy (1969:183-184): "On December 19, 1941, Moholy was appointed to the Mayor's personal staff in charge of camouflage activities in the Chicago area. During blizzards and rainstorms, in fog and in brilliant sunlight, he had to take flights to absorb air views of the city under diverse weather conditions. While he fought air sickness, which he never overcame completely, he pondered how to conceal the vastness of Lake Michigan with a simulated shore line and floating islands. In January, 1942, the School of Design became a certified school for camouflage personnel. As head of the Camouflage Workshop, György Kepes produced a wide range of new techniques and concepts. When they were displayed for the first time in 1943, they aroused wide attention."

Findeli (1991:100-101): "In January 1942 he [Moholy-Nagy] obtained a half-hour audience with President [Franklin D.] Roosevelt and in the autumn of that year, the boss of the Army Education Service granted him the accreditation he had requested [for the School of Design]. The School would give, amongst others, a course on camouflage, train

Above WWI American ship, the *USS Orizaba* (1919), painted in a camouflage scheme designed by Everett L. Warner. Courtesy US Naval Historical Center (NH 44914).

John D. Whiting (1928:80): "They [the German U-boats] have to get pretty darned close to launch a torpedo. That means doping out the course very near, and quickly, while they porpoise. And torpedoes, my dear young fellow, are very expensive. It costs the Kaiser eight thousand bucks every time one is fired."

...

Homer Saint-Gaudens (1941): "I am not at all sure whether camouflage is an art or a science; probably it is both. It certainly started as an art. That part of it has been pretty well explored. Now most camouflage activity is turning scientific, for it concerns itself more and more with an elaboration of camouflage technique which requires scientific training..."

instructors in visual communication, and organize a group of rehabilitation courses—based on the therapeutic characteristics of the preliminary course—for war wounded. Besides this, the product design workshop obtained royalties, from a Chicago firm, for the development of several principles of wooden springs which were intended to counter the effects of the rationing of steel."

Mavigliano (1987:42): "Moholy was appointed to the Mayor of Chicago's Defense Commission and given the task of camouflaging the waterfront of Lake Michigan to fool invading aircraft. He had to endure stormy trips on patrol boats that made him seasick, and wind-blown flights in small reconnaissance planes that made him airsick.

Donald Vogel was an Illinois Project artist at that time and was assigned to Moholy's camouflaging program. 'The group's purpose was to camouflage the oil storage tanks along Chicago's south shoreline. We were assigned drafting desks near the bridge designers at Soldiers' Field stadium. It was bitterly cold and drafty, and there was no supervision. In the weeks that I was there, no one ever appeared to tell us what to do or where to submit what we did. However, we took the project in hand, discussed the ideas we invented and made drawings.'"

•

Émile Mollard
(1895-October 16, 1991)
French military officer

General Mollard was a colonel in the French Army when World War II began. He joined the French Resistance, and, according to the *New York Times* ("Camouflage Specialist," 1991), "founded the Camouflage du Materiel group of officers that specialized in camouflaging armored vehicles, weapons and ammunitions for use by the Resistance."

•

Edward F. Molyneux
(1896-March 24, 1958)
American art director, advertising artist

Molyneux attended Pratt Institute. After serving as a camoufleur during World War I, he worked as a typographer at the Frank Seaman advertising agency, then as art director at Newell Emmett Company—later called Cunningham and Walsh—in Westport CT, a position he continued in for 28 years. In the early 1930s, he was President of the New York Art Directors Club. *New York Times* ("E.F. Molyneux," 1958): "[Molyneux] served in the Camouflage Division of the American Expeditionary Forces in World War I."

Louis de Monard
(January 31, 1873-1939)
French painter and sculptor

The son of a French army officer, Monard was encouraged to follow the same career. He turned instead to painting and sculpture, although he did serve as a camoufleur in World War I, as listed in Countin (1988).

•

Lucien-Hector Monod
(dates undetermined)
French painter

He was the father of Jacques Lucien Monod, a Nobel Prize-winning biologist. He is listed in Coutin (1988) as a World War I French camoufleur.

•

Pierre-Paul Montagnac
(May 9, 1883-1961)
French painter, designer

He is listed in Coutin (1988) as a World War I French camoufleur. A portrait painter and theatrical set painter, he was a knight in the Legion of Honor.

•

Issac Montoya
(1963-)
Spanish artist

In recent years, Montoya has been using large scale photographs to create parodies of contemporary advertising images. When viewed in the usual manner, they are not unduly surprising, but looked at through a filter, it becomes apparent that additional content is "camouflaged" within the work, the nature of which is disturbing. In her discussions of Montoya's work, Maite Méndez

Brian Lavery (2006:104): "[During World War II] There was no set scheme for landing craft at first and [British actor] Alec Guinness [who was serving in the Navy] 'devised a pattern of pale blue and white rectangles, rather like a Braque. My [Guiness'] fellow officers complained that I had made mine like a hospital ship and far too conspicuous, but we went out on the ocean they equally complained that they couldn't see me.'"

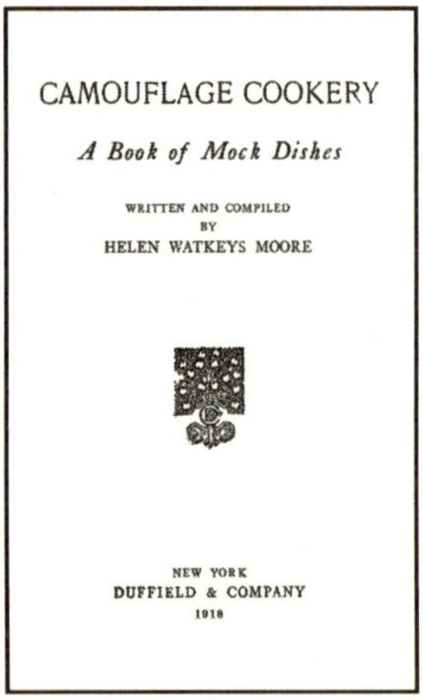
CAMOUFLAGE COOKERY
A Book of Mock Dishes
WRITTEN AND COMPILED
BY
HELEN WATKEYS MOORE

NEW YORK
DUFFIELD & COMPANY
1918

Above Title page of Helen Watkeys Moore's collection of camouflaged dishes (1918).

Baiges (2007) observes that Montoya's work is a comment about the concealment of social reality in duplicitous advertising.

•

Helen Watkeys Moore
(dates undetermined)
Author

The wife of the head salesman at Putnam, Moore was the author of *Camouflage Cookery: A Book of Mock Dishes* (1918). In her foreword, she explains: "In these times, when the housewives of the country are concerned about the conservation of food, I hope that this little book of substitute or mock dishes will prove helpful. Its object is to bring together palatable and economical recipes, by the best known American cooks, for those who wish to prepare wholesome and appetizing dishes..."

Among the featured recipes are: Mock Bisque Soup. Mock Terrapin Soup. Mock Fried Oysters. Mock Meat Hash. Mock Meat Pie. Mock Chili Con Carné. Mock Sausages. Mock Chicken Salad. Mock Cherry Pie. Mock Plum Pudding. Mock Pumpkin Pie. Mock Pate de Foie Gras Sandwiches. Mock Champagne Punch. Mock Bar-Le-Duc Currants. Mock Orange Marmalade. Mock Champagne Punch. Mock Figs. Mock Ice Cream. And many others.

•

John D(rummond Macpherson) Moore
(September 6, 1888-December 9, 1958)
Australian artist, architect

Born in Waverly, Sydney, Australia, Moore was a New York architect in in the firm of B.G. Goodhue (1914-15), along with several others who would later serve as camoufleurs, among them Americans Burnham Hoyt, Ralph T. Walker and Austin Wittlesey. According to Flower (1986): "Moving to London [during World War I] he enlisted in the Royal Engineers in 1915, was commissioned and served in France... In World War II he was deputy director of camouflage for New South Wales in 1942-45."

•

Patrick (Alfred Pierce) Moran
(July 14, 1917-1988)
Australian scientist

Born in Sydney, Australia, Moran studied mathematics at St. Stanislaus College in New South Wales, and at Sydney University (beginning at age 16), where he earned High Distinction in both mathematics and physics.

He then moved to England, where he studied at Cambridge University until World War II began. During the war he worked on rocket development for the British government, then joined the Australian Scientific Liaison Office, where, according to Hall (1989:687), "he was responsible for reporting on all manner of wartime developments in a vast variety of subjects ranging from camouflage to radar."

•

Richmond Knapp Fletcher (1920): "...[During WWI] I joined the Marine Camouflage service of the US Shipping Board, and for the duration of the war experienced the most interesting and strenuous period of my existence to date. The camoufleur's principle job was to superintend the marking off and painting of the ships, following designs sent out by the Navy Department. When not thus engaged, we worked at new designs on small ship models, studying them in a marine theatre equipped with a periscope lens. A complete knowledge of the theory of these designs was essential since the design supplied for any given ship seldom fitted the ship exactly, which meant that a certain amount of originality and initiative was necessary in marking off a vessel. This work was generally done in a hurry, between the docking and sailing of a ship. We often worked from 7 am until 9 pm with perhaps five or ten minutes for meals. It was a continuous round of exercise on rope ladders, 'bosun's chair,' swinging stages, scaffolds, and up and down ropes hand over hand. We lived in an atmosphere of grease, steam, coal and dust, and everywhere, wet paint. It was exciting while it lasted, but I doubt if there were any camoufleurs who were sorry when it was over. Our 'periscope' and theatre, together with a large collection of camouflage models, are now in the department of marine architecture at the Massachusetts Institute of Technology, and form a very interesting exhibit of one of the strangest developments of modern war."

Luc-Albert Moreau
(December 9, 1882-April 25, 1948)
French painter, printmaker

Born Albert Lucien Moreau, he is listed in Coutin (1988) as a World War I French camoufleur. Prior to the war, while studying with Jean-Paul Laurens at the Académie Julian in Paris, he became friends with various artists who would later serve as camoufleurs, including André Dunoyer de Segonzac, Jean-Louis Boussingault, and Pierre Laprade. He was also closely acquainted with French writer Collette and composer Maurice Ravel.

•

Richard C. Morrison
(dates undetermined)
American arts administrator

According to the Archives of American Art (1995:112), Morrison was associated with the Federal Art Project in New England in the 1930s. In an interview in 1965, there is an "extensive discussion of the WPA's work with the Camouflage Laboratory and the way the FAP provided artists, architects, and landscape architects to help develop designs for protective concealment." In 1965, Morrison (living then in Manhattan KS) wrote in a letter to the editor of the *New York Review of Books* that, a year or so prior to the bombing of Pearl Harbor, he "was conducting laboratories in Boston with the aim of mondernizing concealment and camouflage techniques for military services which were reluctant to accept change." He is also described in Knight (1965) as a New England regional WPA supervisor who "began the establishment of a camouflage unit using paints and so on."

•

Edward Morton
(dates undetermined)
American painter

Born in Milwaukee WI, he was a teacher at various times at the Layton School of Art, and the University of Wisconsin-Milwaukee. In the 1930s, he completed a post office mural at Oconomowoc WI, titled *Winter Sports and Rabbit Hunters*, in connection with his service in the Civilian Conservation Corps. He exhibited at the Pennsylvania Academy of the Fine Arts in 1939, and served as a US Army camoufleur during World War II.

•

Richard H. Morton
(1921-?)
American artist

Born in Dallas TX, Morton studied illustration at Pratt Institute in New York. After enlisting in the US Army in 1942, he was assigned to the 603rd Engineer Camouflage Battalion. Following camouflage training at Fort Meade, he was attached to the Ghost Army, a secret unit in Europe, the 23rd Headquarters Special Troops.

•

Irving G(erald) Moshinsky
(1917-April 6, 1998)
American artist, educator

Born in Brooklyn NY, he studied art at Pratt Institute and New York University. According to Obitcentral ("Irving Gerald Moshinsky," n.d.): "During World War II he was a sergeant and an Army engineer in the 606th Camouflage Battalion in France." While serving in France, he painted portraits of the local residents. In the years after the war, he traveled to Europe each year to paint landscapes, in the process of which he became fluent in five languages. He taught WPA artists in New York, and was an art teacher at Roslyn High School. To his students, he was known as "Mr. Mo."

•

Colin (William) Moss

(April 30, 1914-December 16, 2005)
British painter

Moss' father was a grocery owner who died in World War I. He himself studied at Plymouth College of Art and the Royal College of Art, where he was influenced by Gilbert Spencer. During World War II, he worked as a camoufleur under L.M. Glasson. Collins (2006): "War saw him posted to a Leamington Spa skating rink where, along with 150 fellow artists, he devised schemes to disguise buildings and made plans to brighten factory hostels, barracks, Naafi canteens and state restaurants with color and murals." Some of his camouflage-related paintings are reproduced in color in Goodden (2007), and in Newark (2007).

In 1961, he studied in Salzburg with Oscar Kokoschka, who rewarded his students by giving them sweets. Moss recalled that "Anybody who got a sweet was going to take it home and keep it on the mantelpiece for ever, so I ate mine straight away" (Collins 2006).

•

Georges Mouveau

(dates undetermined)
French theatre designer

Mouveau is cited in Coutin (1988) as a World War I French camoufleur. According to Kahn (1984:166), there is an early memo in which he is listed as one of the first French camoufleurs in World War I, along with five others. He was a member of a group of advocates of experimental art, called The Apaches, and a stage designer for the Theatre des Arts and the Paris Opera. In the late 1920s, one of his students was Robert Jay Wolff, who later taught at the New Bauhaus in Chicago.

•

New York Times ("Chaplin as Soldier..." 1918): "In *Shoulder Arms*, [Charlie] Chaplin is as funny as ever...camouflaged as a tree trunk, [he] plays destructively with one of the tree's branches."

Above World War I photo of a German frontline observation post, disguised as a dead tree.

J. André Smith (1928:477): "Just what the American camoufleur will bring to this new art [of camouflage] is still too early to predict; he will approach it at first in true humility, asking of his French brother the benefit of his experience, inspecting carefully the work that has been done on the English as well as the French fronts, and after that–who knows? This art is still in an early stage of development. If the French were ingenious enough to invent it and the Germans to copy it, it is safe to say that we Americans shall first of all systemitize it; not a cut-and-dried business, but one directed with level reasoning and touched by American humor and inventiveness."

Above Among the Allies as well as the Germans, during World War I, there was considerable experimentation in the use of dead trees as frontline observation posts. As a replacement for the actual tree, a full-scale hollow replica was constructed in a camouflage workshop, lined with armor, and transported to the front, where it was clandestinely switched with the actual tree during the night. The two lower images on this page are portrayals of that. Adapted from a series of lithographs in a French publication titled *La Guerre Documentée* (Paris c1920).

Rodrigo Moynihan
(1910-1990)
British portrait painter

Moynihan was educated at the Royal Academy. According to the Tate Archives, after training for artillery in 1940, he was assigned to camouflage. Three years later, he was designated an Official War Artist. Also see "Shared Experience" (n.d.).

•

Claude Muncaster
(1903-1974)
British marine painter

Muncaster's birth name was Oliver Graham Hall. According to the National Maritime Museum website, "The artist later became a lieutenant commander in the Royal Naval Volunteer Reserve and an advisor on camouflage." In Williams (1989:155) and Williams (2002), he is described as having conducted camouflage research with Alphonse E. Schuil (in which he is listed as Oliver Grahame-Hall).

•

Hermann [Dudley] Murphy
(1868-April 16, 1945)
American landscape artist

Born in Marlboro MA, he studied in Paris at the Académie Julian. He was also a canoe sailing champion. He served as the artist with the expedition of the Nicaragua Canal Construction Company in 1887-88. From 1902-37, he was a member of faculty of the School of Architecture at Harvard University. *New York Times*: ("Hermann Murphy," 1945:23): "In the first World War he was an inspector of camouflage for the US Shipping Board."

In the same article, there is the following note: "In 1928 the Boston Art Club 'fired' its art committee on the ground[s] that it leaned too much toward

Above German airplane camouflage during World War I.

Facing page These two photographs were published with an article on military entrenchments In *Golf Illustrated* by British-born golf course designer Alister Mackenzie (1915). The captions read as follows: "[top] Interior of trench. Note the complete absence of straight lines and angles, the overhanging lips of the recesses, the depressions opposite the traverses, the use of the traverse as an elbow-rest while firing, the long serpentine line of trench shown by the men's caps held up on rifles. [bottom] The same trench. Photograph taken from 10 yards away. The man standing up is behind the trench. The men's heads while firing were completely concealed at 40 yards away."

modernism and replaced it with a committee headed by Mr. Murphy. At the time he denounced 'crazy stuff' in art."

•

John J.A. Murphy
(1888-January, 1967)
American wood engraver, book designer

Murphy studied at the Boston Museum of Fine Arts School and the Art Students League. He initially worked as a portrait painter, then moved to London where he became chief assistant to the well-known Welsh engraver and designer Frank Brangwyn. *New York Times* ("John J.A. Murphy," 1967:42): "During World War I Mr. Murphy served in the camouflage section of the Army Corps of Engineers in France and illustrated posters."

•

N

Jacques (Lehman) Nam
(1881-1974)
French painter, illustrator

He is listed in Coutin (1988) as a World War I French camoufleur. His illustrations appeared in various French magazines, including *La Vie Parisienne*, *Le Petit Journal Illustré de la Jeunesse*, and *La Baïonette*. He specialized in animal subjects, especially cats.

•

M. Nash
(dates undetermined)
American painter

He was a member of Everett L. Warner's ship camouflage team during World War I.

•

William Nell
(dates undetermined)
American painter

He is mentioned in *Literary Digest* ("Company A" 1917) as having been a camoufleur, and is also listed by Faulkner (1957) and (1973).

•

Facing page Drawings submitted for US Patent No. 285,951 titled "Stuffed Toy Bear," invented by S.J. Raymo (1986).

New York Camouflage Society

Founded in 1918 by Barry Faulkner and Sherry Fry, also called the American Committee, and the American Camouflage

Faulkner (1957): "[When the US entered WWI] The New York Camouflage Society didn't enlist heavily. Belonging to the society had allayed the patriotic unrest of most of the members."

Peixotto (1917:9-10): "On the very day of America's entry into the world war a group of New York artists wanted to organize for war work...

When the idea of the 'camouflage' was suggested it was received with enthusiasm. Several of the men had even then made careful researches in the domain of protective coloration, and one at least had been working with the full knowledge and approval of naval officers."

"The American 'Camouflage'" (1917:1411-1412): "To carry out the plans of the American Committee [New York Camouflage Society], there will be needed the cooperation of scene-painters and house-painters. The organization might ultimately number as high as 7,500. Such as are already enrolled number well-known names like E.H. Blashfield, George de Forest Brush, Barry Faulkner, William A. Mackay, Abbott Thayer, Daniel French, J. Alden Weir, Herbert Adams, Paul Dougherty, Cass Gilbert, and Bertram Goodhue."

US Committee on Public Information (1918:48): "A special contingent of camouflage is being organized by the War Department, and a private society of artists, the American Camouflage [New York Camouflage Society], has an office at 215 West Fifty-seventh Street, New York City."

Above Photograph of Kimon Nicolaides by Peter A. Juley & Son. Courtesy Photograph Archives, Smithsonian American Art Museum (J0002016).

Below Dust jacket of *Camouflage* by Tim Newark (2007).

John Winsor (1996): "...the British army's adoption of khaki instead of scarlet in 1902 was a response more to hot weather than the need for concealment."

•

David Livingston Smith (2004:21): "As a species, we are so well practiced in the art of deception that it comes to us almost as naturally and effortlessly as breathing. Just try keeping track of the countless deceptions that you, and all of us, perpetrate on an ordinary day and you will quickly discover that you rarely have to work at deception."

Tim Newark
(1961-)
British historian and journalist

He studied art history at University College in London and the Warburg Institute. Since 1993, he has been the editor of *Military Illustrated*, a military history journal, and has authored several valuable books about historic and cultural aspects of camouflage, including *Brassey's Book of Camouflage* (Newark 1996), *Brassey's Book of Uniforms* (1998), *The Future of Camouflage* (2003) and, of particular interest, *Camouflage* (2007), introduced by Jonathan Miller, printed in conjunction with a major exhibition that year at the Imperial War Museum in London.

•

Kimon Nicolaides
(1891-1938)
American artist, author, educator

Born in Washington DC. Nicolaides' father was a Greek-born importer of Oriental artifacts; while his mother's ancestors can be traced to the US Colonial period. As a young man, he earned money by framing pictures, writing newspaper articles, and playing the role of an art student as an extra in a film. He studied drawing at the Art Students League with George Bridgman and John Sloan.

When the US entered World War I, Nicolaides volunteered for the American Camouflage Corps, for which he served in France for over a year, and for which he received a citation. One of his military duties required the study of geographical contour maps, an experience that may have influenced his use of 'contour drawing' as Exercise One in his famous, widely used textbook, *The Natural Way to Draw* (Boston: Houghton Mifflin, 1941). Nicolaides died before the book was finished. Its completion and publication were overseen by a devoted former student, Mamie Harmon, whose papers are in the Archives of American Art.

•

Martin Nixon-Miller
(dates undetermined)
American civil engineer

Nixon-Miller graduated from Haverford College in Haverford PA in 1894. In 1917, he was briefly placed in charge of Company A of the American Camouflage Corps.

Literary Digest ("Company A" 1917:32-33): "The company is under the capable command of Capt. Martin Nixon-Miller, USR, from whom they get daily a grueling course in military instruction. Captain Nixon-Miller was for more than seven years a member of the Pennsylvania National Guard. He is a civil engineer by profession and served as assistant engineer under Goethals on the Panama Canal... Incidentally, it was Captain Nixon-Miller who, as a Government agent, uncovered the garbage scandals of the reform administration in Philadelphia. The men in his command are good sports and are taking their medicine cheerfully..."

Gertrude Stein (Stein 1938): "Really the composition of this war [WWI]...was not the composition of all previous wars, in which there was one man in the center surrounded by a lot of other men, but a composition in which one corner was as important as another corner, in fact the composition of Cubism."

Above Two views of disruptively painted railway cars in France during World War I.

Facing page A painting of a copperhead snake from *Concealing Coloration in the Animal Kingdom* (Thayer 1909). The full-page painting (*top*) can be overlayed by a cut-out silhouette of the snake, revealing its location in the setting of dead leaves (*bottom*). The caption states that it was "painted from life by Rockwell Kent [Thayer's student] and A.H. Thayer," while the book's preface includes a note of gratitude to "Mr. R.L. Ditmars, Curator of Reptiles at the Bronx Zoological Park, New York, for the loan of a live Copperhead snake, and other favors."

•

Percy E(rskine) Nobbs
(1875-1964)
Scottish-born Canadian architect

Nobbs was a prominent architect, and professor of architecture at McGill University, where he designed or remodeled a number of the campus buildings. He was an active proponent of the Arts and the Crafts Movement.

According to Schoenauer (n.d.): "During World War I Nobbs enlisted and was engaged in military training, teaching young Canadian recruits bayonet fighting. [He had been an Olympic fencer in 1908.] Later he joined the Royal Engineers and served as a camouflage expert in France, attaining the rank of major by the end of the war."

•

B[ror] J[ulius] O[lsson] Nordfeldt
(April 13, 1878-April 21, 1955)
Swedish-born American painter

Born in Tullstorp, Sweden, Nordfeldt came to the US with his family in 1891, and settled in Chicago, where he later studied at the Art Institute of Chicago. He was an assistant muralist to Albert Herter (the father of camoufleur Everett Herter). Returning to Europe in 1900, he studied in Paris at the Académie Julian, and in Oxford, England, with

woodblock artist F. Morley Fletcher. He moved back to Chicago in 1903. According to the website of Luther College ("B.J.O. Nordfeldt"), the collection of which includes one of his artworks: "During the war [WWI], Nordfeldt was assigned to San Francisco to supervise the camouflage of merchant ships while his wife, a physician, worked for the War Department."

Online bio ("B.J.O. Nordfeldt," n.d.): " During World War I, Nordfeldt was employed as Assistant District Camoufleur for the US Shipping Board in San Francisco, specifically assigned to camouflage merchant ships. After the war, Nordfeldt visited his friend and fellow artist William Penhallow Henderson (1877-1943) [with whom he had worked on ship camouflage] in Santa Fe and made the decision to relocate to the Southwest, where he eventually settled for the next twenty years." He was a WPA artist in the 1930s, and a friend of artist Russell Cowles.

•

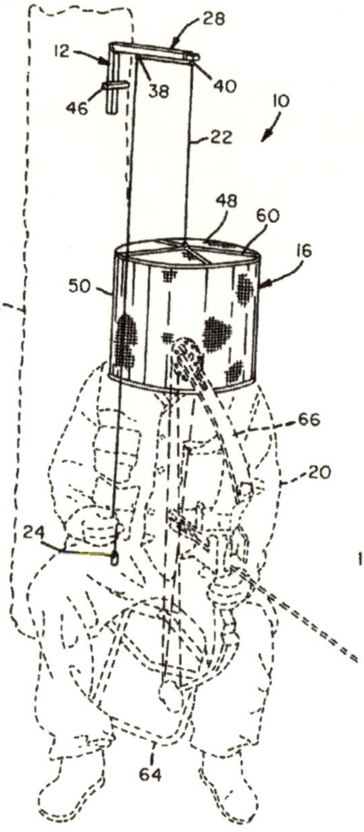

28
12
38
40
46
10
22
48
60
16
50
14
66
20
24
64

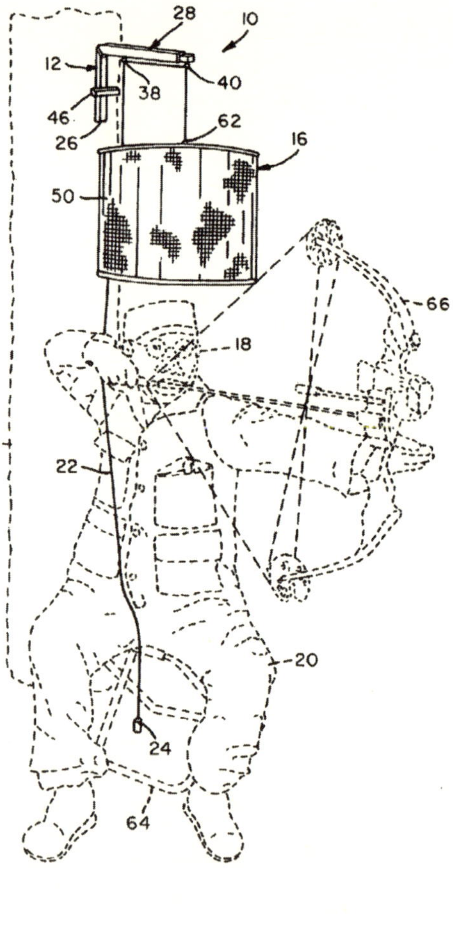

28
10
12
38
40
46
26
62
16
50
66
18
14
22
20
24
64

OPQ

Objections to camouflage

In Reit (1978:65-66), there is a discussion of the initial reluctance of military officers to support the use of camouflage. Their reactions to it, he states, "ranged from casual indifference to mild suspicion to active dislike." So-called "protective concealment," he continues, "had a taint of passivity or timidness and, like the loss of the vivid *pantalons rouges*, was looked on as somehow bad for morale." Finally, "it went against the military obsession with neatness, spit-and-polish, drill-field regularity... It made a virtue of dispersion, irregularity, and improvisation; and like modern ecologists, camoufleurs tried to blend with the natural surrounding instead of exploiting and dominating it."

•

M. O'Connell
(dates undetermined)
American artist?

He was a member of Everett L. Warner's World War I ship camouflage team.

•

Donald M. Oenslager
(March 7, 1902-June 21, 1975)
American theatre designer

Born in Harrisburg PA, he studied theatre at Harvard University. In 1925, he joined the faculty at Yale University, where he taught for 46 years, while also

Facing page Drawings submitted for US Patent No. 0081272 titled "Camouflage Hood Assembly," invented by Richard Dean Shaklee (2005).

working as a set designer on Broadway. Among the productions he designed were *The Emperor Jones*, *You Can't Take It With You*, *Of Mice and Men*, *Pygmalion*, *Major Barbara*, and *A Doll's House*. He was also an advisor for theatre architectural projects (including Lincoln Center Philharmonic Hall, and the John F. Kennedy Center for the Performing Arts), as well as the author of several books on theatrical design.

New York Times ("Donald M. Oenslager," 1975:29): "In World War II, he served as an Army major in the South Pacific. He specialized in camouflage and intelligence work, using his technical experience to help B-29 flight crews prepare for bombing missions over enemy territory."

•

Northrup Frye (2004): "[Beyond camouflage] Another form of puzzle is provided by the painter who uses darkness as a positive basis and light as a sporadic scattering of it. The Dutch pictures with a candle in the center and the rest of the picture shading off to blackness at the edges are typical of this, and even Rembrandt and Vermeer do the same thing in subtler ways."

...

Alexander Nemerov (1997:79): "Camouflage, as both [Theodore] Roosevelt and [Friedrich] Nietzsche independently argued, was a form of effeminate cowardice, a mere defensive strategy. The marks of concealment all but announced an unmanly desire to hide instead of fight."

Eliot O'Hara
(1890-1969)
American painter and author

Above Photograph of Eliot O'Hara (1964) by Peter A. Juley & Son. Courtesy Photograph Archives, Smithsonian American Art Museum (J0002028).

O'Hara was born in Waltham MA, where his family owned a business called the O'Hara Dial Company. He briefly attended Norwich University in Vermont, then moved to Paris to enroll in a hotel management school. Returning to the US in 1912, he managed the family business for fifteen years, then returned to Paris to focus intently on painting. In 1931, he founded his own art school, and, a year later, began to publish instructional books on watercolor painting.

During World War II, O'Hara worked in the US Navy camouflage section with Everett L. Warner. As recalled by one of his co-workers, Robert Hays (1999), "Eliot O'Hara [was] a well-known watercolorist in the Washington DC circuit, who berated me one day for allowing people to come in to see him—it annoyed him. Even though everything was restricted and we all had badges permitting us to enter the area, only those with badges could enter and some were curious about his presence—ha! I felt like I was a baby sitter or something."

•

John F. Ohmer
(dates undetermined)
American engineer

Ohmer was originally from Ohio, and attended Cornell University, graduating in 1913. He apparently served in World War I with the US Corps of Engineers, then served again in World War II. He is described in Reit as "a small, peppery, enthusiastic officer" who was attached early in that war to the Camouflage Board at Fort Belvoir. He was an early advocate of the camouflage of vital domestic landmarks, such as air bases and aircraft factories.

Casey (n.d.): "Colonel John F. Ohmer was stationed at March Field, 60 miles east of Los Angeles. He was a pioneer in camouflage techniques. In Britain in 1940, carefully-made and positioned camouflage caused the Luftwaffe to waste thousands of tons of bombs on empty fields. Months prior to Pearl Harbor, Colonel Ohmer campaigned for the protective cover of Wheeler Field, a major air base near Pearl Harbor. His plan was rejected for being too costly, $56,210.

Weeks later Wheeler and its aircraft were destroyed.

On 7 December 1940, Goodyear Tire and Rubber agreed to manufacture rubber decoy aircraft similar to those used by the British for $1,000 a copy. But the War Department again turned down Ohmer's plans. He had hoped to scatter the dummy planes around the American air bases in Hawaii. But now, Ohmer was charged with disguising California."

•

Julius Olsson
(February 1, 1864-September 7, 1942)
British Marine painter

According to Minnie Churchill (2004:242): "Then, out of the blue I received via Mary Soames a letter from Austin Wormleighton enquiring about the connection between [her father Winston] Churchill and the sea painter Julius Olsson (1864-1942), whose biography he was writing. Mr. Wormleighton stated, according to family tradition, 'Olsson was one of several painters who gave lessons to your father and accompanied him to the South of France and to Cornwall on painting excursions.' Austin Wormleighton suggested that Olsson may have known Churchill from the time of the First World War, when the former worked on the camouflage of merchant ships by dazzle painting."

This is supported by information on the National Maritime Museum website: "Olsson was in the Royal Naval Reserve during World War I and directed the camouflage painting of warships in Plymouth and Portsmouth dockyards."

Francois Omer
(August 14, 1885-?)
French painter, printmaker

Omer was born in Avignon, France, and was a student of (camoufleur) Victor Lucien Guirand de Scévola. He appears to have been primarily a landscape painter, and is listed in Coutin (1988) as a World War I French camoufleur.

If camouflage was considered unmanly, it was also regarded as inappropriate for women, especially during World War I. Nevertheless, a Women's Reserve Camouflage Corps was formed during that war in New York by the National League for Woman's Service. According to Smith (1921:164-167), the female members of the group were "artists, writers, stenographers, school teachers, debutantes, many sorts of women." Initially led by an artist named Anne Furman Goldsmith, then later by Myrta M. Hanford, a New York State official, its courses were taught by H. Ledyard Towle, and later by the US Shipping Board. The training took place on the Billings estate, loaned to them by John D. Rockefeller, Jr., at the conclusion of which about a dozen women were hired by the US Navy to work on ship camouflage in Washington DC, or on ship painting at the Philadelphia shipyards. One of their projects was to camouflage the *USS Recruit*, a non-ship built in Union Square in New York City for use as a recruiting station. Writes Smith: "This was a night's work for the women and was done at the request of the Navy to further recruiting. The camouflage design was worked out in the classrooms of the Corps. One day at sundown New Yorkers saw the ship a tame, neutral gray. The next morning it wore a wild, fantastic design of many colors."

P

Henry (Marriott) Paget
(December 31, 1856-March 27, 1936)
British painter

Paget was the oldest of three brothers, all of whom were artists, the others being Sidney and Walter. Henry was born in London, where he studied at the Royal Academy Schools. His main achievements were in illustration, portraiture, and historical painting. In Hartcup (1980:20), he is listed as having been a participant in the World War I British camouflage section, called Special Works Park. He is described as "a black and white artist [magazine illustrator], who went on to specialize in tank camouflage."

•

William C. Pahlmann
(December 12, 1900-November 6, 1987)
American interior designer

Pahlmann was born in Pleasant Mound IL, then moved to Texas, where he grew up in San Antonio. After completing a correspondence course with *Arts and Decoration* magazine, he studied at Parsons School of Design in New York, and in Paris. In 1936, he became head of the decorating department at Lord and Taylor, where he remained until 1942, when he was appointed director of the camouflage training facility at Jefferson Barracks in St. Louis MO. See Warren (1965); Reit (1978); and *New York Times* ("William C. Pahlmann," 1987).

According to Reit (1978:78-79), among those who worked with Pahlmann in developing camouflage were sculptor Julian Harris, painter Gardiner Cox, textile designer Claude Herndon, architects Nathaniel Saltonstall and Hampden Robb, and stage designer Donald Oenslager.

•

Panthéon de la Guerre
French World War I diorama

In 1916, seventeen French artists collaborated on a huge circular diorama, as a tribute to the heroes on the Allied side of World War I. Called the *Panthéon de la Guerre*, it contained about five thousand full-length figures, with a circular expanse of four hundred feet. It was completed in 1918, and displayed in a custom-made building until 1927. As documented in Levitch (2006), of relevance to camouflage is the fact that among those depicted on its "staircase of heroes" are a handful of identifiable World War I French camoufleurs. In 1927, the panorama was sold to US businessmen and shipped in a ten-ton crate to New York, where it was displayed at Madison Square Garden and at subsequent expositions. In 1956, it was donated to the Liberty Memorial in Kansas City MO, where, two years later, selected sections were cut up and reassembled to make a smaller mural, only sixty-nine feet wide.

•

Victor Papanek
(1927-1999)
Austrian-born American designer

Born in Vienna, Papanek attended public school in England. He moved to the US to study design and architecture, earning a Bachelor of Arts degree in Industrial Design and Architecture at Cooper Union (1942-48). In 1949, he studied with Frank Lloyd Wright at Taliesin and Taliesin West. He was the author of *Design for the Real World* and other influential books. According to his biography in Morgan (1984:471), he "served in the US Army as a Warrant Officer in camouflage, 417th Combat Engineers Assault Unit, Alaska and the Aleutian Islands (1944-45)."

•

Albert L. Parella
(1909-1999)
American painter, designer

Educated at the Cleveland School of Art, Parella taught advertising design at the Cleveland Museum School, the Butler Institute of American Art, and

Youngstown State University. He was also art director at WKBN-TV in Youngstown OH. According to a biographical note on AskArt online, "He served in the Air Force as a graphic artist and camouflage expert."

•

Cobden Parkes
(August 2, 1892-August 15, 1978)
Australian architect

Born in Sydney, Australia, Parkes served in World War I, during which he fought at Gallipoli. He was wounded in both hands, with the result that his capabilities as an architectural draughtsman were limited. He worked as a government architect. According to Reynolds (2000): "Following Japan's entry into the war in 1941, he provided support for the Air Raid Precautions and New South Wales Camouflage committees."

•

James Hollins Patrick
(1911-1944)
American painter, printmaker

Patrick studied at the Chouinard Art Institute in Los Angeles, where he married costume and dress designer Eleanor Jackson Rigby. Patrick was associated with Millard Sheets, with whom he collaborated on several murals. During World War II, he was a camoufleur in the US Army.

•

Elliot (Harold) Paul
(February 10, 1891-April 7, 1958)
American journalist and author

Born in Malden MA, he worked in Boston as a political reporter, then served in the US Army in France during World War I. He turned to fiction writing, and moved to Paris in 1925. He worked

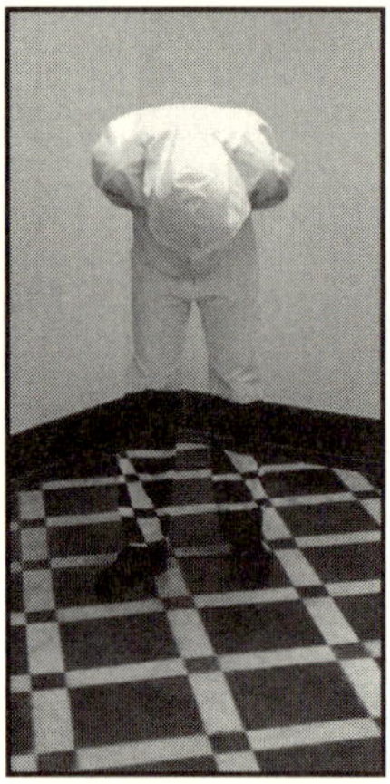

Above Since the late 1990s, Dutch artist Désirée Palmen has produced a series of photographic performances in which human figures are partially merged with their surroundings. Reproduced here is *Floor* (1999). Courtesy the artist.

Charles Burchfield (University of Arizona Art Gallery 1965:23): "[During WWI] Like many others, I was not temperamentaly suited for army life; but after the usual rookie training, I was fortunate in that I was transferred to the camouflage corps, where I was with kindred spirits doing interesting work."

...

William Pahlmann, quoted in Reit (1978:79): "At that point, early in '42, anyone in the arts was invited to apply for camouflage work—and we all rushed out there very green; not even knowing how to salute."

for the *International Herald Tribune*, and was a co-editor of the innovative magazine *transition*. He was a friend of Gertrude Stein, as mentioned in *The Autobiography of Alice B. Toklas*. His most famous book is *The Last Time I Saw Paris*. He returned to the US to write screenplays, including *Rhapsody in Blue*.

In 1942, an essay by Paul, titled "Art and Camouflage," published earlier in *Vogue* magazine, was reprinted in an anthology called *Vogue's First Reader* (Crowninshield, 1942).

•

Charles Johnson Payne
(January 17, 1884-December 30, 1967)
British illustrator, cartoonist

A popular book illustrator, Payne was known as "Snaffles." He contributed to ship camouflage in the final years of World War I, when he was assigned to the dazzle painting team headed by Norman Wilkinson. During World War II, he contributed to the camouflage of aerodomes, and then served in the Home Guard. His views on ship camouflage are quoted in Hartcup (1980:45).

•

Pierre Stephen Robert Payne
(December 4, 1911-March 3, 1983)
British novelist, biographer and historian

Born in Cornwall, England, he was the son of an English naval architect. In 1946, he conducted an important, early interview with Mao Zedong in Yenan. He is best known for his biographies of Charles Chaplin, André Malraux, Greta Garbo, Albert Schweitzer and others. According to the website of the Stony Brook University Special Collections: "From 1939 to 1941 Payne worked as a shipwright at the Singapore Naval Base and in 1941 he became an armament officer and chief camouflage office for the British Army Intelligence there." See also Grant (1969).

•

Joseph Thurman Pearson, Jr.
(1876-1951)
American painter

Pearson taught at the Pennsylvania Academy of Fine Arts. According to Roy Wood, Jr. on the Woodmere Art Museum website, "Pearson volunteered during the First World War to assist with camouflage." As a source, he cites a clipping from an unidentified newspaper, with the headline "Noted Philadelphia Artist Enlists to Camouflage Ships."

•

Above Photograph of A. Sheldon Pennoyer (1932), by Peter A. Juley & Son. Courtesy Photograph Archives, Smithsonian American Art Museum (J0002047).

Horace C. Pearsons
(1914-September 2001)
American advertising designer

Born in Cleveland OH, Pearsons studied advertising art at the New England School of Practical Art in Boston. According to the *Cape Cod Times* ("Horace C. Pearsons," 2001), during World War II "he worked for the Air Force at Fort Belvoir VA as a specialist and instructor in camouflage and booby traps. He traveled to regional air bases to provide instruction to units heading for the Pacific theatre." After the war, he worked for a New York advertising firm as its production manager. After his retirement, he moved to Cape Cod, where he worked as art director of the *Cape Cod Guide*.

•

Above *Ernest Peixotto, c1915*. 1 photographic print: b&w; 22 x 16 cm. Courtesy of the Charles Scribner's Sons Art Reference Department records, 1839-1962, Archives of American Art, Smithsonian Institution.

Gordon Peers
(1909-1988)
American painter

Peers studied art with John Robinson Frazier at the Rhode Island School of Design, and at the Beaux Arts Academy in New York, where he worked with World War I camoufleur Thomas Hart Benton. In 1941, when the US entered World War II, Peers was on the faculty at RISD, and was granted a leave of absence to serve as a camoufleur in the US Army.

•

Ernest (Clifford) Peixotto
(1869-1940)
American painter, illustrator

Peixotto was originally from San Francisco. He studied art in Paris, then embarked on an active double career as a writer and artist. He often illustrated his own travel articles, in *Harper's*

Magazine, *Scribner's Magazine* and so on. He wrote an early, important article on the role of French artists in World War I (Peixotto, 1917) and, in addition to other books, an illustrated report on his battlefield experiences during World War I, titled *The American Front* (Peixotto, 1919), including accounts of his contacts with American camoufleurs. Krass (2007) provides a vivid narrative of the experiences of Peixotto and other artists who served as wartime artists for the American Expeditionary Force.

Above Photograph of the cardboard replica of a railroad field gun made by A. Sheldon Pennoyer as a centerpiece for a Camoufleurs' Ball in New York in 1917.

A(lbert) Sheldon Pennoyer
(1888-1957)
American painter, printmaker

Pennoyer was originally from California. In 1913, he was living in Paris, where he was a student in the Architectural Section of the École des Beaux Arts. When his interests turned to art, he enrolled at the Académie Julian. Soon after, World War I broke out, and he returned to California. In Pennoyer (1954:26-34), he writes: "World War I found me in the 40th Engineers Camouflage Corps that was forming at Camp American University, Washington DC." He recalls that one of the major events during the winter of 1917 was "the famous Camoufleurs' Ball at the New Willard Hotel," for which Pennoyer was assigned by his commanding officer, Captain Aymar Embury II, "to build a model of a railroad field gun [pictured above] which he wished to use as a centerpiece for his dinner table when he entertained a number of his officers and men before the opening of the ball... Sash pulleys for wheels, heavy cardboard mailing tubes for the gun barrel, plenty of quick-drying camouflage paint and moss representing the grass along the roadbed, tied in fairly closely with all we could see in a photograph we had of a railway gun in France. It had a fake realism quite in keeping with all the deception we, as camoufleurs, were attempting at the time." That same winter,

Pennoyer's unit sailed to France, then traveled by train to Dijon. There, he and several others (with Sherry Edmundson Fry in the lead) were assigned by Embury to function as a liaison to the French camouflage section. In that capacity, one of his most challenging requests was to camouflage a massive fourteen-inch naval railway gun, with a firing range of up to thirty miles. A lifelong railroad enthusiast, Pennoyer was especially known for his depictions of subjects related to railroads.

Lamb (1995:167): "In general, if you are distasteful, the last thing you want is for a predator to learn on you. Much better if it learns on another species that you resemble. By the time it comes to you, with luck, it will have learned not to peck at you. Such mimicry occurs in New World heliconiid butterflies that mimic very distasteful ithomiid butterflies: both are aposomatically arrayed in orange, white-and-black ornament—basic warning colors. It is interesting that humans have also adopted these colors for warning on vehicles, waste tips and office signs."

•

Roland Penrose
(October 14, 1900-April 23, 1984)
British artist, art historian and art collector

Born in England to a wealthy Quaker family, Penrose initially studied architecture at Cambridge University. His interests shifted to painting, and he moved to France, where he became friends with various prominent artists, among them Pablo Picasso, Joan Miro, Man Ray and Max Ernst. Moving back to London in 1936, he became a leading proponent of Modernism, and surrealism in particular. His second marriage was to American photographer Lee Miller.

Above Cover of British Surrealist Roland Penrose's *Home Guard Manual of Camouflage*, published during World War II.

During World War II, although a pacifist, Penrose was a participant in the Industrial Camouflage Research Unit (with Stanley William Hayter, John Buckland Wright, Julian Trevelyan and Dennis Clark Hall), taught military camouflage at the Home Guard training center at Osterley Park, and produced a printed guidebook on civilian camouflage called the *Home Guard Manual of Camouflage*. Subsequently, he became an officer in the Royal Engineering Corps, and a senior lecturer at the Eastern Command Camouflage School in Norwich. He was also associated with the camouflage training center at Farnham Castle.

In his own words (Penrose, 1981:131): "...owing to my close links with the army camouflage unit that had absorbed a number of artists, actors, fashion designers and even a conjurer into its ranks, as soon as the menace of invasion had evaporated, I was

Above A World War I American tugboat called *Narkeeta*, with experimental camouflage by Gerome Brush. As described by Everett L. Warner in "Marine Camouflage" (1919c), "the white ground is covered with black horizontal stripes two inches wide by twelve or fourteen inches long with a few large splotches of black. The Commandant [of the New York Yard] states that this method is not original as the British are reported to be using similar methods."

given the rank of captain and, to my surprise and embarrassment, overnight I found myself in command of the Eastern Command Camouflage School, which had its headquarters in the elegant eighteenth century Assembly Rooms in Norwich."

•

Carl W[illiam] Peters
(November 14, 1897-July 7, 1980)
American painter

Peters was from Rochester, New York, and studied art at the Art Students League in New York. He received national recognition for his work, and created murals for the WPA, but, like other Regionalist and American Scene painters, was displaced and soon forgotten in the tide of Modernism.

There is a massive, astonishing book about Peter's life titled *Carl W. Peters: American Scene Painter from Rochester to Rockport* (Love 1999) that provides a compelling account of the shifts of influence in American art in the first half of the 20th century. It also makes frequent references to camouflage, since Peters was a camoufleur in France during World War I.

Love (1999:316): "Ironically, Private Peters, the soldier, learned that he, too, could spend his time in France as a painter—however, not as a cavalryman making quick sketches along his route, and not as he had dreamed, as one making *plein air* landscapes or genre, but, like [Abraham] Rattner and others, as an American soldier who painted camouflage patterns on military equipment."

•

Above Photograph of Waldo Pierce (n.d.) by Peter A. Juley & Son in the Photograph Archives, Smithsonian American Art Museum (J0002063). As reproduced in Stahl (1995), Dover Pictorial Archives.

Archibald Phillips
(dates undetermined)
British art dealer

Phillips was a British art dealer, who in 1893 established a company, in partnership with Rayner MacConnal, called Phillips and MacConnal (now known as MacConnal-Mason). According to Williams (1989:80), in 1922 Phillips claimed to have invented dazzle painting. He "sent a veritable catalog of camouflage designs, some described as having a 'dazzle effect,' to the [British] Admiralty on 9 and 28 May 1915 [far in advance of its proposal by Norman Wilkinson in 1917]. One of these designs was based on a peculiar mosaic pattern system which produced a strongly three-dimensional stepped effect. Phillips went to considerable expense in the promotion of his claim, publishing an elaborate booklet entitled 'Suggestions for the Camouflage or Dazzle of British Merchant Ships in the Great War.'" A camouflaged ship bearing Phillips' proposal of a pattern of large, colorful hexagons is reproduced in Williams (2002:63).

Mary Chatton (1993:65): "Poets often see patterns in nature. One object will be compared to another object that has a similar pattern. Photographer Tana Hoban's *Take Another Look* and *Look Again* use pictures with peepholes in them to show how patterns and texture in nature can be deceiving; some things look like other things. Her photograph of a fish's scales, for example, could be a snake's skin."

•

P.E. Phillips
(dates undetermined)
British painter

In Barkas (1952), he is mentioned as having been a camouflage officer during World War II.

•

Waldo Pierce
(December 17, 1884-March 8, 1970)
American painter

Pierce was born in Bangor ME, to a family of properous lumber mill owners. He went to school at Harvard, then joined the American Field Service in 1915, for which he drove an ambulance on the battlefield in France, and for which he was awarded the Croix de Guerre. Later, when the US entered the war, as noted in Wolanin (1983:64), he worked with Arthur B. Carles as a camoufleur in the Philadelphia Navy Yard. Following the war, he lived in Paris, where he became associated with the "Lost Generation" of American expatriates, among them Ernest Hemingway. It was Pierce who painted a portrait of Hemingway for the October 18, 1937 cover of *Time* magazine.

•

Émile Pinchon
(1872-1931)
French sculptor

The brother of camoufleur Joseph Pincon, he is listed in Coutin (1988) as a World War I French camoufleur.

•

Joseph Pinchon
(April 17, 1871-June 20, 1953)
French painter, designer

He is listed in Coutin (1988), along with his brother Émile Pinchon, as a World War I French camoufleur. He was a portrait painter and a costume designer for the Paris Opera. According to Kahn (1984:166), an early memo lists him as one of the first French camoufleurs in World War I, with five others. He was a knight in the Legion of Honor.

•

Joseph B[rereton] Platt
(1895-February 6, 1968)
American industrial and stage designer

Born in Plainfield NJ, Platt studied at Parson's School of Design in New York. Following World War I, he worked as a mural painter, then as the Paris correspondent for Condé Nast publications, including *Vanity Fair*, *Vogue* and *House and Garden*. His interests shifted to interior design, and he became head of design at the Marshall Field Department Store. He was responsible for the popularization of Parsons tables, and the design of the Whitman's

Chocolates Sampler Box, a Parker Pen, and packaging for Elizabeth Arden cosmetics. In 1938, he was hired by David O. Selznick to design the sets for the film *Gone With the Wind*, which led to other set designs for such films as *Rebecca*, *The Parradine Case*, and *Lady of Burlesque*. His father-in-law was Rudolph Evans, sculptor for the Jefferson Memorial. *New York Times* ("Joseph B. Platt," 1968): "In 1917 Mr Platt when to France with the Army Camouflage Corps."

•

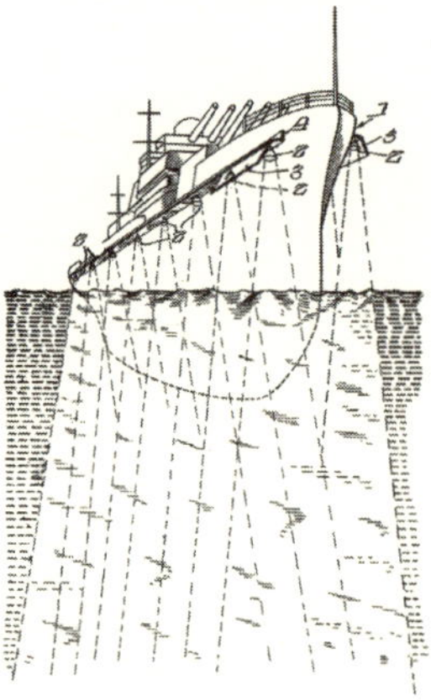

Above Drawing submitted for US Patent No. 2,300,067 titled "Ship," invented by Martin C. Schwab (1941). In this proposal, a ship is fitted with downward-pointing lights for the purpose of reducing its shadow and preventing detection by U-boats.

Facing page Paintings of World War I camouflaged ships by American artist Burnell Poole (clockwise, beginning at the top): *A Distorted Target for U-Boats* [*Mauretania*] (1924); *A Fast Convoy* [*USS Allen* escorting *USS Leviathan*] (n.d.); *A Critical Situation* [*USS Stockton*] (n.d.); *USS Kimberly Escorting USS Leviathan* (1926); and *USS Stevens escorting USS Leviathan* (1918). Courtesy US Naval Historical Center.

Walter C. Pleuthner
(1885-?)
American artist, architect

In Williams (2002:68), there are reproductions of two drawings for a ship camouflage proposal that was submitted by Pleuthner during World War I, then abandoned after testing by Loyd A. Jones and his associates at the Eastman Kodak Research Laboratories. The pattern was comprised of circles and triangles, in random distribution on the ship's surface.

Pleuthner was an architect and designer from New York. As a painter, he had exhibited at the Armory Show in 1913.

•

Pierre-Marie Poisson
(1876-1953)
French sculptor and metals artist

Poisson received his training at the École des Beaux-Arts in Paris. He is listed in Coutin (1988) as a World War I French camoufleur. One of his sculptures, titled *Allegory of Maritime Commerce*, was commissioned for *Le Normandie*, a French ocean liner that sank in 1940.

•

Burnell Poole
(August 30, 1884-1933)
American painter

Poole was born in Boston MA. He was not a camoufleur, but as a US government artist and a wartime correspondent for *Everybody's Magazine*, he made a number of paintings of World War I camouflaged ships, as shown on this page.

Henry Poole
(January 28, 1873-August 15, 1928))
British sculptor

Born in London, he was the son of sculptor Samuel Poole. He was educated at Lambeth School of Art and the Royal Academy (1892-97). According to the Tate Archives Online, he served in the Camouflage School in 1915-1918.

Gladys Thayer (Abbott H. Thayer's daughter, quoted in White 1951:109): "Our fourth trip was to the West Indies...My father's special mission was tasting butterflies! This was in order to disprove what his very dear friend, Professor Poulton of Oxford has written many lengthy books to *prove*, the theory of mimicry, trying to show that harmless butterflies or other insects had through natural selection acquired similar patterning and coloring to those of bad tasting butterflies for their protection. He actually tasted them and could find no difference in the flavor."

•

Louis Dalton Porter
(1919-June 28, 2006)
American artist

Born in Kaplan LA, Porter grew up in Mississippi and Louisiana. He served in the Ghost Army during World War II. See Holley (2006).

•

Facing page An interior page spread from Adolf Portmann's excellent book on *Animal Forms and Patterns* (1967), made all the more wonderful by Sabine Baur's exquisite pen and ink illustrations.

Adolf Portmann
(May 27, 1897-June 28, 1982)
Swiss zoologist

Portmann was born in Basel, Switzerland, and studied zoology at the University of Basel. In 1931, he became a professor of zoology at the same university, where he specialized in marine biology and comparative morphology. He wrote two important books on animal coloration: *Animal Forms and Patterns: A Study of the Appearance of Animals* (Portmann 1967), and *Animal Camouflage* (Portmann 1959).

•

Above Edward B. Poulton, as photographed by James Lafayette, date unknown. Public domain.

Edward B(agnall) Poulton
(January 27, 1856-November 20, 1943)
British zoologist

A professor of zoology at Oxford University, Poulton was a fervent Darwinian. He was a loyal supporter of Abbott H. Thayer's theories of protective coloration,

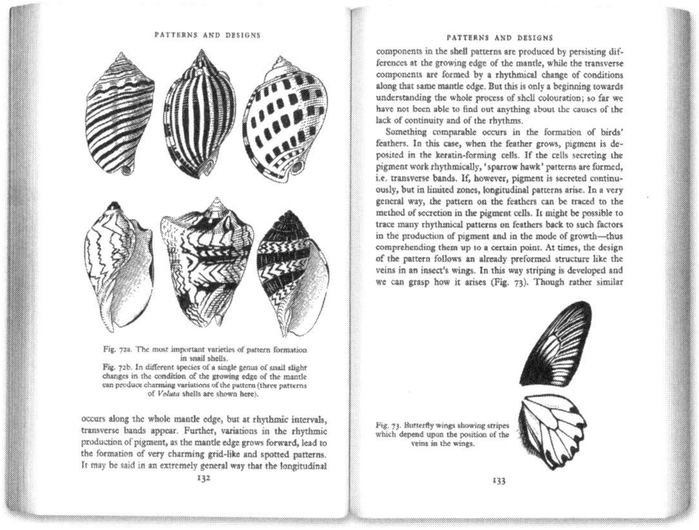
PATTERNS AND DESIGNS

Fig. 72a. The most important varieties of pattern formation in snail shells.
Fig. 72b. In different species of a single genus of snail slight changes in the condition of the growing edge of the mantle can produce charming variations of the pattern (three patterns of *Voluta* shells are shown here).

occurs along the whole mantle edge, but at rhythmic intervals, transverse bands appear. Further, variations in the rhythmic production of pigment, as the mantle edge grows forward, lead to the formation of very charming grid-like and spotted patterns. It may be said in an extremely general way that the longitudinal

132

PATTERNS AND DESIGNS

components in the shell patterns are produced by persisting differences at the growing edge of the mantle, while the transverse components are formed by a rhythmical change of conditions along that same mantle edge. But this is only a beginning towards understanding the whole process of shell colouration; so far we have not been able to find out anything about the causes of the lack of continuity and of the rhythms.

Something comparable occurs in the formation of birds' feathers. In this case, when the feather grows, pigment is deposited in the keratin-forming cells. If the cells secreting the pigment work rhythmically, 'sparrow hawk' patterns are formed, i.e. transverse bands. If, however, pigment is secreted continuously, but in limited zones, longitudinal patterns arise. In a very general way, the pattern on the feathers can be traced to the method of secretion in the pigment cells. It might be possible to trace many rhythmical patterns on feathers back to such factors in the production of pigment and in the mode of growth—thus comprehending them up to a certain point. At times, the design of the pattern follows an already preformed structure like the veins in an insect's wings. In this way striping is developed and we can grasp how it arises (Fig. 73). Though rather similar

Fig. 73. Butterfly wings showing stripes which depend upon the position of the veins in the wings.

133

particularly the phenomenon of countershading, which, as it turned out, Poulton had discovered in advance of Thayer.

•

Francisque Poulbot
(1879-September 16, 1946)
French designer, illustrator

Poulbot is best known for his comic illustrations (in such magazines as *Le Pele-Mele* and *La Baionnette*) as well as for a book about the impact of World War I on French children. In addition, he and his partner Leona were known for their tradition of holding an annual mock wedding. He served as a French camoufleur in World War I, and was awarded the Legion of Honor.

•

Pow
(full name and dates undetermined)
American soldier

With regard to the World War I American Camouflage Corps, Faulkner (1957) mentions a Sergeant Major Pow "who was to be the backbone of the regiment [and who] made his appearance early and was an exception to the general artistic trend."

•

Robert O. Preusser

(1919-1992)
American artist, design educator

Preusser was a teacher of visual design at the Chicago Institute of Design in the 1940s. In 1954, he was recruited by György Kepes to teach at MIT at the Center for Advanced Visual Studies, where he remained until 1985. According to the Archives of American Art, he served "with a camouflage unit in the US Army, 1942-45".

•

Robert Chenault Givler (1922:299-300): "The dream is a *psychic camouflage*...[The deceptions of dreams] are one and all intentional, for the subconscious uses all these dramatic devices in order to release the long pent-up energies of its fears and desires without rousing the sleeper and spoiling his rest. Sometimes, of course, the camouflage fails, and then we have the nightmare or some other unlovely awakening..."

Peter Proud

(1913-1989)
British film director

Born in Glasgow, Scotland, his birth name was Ralph Priestman Proud. He began working in the film industry at an early age, first as a sound recorder, then as an assistant art director, for Alfred Hitchcock, among others. By the end of his career, he was credited with more than seventy films. He is described in Reit (1978:138) as having served as a camoufleur in North Africa with Geoffrey Barkas in WWII. Barkas (1952) recalls that he did "as much as anyone to give 'battle camouflage' its start in life at Tobruk."

•

Above French "Quaker gun," a decoy made entirely of *papier maché*. WWI news photo.

Mary Pukui

(1865-1956)
Hawaiian folklore scholar

Born in Hawaii, Pukui worked for the Bishop Museum in Honolulu as a Hawaiian language translator, folklore recorder, and consultant. According to Peterson (2004): "During World War II she was consulted as a Hawaiian expert and worked as a foreperson in a US Army Corps of Engineers unit to camouflage military installations."

•

John D. Whiting (Whiting 1928:226-227): "She [*HMS Stockforce*] was, to the naked eye, a common cargo tramp, as shiftless-looking a tramp as ever racketed time-loosened rivets to the rough buck of a sea. Her decks, littered with scraping of cargo, were ill kept and foul, even had chicken coops, and she displayed washing on her cargo booms. But through the glass [telescope] Hugh could make out the wicked muzzles of guns peering darkly over the wooden bulkheads which had masked them.

'A Q-ship, by thunder,' he thought, 'a mystery ship. The censors keep those things secret as fate; I never thought I'd really see one. I wonder what she's been through...'"

Jean Puy
(November 8, 1876-March 7, 1960)
French painter

Puy was born in Roanne, France, and studied at the École des Beaux-Arts in Lyons, and at the Académie Julian and the Académie Carriere in Paris. He was closely associated with Henri Matisse and other Fauvists. During World War I, he was a French camoufleur. See Gribaudo (1977:9 and 15).

Q

Q-Ships
Also called Q-boats or mystery ships

Q-ships were armed British ships that appeared to be harmless, vulnerable targets for German U-boats. As Bond (1918-1919:502) describes them: "The Germans had some unhealthy experiences with the 'Q-boats' or 'mystery' ships of the British. These were vessels rigged up much like ordinary tramp steamers, but they were loaded with wood, so that they would not sink, and their hatches were arranged to fall open at the touch of a button, exposing powerful guns. They were also equipped with torpedo-tubes, so that they could give the U-boat a dose of its own medicine. These ships would travel along the lanes frequented by submarines, and invite attack."

Q-Ships are described at length in Hartcup (1980). See also Gordon Campbell.

•

Quaker Guns
Dummy artillery decoys

In contrast to Q-ships, so-called Quaker guns only appeared to be weapons. They were instead dummy installations of troops, rifles and so on, for the purpose of causing the enemy to miscalculate the size of the Allied force or its location. They might also draw fire from the enemy, causing them to waste ammunition while enabling their position to be located.

•

Jean-Paul Quint
(1889-1953)
French illustrator

Quint is listed in Coutin (1988) as a World War I French camoufleur. Following the war, he illustrated a 1922 edition of Stendhal's *The Red and the Black*.

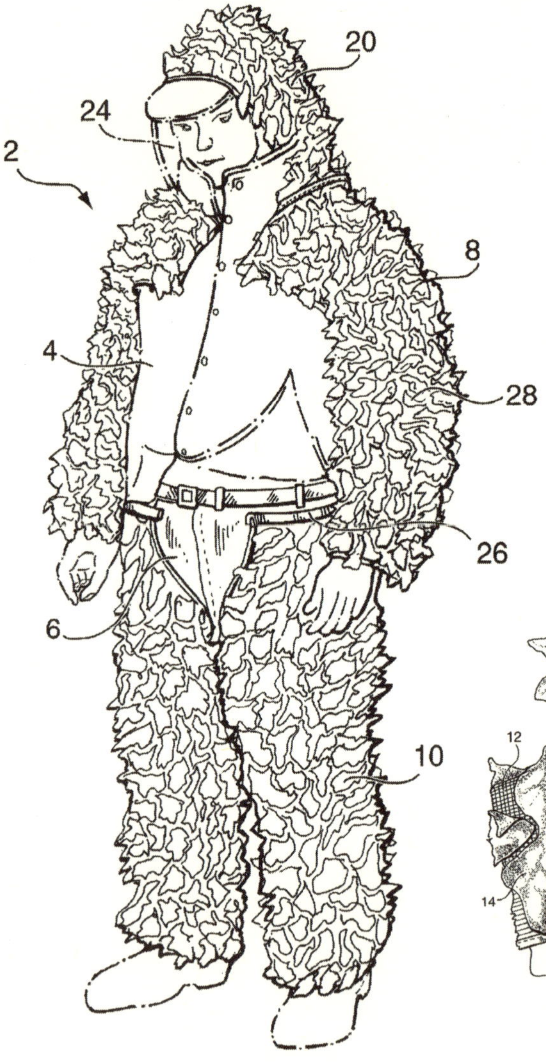
20
24
2
8
4
28
26
6
10

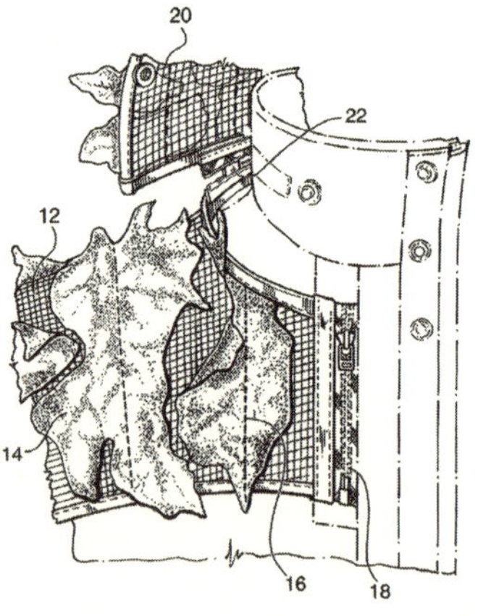
20
22
12
14
16
18

R

(Wilmer) Bruce Rabenold
(dates undetermined)
American architect

Rabenold studied architecture at the University of Pennsylvania, and was later on the faculty at Cooper Union in New York. He is listed as a World War I American camoufleur in Embury (1917:138), who describes him as "of W.W. Bosworth's office."

•

Charles (Frederic) Ramsey
(September 23, 1875-May 21, 1951)
American artist

Born in France, he was the son of the painter Milne Ramsey. He studied at the Pennsylvania Academy of the Fine Arts, then taught at the same school until 1912. In 1916, he held a brief appointment as the director of the Minneapolis School of Art. Dismissed for supposedly radical views (he was accused of being a Socialist), he moved to New Hope PA, where he and others formed an association of artists called The New Group. During World War I, he worked for the US as a marine camoufleur. For a full account, see Alterman (2005).

•

Facing page Drawings submitted for US Patent No. 6,460,185 titled "Tree Leaf Camouflage Suit," invented by Phil Hardy (2002).

Abraham Rattner

(1893-1978)
American painter

Richard Wolin: "By dispensing insight under the camouflage of humor, satire makes distasteful revelations palatable."

Above Photograph of camoufleur Abe Rattner (n.d.) by Peter A. Juley & Son in the Photograph Archives, Smithsonian American Art Museum (J0002095). As reproduced in Stahl (1995), Dover Pictorial Archives.

Facing page Camouflaged German airplane during World War I.

National Marine Magazine (March 1919): "[In WWI ship camouflage] Each nation seemed to have a characteristic type of camouflage and after a little practice you could usually spot a ship's nationality by her style of camouflage long before you could make out her design."

Born in Poughkeepsie NY, of a Russian father and Romanian mother, he studied architecture at George Washington University, then went on to study art at the Corcoran School of Art and the Pennsylvania Academy of Fine Arts. He is listed in Embury (1917:138) as a World War I camoufleur, and is described as "one of the cleverest colorists in the first company."

Culkin (1980): "Rattner served in the US Army during World War I. He was recruited by Homer Saint-Gaudens for service in the emerging work in camouflage. He was promoted to sergeant, put in charge of camouflage research, and served at the front in the second Battle of the Marne, Chateau-Thierry, Belleau Wood, and on the Hindenberg Line. He received a severe back wound which troubled him for the rest of his life.

...He volunteered for camouflage service in the second World War but had to be content with delivering a few lectures and doing some research at home."

Rattner, as quoted by Culkin (1980): "The camoufleur seeks to conceal. The perfect camouflage is a job where complete concealment is achieved...Where complete camouflage is not possible, the camoufleur must a) *reduce visibility*, b) *create doubt as to the nature of the object so as to confuse the enemy* or c) *distract attention of enemy to a decoy*. Forty-five seconds of hesitation on the part of the bombardier is enough to cause him to miss his target."

•

Georges Redon

(November 16, 1869-1943)
French printmaker, draughtsman

Born in Paris, Redon was a restaurant designer and illustrator. He is listed in Coutin (1988) as a World War I French camoufleur. Following the war, he

Popular Mechanics (1918:884): "Airplanes captured in recent months from the enemy [the Germans] show that hexagons of various colors are being frequently used in the enemy's camouflage scheme. The pattern unit consists of a purple hexagon surrounded by six others, two black, two green, and two grayish blue..."

wrote and illustrated a large format book for children titled *To the Youngsters Who Will Constitute the France of Tomorrow.*

•

Paul Regusis
(?-November 25, 2007)
American photographer, graphic designer

Regusis was a member of the Ghost Army. After World War II, he studied art in Paris, then returned to New York to work as a photographer and designer. He settled in Albuquerque NM in 1954.

Regusis (1990a): "I was in the 603rd Engineers Camouflage Battalion, 21st Special Troops, US Army in World War II. My unit not only built and placed dummy tanks and guns...but we also used sound tapes (records at that time), simulated artillery fire with colored powders in empty shell casings, painted showcard signs of the simulated infantry division, worked with real units at times, faked troop movements with our own vehicles, etc., etc."

•

Theodor Reik
(May 12, 1888-December 31, 1969)
Austrian-born American psychoanalyst

A student of Sigmund Freud in Vienna, Reik had no formal connection with camouflage, military or otherwise. However, he wrote a provocative essay titled "Neurotic Camouflage," in *Listening With the Third Ear*, in which he described the methods by which people with neuroses often conceal their symptoms from others. In *social* or *neurotic camouflage*, said Reik, "The patient has to be obedient to the complusion imposed by the neurosis, but at the same time he feels impelled to heed the psychic necessity of making his behavior conform to the requirements of his social milieu" (Reik 1949:293).

Roland Reiss
(May 15, 1929-)
American artist

Born and raised in Chicago, he studied at the American Academy of Arts in Chicago, then earned BA and MA degrees at the University of California at Los Angeles. He taught for many years at the University of Colorado in Boulder, and at Claremont in California. During the Korean War, he was not assigned to camouflage, but he did serve as the director of about fifty US Army artists, who produced instructional materials (charts, graphs, posters and so on), including camouflage training aids. See Reiss (1997/1999).

•

Seymour (Victory) Reit
(November 11, 1918-November 21, 2001)
American author, illustrator

Above The dust jacket of *Masquerade* by Seymour Reit, who was himself a WWII camoufleur.

Facing page A selection of humorous drawings (c1919) about disruptive camouflage from WWI-era publications. They include (moving clockwise, beginning at the top): Two boys dazzle painting a row boat; a cartoon from *Punch* in which a British sailor with paint-splattered clothes is compared to the camouflaged ship in the background; a *Punch* cartoon in which street workers filling cracks in the road with tar are making patterns that resemble dazzle painting; and British artist Bruce Bairnsfather's characterization of a British Tommy in a disruptively patterned uniform.

Born in New York on Armistice Day, he was given the middle name Victory. While attending New York University, he produced cartoons for the school's humor magazine. Later he worked for Fleischer Studios, making drawings and writing jokes for animated cartoons, including *Popeye* and *Betty Boop*. In 1940, he created the character *Casper the Friendly Ghost*.

Nash (2001`): "During World War II he used his artistic skills in a camouflage unit. 'They began by camouflaging the West Coast, because they were afraid of the Japanese,' recalled his widow, Edmée Reit. 'It was a unit full of wonderful people like Billy Baldwin who had been scenic designers and window decorators.'"

Following World War II, Reit returned to writing humorous dialog for comic books and animated cartoons, among them *Little Lulu*, *Archie* and *Casper*. He was also the author of more than sixty satirical pieces for *Mad* magazine, and more than eighty children's books.

SPREAD OF THE "DAZZLE" CULT

He also wrote a number of books for adult readers, one of which was *Masquerade: The Amazing Camouflage Deceptions of World War II* (Reit 1978).

Theodor Reik (1949:296): "Where would a clever man hid a particular leaf? In the forest."

Colin Watson: "A needle is much simpler to find in a haystack than in a bin of other needles."

•

Henry Reuterdahl
(August 12, 1871-December 22, 1925)
Swedish-born American painter

Above Photograph (detail) of American camoufleur Henry Reuterdahl (1908) by George Grantham Bain. Public domain.

Born in Sweden, in 1893 he immigrated to the US where he taught at the Art Students League and gained prominence as a marine painter during the Spanish-American War and World War I. Of particular note are his paintings of dazzle painted ships.

Williams (1989:75): "Another [US] Camouflage Section under Henry Reuterdahl and quite separate from the Eastman Kodak Research Center was attached to the Emergency Fleet Corporation, an organization which had been created to manage and operate vessels acquired by the [US] Shipping Board specifically for the war effort. The painting of merchant vessels in camouflage colors was the exclusive concern of this corporation."

Rodyenko (1941:39): "To most people, the art of camouflage means 'painting things up.' They are, of course, very wrong. In general, they get this impression from seeing pictures of guns and tanks 'camouflaged' with crisscross and zigzag designs. In their newspapers they read of snipers who wear strange-colored costumes to 'confuse' the enemy. They hear and believe stories about industrial buildings which have been so decorated with dazzle paint that they are invisible from the air. What they do not realize is that most of such stuff is utter nonsense."

•

Norm(an) Rice
(1913-December 31, 1956)
Australian cartoonist

Born in Newtown, New South Wales, he studied at Sydney Art School, then contributed cartoons to various Sydney-area publications. Kerr ("Norm Rice" n.d.): "At the beginning of World War II he joined an army camouflage unit, later transferring to a survey unit when he served in New Guinea."

•

Bart Whaley (2000:76): "When among nudists, going naked is the best disguise."

Grenville Rickard
(1890-May 3, 1956)
American architect

Rickard was born in Denver CO. He graduated from Yale University Architectural School, then worked as an architect in Denver and New York. *New York Times* ("Grenville Rickard, Architect Here" 1956): "In World War I, Mr. Rickard served with the Army Engineers. He was the Chief of the Camouflage Unit of the Office of Civilian Defense in World War II. In 1941 he was chairman of the New York Civilian Camouflage Council, and for the two years following, secretary of the Camouflage Commission, Washington." He is listed in Embury (1917:138) as "of Cass Gilbert's office." See also Rickard (1942).

•

André Rivaud
(April 6, 1892-?)
French sculptor, designer

Born in Paris, he is listed in Coutin (1988) as a World War I French camoufleur. A medals artist, he is credited with designing the War Cross, a Greek military decoration, in 1917, as well as the winners' medals for the 1924 Olympic Games in Paris.

•

Left Installation of a screen to conceal activity on a road in France during World War I.

Roads, Camouflaged

Bairnsfather (1919:226): "We drove down an elaborately camouflaged road. These are just ordinary roads, with a screen constructed from a kind of rush matting fixed up on the side nearest the enemy.

The appearance of these roads from a distance is just like the rest of the country. Of course this doesn't prevent the enemy from firing at such roads which they know exist, but it prevents deliberate aim at a definite object and therefore it would probably be a sheer waste of shells to fire on the off-chance of hitting something. It's not a nice sensation driving along these camouflaged roads, but there it is, and the danger is not really great."

J(ames) Hampden Robb

(1898-1988)
American architect

A Boston architect, he was educated at Harvard (1922) and Columbia University School of Architecture. He is listed in Reit (1978:79) as having worked with William Pahlmann during WWII at Jefferson Barracks, St Louis MO.

•

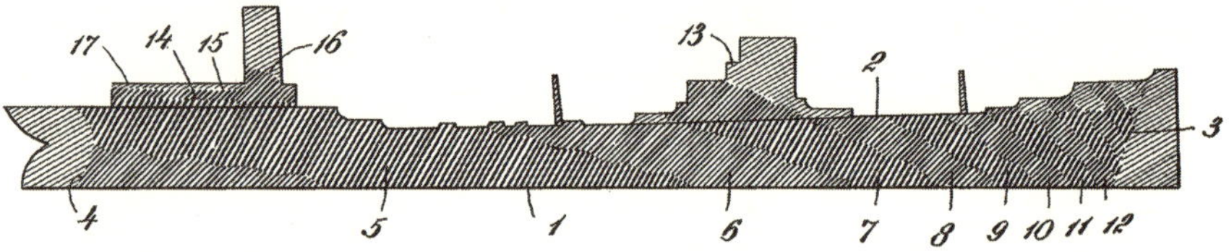

Above Drawing submitted for US Patent No. 2,292,848 titled "Camouflage," invented by Arthur Robson (1942). He describes it as a method that "provides, in addition to low visibility at a distance, effective distortion within the range of visibility."

Frothington (1971:163): "What would a deep water sailor of the old type have thought, if he had been dropped into the midst of the parti-colored [dazzle painted] ships of 1918? Concealment camouflage had been practically given up, and the British had led the way in daring designs of stripes and blocks of color to deceive as to sizes and courses of ships."

Arthur D. Robson

(dates unknown)
American chemist

The son-in-law of (camoufleur) Maximilian Toch, Robson was a First Lieutenant in Company I of the 40th Engineers, stationed at Dijon during World War I.

Tracy (1919:177): "On October 3rd, 1917, the officer in charge of organization at Washington and three lieutenants, two of whom were chemists, skilled in dyes, painting and colors, and had been asked for by General Pershing, embarked for France, and reported on arrival to the Commanding Officer, Camouflage Section, at Paris."

Toch (1931:309): "The first chemist to go over in October 1917 was my own son-in-law, Arthur D. Robson, a Cornell graduate in chemistry and a man who had several years' experience in factory management."

•

Norman (Percival) Rockwell
(February 3, 1894-November 8, 1978)
American illustrator

Born in New York, Rockwell studied at the National Academy of Design and the Art Students League, where he worked with George Bridgman and others. By age 19, he had become the art editor of *Boy's Life* magazine. Several years later, with the advent of World War I, he tried to enlist in the Navy, at first unsuccessfully because he was underweight. When he was accepted later, according to McCloskey (2005), "He was assigned to paint camouflage on airplanes. When his talent became evident to his superiors, he was returned to the home front and commissioned to paint portraits of army officers." Following the war, he became famous for his covers for *The Saturday Evening Post*.

•

Guy (Lindsay) Roddon
(December 6, 1919-July 2, 2006)
British painter

Beechey (2006): "Along with many artists during the second world war—including William Coldstream, Claude Rogers and Julian Trevelyan—Roddon found a comfortable billet, diverting work and genial company in the Camouflage Corps. In his first posting, to Farnham castle, Surrey, he was under the command of Colonel Jack [sic Frederick] Bennington, the suave director of Wildenstein's, and the gallery owner Freddy Mayor. Following a transfer to Norwich, he was one of a band of camouflage officers working for the theatre designer Oliver Messel.

The headquarters of the Eastern Command camouflage school were the city's 18th-century Assembly Rooms, which Messel had lavishly restored by the troops. Messel was replaced by the painter and critic Roland Penrose, who hung the officers' mess with examples from his impressive Surrealist art collection."

•

Peter Rodyenko
(1891-?)
Austrian-born American designer

Born in Austria, Rodyenko was educated at the University of Vienna. After moving to the UK, he served in the British merchant marine, then worked as a war correspondent in India and China. From 1912 to 1919, he was a military and legal advisor to the Chinese government. He then immigrated to the US, and in 1923 became a lieutenant in the National Guard. During World

War II, he served in the US Corps of Engineers as a camouflage instructor. See Rodyenko (1941) and the *New York Times* ("Major, 'Over 60,'" 1951).

•

Grenville Rickard (1942:190-191): "[During World War I] We heard much of the theories of Abbott Thayer concerning protective coloration in animals and birds. So we painted the bottoms of wooden gun barrels light and the highlights dark and broke the whole into jagged patterns of color. We made snipers' uniforms, hollow dead cows for observers, and other such contrivances."

...

Jane Toombs (2002:40):

"'Would you take a gander at that, Alfie,' Sid said. She's painted all in stripes like a brown and green zebra.'

'Camouflage, Sid. I read about it. The U-boats can't see her good when a ship's been camouflaged. It's supposed to make them look like clouds or something."

Louis Roger
(August 26, 1874-June 12, 1953)
French artist

Roger was a student of Jean-Paul Laurens and Jean-Joseph Benjamin-Constant. According to Kahn (1984:166), there is an early memo in which he is listed as one of the first French camoufleurs in World War I, along with five others. He was awarded the Legion of Honor.

•

Stanley R.H. Rogers
(1888-January 21, 1961)
British-born Marine artist and author

Born in Nottingham, England, he attended art school in London. As a youth, he also lived in Walla Walla WA for a period. He wrote and illustrated more than thirty books pertaining to the sea and ships. He returned to live in the US in 1945. *New York Times* ("Stanley Rogers" 1961): "From 1920 to 1945 he lived in England and during World War II he served with a Royal Navy camouflage unit as a civilian."

•

Eugène Ronsin
(dates undetermined)
French theatre designer

He is listed in Coutin (1988) as a World War I French camoufleur. According to Kahn (1984:166), there is an early memo in which he is listed as one of the first French camoufleurs in World War I, along with five others.

•

John W(elborn) Root, Jr.
(1887-1963)
American architect

Born in Chicago, he was the son of a prominent architect. He studied architecture at the École des Beaux-Arts in Paris. He became friends with John Augur Holabird (the son of Chicago architect William Holabird), with whom he would later work in the firm of Holabird and Roche, later Holabird and Root. He designed the Chicago Daily News Building (1929), Chicago Board of Trade Building (1930), and the Chrysler Building at the Century of Progress World's Fair (1933-34).

During World War I, he served as a captain in France in the American Camouflage Corps.

Faulkner (1957): "Captain [John] Root was full of ideas and enthusiasm and it was from his lips that we first heard the magic words of 'field lines' and 'trip wires', 'camouflage discipline' and 'duck boards'. We all made lots of mistakes to begin with but he was fertile in new devices and pulled us out of the hole with quite amazing dexterity. He was a good man to work for, cheerful and patient. Root was quite a figure as he used to come to us across the fields with his big horn spectacles, his famous leather overcoat upon his back, and swinging his walking stick with the black and white woven handle."

In an interview (Benjamin 2003), Holabird explains that he was transferred to a camouflage unit during World War II at the suggestion of his father's partner, John W. Root, Jr., who had served as a camouflage officer in World War I. However, when Holabird found that the camouflage work consisted of weaving burlap netting, he transferred instead to a parachute unit.

•

Louis C(onrad) Rosenberg
(May 6, 1890-1983)
American architect and printmaker

Born in Portland OR, he received private instruction in architecture as a child. He attended MIT, where he won the Traveling Fellowship in Architecture. Rosenberg (n.d.): "In 1917 Rosenberg joined the US Army and served in the Camouflage Corps, 40th Engineers, AEF, under the command of an architect, Aymar Embury." In his radio talks, Faulkner (1957) describes him as a "paint manufacturer." In 1919, he returned to the US to teach architectural design at the University of Oregon. As his involvement in printmaking increased, he switched from architecture to art, in the process of which he enrolled at the Royal College of Art in 1924. Rosenberg (n.d.): "With the advent of World War II, Rosenberg enlisted in the US Army and was again assigned to a camouflage unit under the command of Aymar Embury, the New York architect. His company was sent to North Africa where Rosenberg sketched and water-colored. These are included in the book *Middle East War Projects of Johnson, Drake, and Piper, Inc., for the Corps of Engineers, U.S. Army, 1942-43.*"

•

(James) Milford Zornes (1965): "The amusing thing about our battalion up there [at March Field in California during WWII] was that I think they didn't know what to do with stagehands, stage designers, window dressers, painters, and this whole crew of creative people in one field or another, so they put us all in camouflage."

...

Hugh B. Cott (1940:2): "So accustomed are we to reject what the eye sees in nature, so dull and dead have we become as a result of visual experience, that to appreciate the wonder and wealth of color around us we must be shown our surroundings in some novel or unusual manner—in a picture, for instance, or as they appear when we stand on our heads, or when seen inverted in the focusing screen of a camera. Indeed, so largely does experience enter into and modify our perception of objects, that many people are quite unable to accept what the eye gives them, but only what they have learned to expect it to give them: they *see* only what they *know*. They have lost that power which artists by patient striving have recovered, and which Ruskin calls the 'innocence of the eye.'"

Vladimir Baranoff Rossine
(1888-1944)
Russian painter and sculptor

Born in Kherson, Ukraine, he studied at the Academy of Fine Arts in St. Petersburg, c. 1903-1907. He moved to France, became associated with the School of Paris, and adopted the French-sounding name Daniel Rossiné. He also became friends with Robert and Sonia Delaunay. He traveled to Norway, where he met his first wife, then returned to Russia to take part in the Russian Revolution. There, he experimented with the Moebius strip, and with color, combining sound with vision. He devised an optophonic piano, in which colored disks are moved when piano keys are struck. He left Russia in 1925 to return to Europe, where he gave optophonic concerts at Paris, Riga and Berlin. In the early years of World War II, he is said to have researched the use of Pointillism in camouflage, which he and Robert Delaunay (with whom he formed a business) called Chameleon. He disappeared during World War II.

•

Ernest-Henri Rouart
(August 24, 1874-February 27, 1942)
French painter

Born in Paris, he was initially interested in mathematics, but at some point turned to art. He studied with Edgar Dégas, who guided his copying sessions at the Louvre. He is listed in Coutin (1988) as a World War I French camoufleur. He organized centenary exhibitions for Édouard Manet, Dégas and Berthe Morisot (Manet's sister-in-law).

•

Above Camouflaged World War II fleet oiler, the *USS Sebec* (n.d.). US Navy photograph.

Auguste Jean-Baptiste Roubille
(December 15, 1872-1955)
French painter, designer

Roubille, a native of Paris, was particularly interested in humorous illustration. He worked for a variety of French humor magzines, including *Rire*, *Cri de Paris*, and *Cocorico* [French for "cock-a-doodle-doo"]. He created a mural for the House of Laughter at the Paris Éxposition Universelle in 1900. He is listed in Coutin (1988) as a World War I French camoufleur.

•

Lewis S. Rubenstein
(1908-2003)
American painter and printmaker

He attended Harvard University, graduating in 1930. He studied fresco painting, initially in France and Italy, and then in the US with Rico Lebrun. He was awarded a Fulbright fellowship to study Japanese ink painting. He was on the faculty at Vassar College in 1946-74. "Lewis S. Rubenstein" (2008): "He also painted camouflage during his four-year stint in the United States Navy." Rubenstein worked under Everett L. Warner in the World War II US Navy Camouflage Section. Hays (1999): "he was assigned to put out technical and administrative 'fires'—in and out most of the time."

•

Charles C(ary) Rumsey

(August 29, 1879-September 21, 1922)
American sculptor

Darwin Porter (*The Secret Life of Humphrey Bogart*): "[In November 1918, Bogart] was pronounced fit for duty and assigned to the *USS Leviathan*, a transport ship with a trio of gigantic funnels and zebra-like camouflage stripes."

...

Christopher Morley [British actor] (1920:73): "We hurried down toward the waterfront [in Philadelphia during WWI] and peeped through the high board fence to see a steamer in drydock for a coat of camouflage. Great stripes of black and blue and white were being laid across her hull."

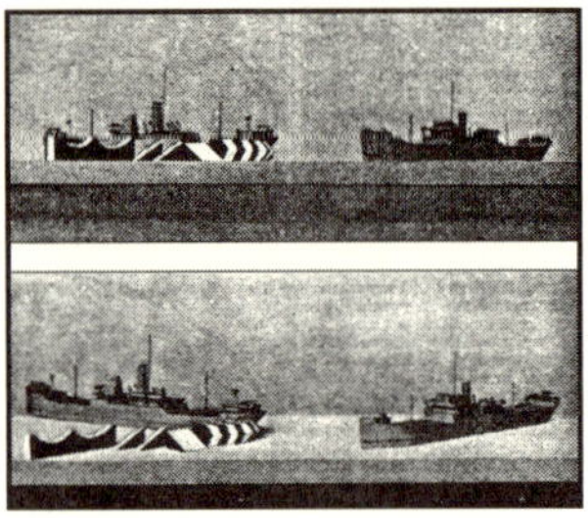

Above Everett L. Warner's demonstration of the effectiveness of dazzle painting for course deception, in contrast to a ship painted only in gray. It also demonstrates an ingenious method that Warner devised for generating new dazzle designs, simply by arranging painted portions of wooden blocks in front of an unpainted model, then applying that arrangement to the side of the ship model, regardless of its actual course.

Born In Buffalo NY, Rumsey studied at Harvard University, then at Boston Art School, and the École des Beaux-Arts in Paris. He specialized in bronze statues of horses, including one of race horse Nancy Hanks. During World War I, he was a captain in the Headquarters Troop, Seventy-seventh Division and Fortieth Engineers. He married Mary Harriman, the sister of W. Averell Harriman. He died in a car accident in New York at age 43.

•

Walter Russell

(May 31, 1867-April 16, 1949)
British painter

Russell was a prominent portrait and landscape painter who taught for many years at the Slade School of Art, and was eventually knighted. During World War I, he served as a camoufleur in the Royal Engineers under Solomon J. Solomon (presumably in the same unit as Alister Hardy). In the Imperial War Museum in London, there is a well-known painting by Solomon, titled *Erecting the First Camouflage Tree*, showing the nighttime installation in 1916 of an armored fake dead tree trunk, to be used as a battlefield observation post. Standing on the right, supervising the event, are Solomon and Russell. He was also instrumental in the camouflage of tanks. As described in Chesney (1943:73), he was put in charge of supervising four hundred French women workers, who garnished nets (for umbrella camouflage) at a factory at Amiens. See also Hartcup (1980).

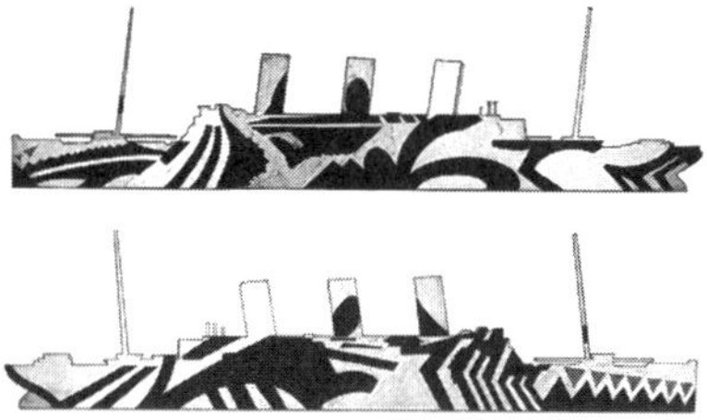

Above More examples from World War I of the astonishing dazzle camouflage schemes produced by Everett L. Warner and his team of navy camoufleurs: (top) The dazzle design for the two sides of the *USS Leviathan*, designed either by Frederick Waugh or Norman Wilkinson; (center) the same design applied to the actual ship; and (bottom) various models, including (first row, left to right) a flush deck destroyer, *USS America*, and *USS Columbia*; and (second row) *USS Proteus*, *USS George Washington*, and (according to Warner's notes) a "Nansemonde type."

S

John Saccaro
(1913-1981)
American painter, writer

Saccaro was born in San Francisco, worked as a WPA artist during the Depression, and later taught at UCLA. According to Benezit (2006), he served in the French Army during World War II, while other sources state that he was a US Army camoufleur. He is sometimes credited with promoting Abstract Expressionism in San Francisco.

•

Homer Saint-Gaudens
(September 28, 1880-December 8, 1958)
American theatre designer, arts administrator

Saint-Gaudens is remembered for several distinctions: He was the son of American sculptor Augustus Saint-Gaudens, who was one of the country's most famous artists at the end of nineteenth century. After studying at Harvard, where his first year roommate was (camoufleur) Barry Faulkner (who soon dropped out to study art), he went into theatre as a career. With the US entry into World War I, Saint-Gaudens was put in charge of the American Camouflage Corps, perhaps for no better reason than his father's celebrity. Following the war, he continued to work in theatre, until his appointment as head of the art department at Carnegie Institute of Technology in Pittsburgh.

Facing page Disruptive patterns in butterflies as portrayed by French designer E.A. Seguy (c1920) from *Seguy's Decorative Butterflies and Insects in Full Color* (New York: Dover Publications, 1977), in the Dover Pictorial Archives.

When World War II broke out, he returned to active duty and was once again assigned to camouflage. For his own views on the subject, see Saint-Gaudens (1919 and 1933).

Rickard (1942:190): "The 'Saint' [was] our generally accepted term for Lieutenant Saint-Gaudens..." Rickard is being sardonic perhaps, because Faulkner (1973:110), who was in the same unit, recalls that Saint-Gaudens was "intensely disliked by the men."

Saint-Gaudens [quoted in *Current Biography 1941*:747]: "I found myself guiding the emotional destinies of 400 tempermentalities in the form of artists, plumbers, carpenters, and other eccentrics, who ultimately won the war by spreading scenery over the gory fields of the AEF [American Expedition Forces]."

Saint-Gaudens (1933:242): "Camouflage, both as we practiced it in France and as we plan it for the future, has suffered from too much loose conversation, newspaper propaganda, and the inability on the part of the average soldier to use his eyes."

Saint-Gaudens (1933:248): "Camouflage then is no vaudeville magic. It requires trouble, horse sense, and an ability to take advantage of the local conditions. It is Indian fighting. Camouflage is not so much to hide an object from the enemy as to fool the enemy as to the object's whereabouts."

•

Madeleine L'Engle (1974:88): "...my parents often spent their summer [in Cornish, New Hampshire] as part of the Saint-Gaudens Artists Colony. Homer Saint-Gaudens was one of my father's [Charles Wadsworth Camp] friends from the war, and he found the quiet and companionship a nourishing climate for his writing."

...

Charles Burchfield (Bauer 1982:151): "[With the bombing of Pearl Harbor] One of my first thoughts was, is there anything concrete I can do to help? I'm too old for the army; and [Homer] Saint-Gaudens let me know pretty decidedly some time ago that I would be useless in camouflage [although Burchfield had been a camoufleur in World War I] (I don't know why he did, for I certainly made no overtures to him...unless it was just to forestall any possible application from me...)"

...

Scientific American ("Seeing But Not Seen" 1918:451): "...it is reported that Homer Saint-Gaudens, son of the great sculptor and a first lieutenant in the company, has invented a machine which converts old newspapers into blankets that can then be tinted like the surrounding grass and used as a cover for a body of men going up to the front."

George Sakier
(1897-1988)
American industrial designer

Born in New York, he studied engineering at Columbia University and Pratt Institute. At age 19, he wrote a book on *Machine Design and Descriptive Geometry*. During World War I, he served as a camouflage technician, during which he developed an interest in art and specifically painting. For many years, he painted both abstract and realistic landscapes. Following World War I, he taught machine design and engineering mathematics, and worked as an art director for *Vogue*, *Harper's Bazaar* and *Modes and Manners*. He designed Art Deco glassware for Fostoria, modern faucets, and pre-fabricated single unit bathrooms. In 1934, when *Fortune* magazine published an article on the fledgling industrial design profession, he was among the featured designers, along with Raymond Loewy, Henry Dreyfuss, Walter Dorwin Teague and others. There is a George Sakier Foundation that offers a prize for excellence in photography at Yale University.

•

Charles Louis Sallee
(1913-2006)
American painter, printmaker

An African-American artist, Sallee was born in Oberlin OH. He studied art at Western Reserve University, Cleveland Art Institute and other schools, then taught art in Cleveland. He was a muralist for WPA, and served as a cartographer and camoufleur during World War II. In his later life, he worked as an interior designer.

•

Evan Salmon
(1971-)
Australian artist

Salmon was born in Sydney, Australia, and studied at the East Sydney Technical College, and the University of South Wales, where he earned an MFA degree in 2002. In 2006, he exhibited a sculptural interpretation of dazzle camouflage, titled *Dazzle Ship*, at the Wollongong City Gallery, in Wollongong, New South Wales. It consisted of a model of a dazzle camouflaged ship (225 inches in length), made of painted plywood, balsa wood, canvas, and wire. See Salmon (2006).

•

Nathaniel Saltonstall
(1903-1968)
American architect

Saltonstall is listed in Reit (1978:79) as having worked with William Pahlmann during WWII at Jefferson Barracks, St Louis MO. Among his ancestors is a famous Puritan of the same name, who served as a judge during the Salem witch trials.

William Saltzman
(1917-2006)
American artist, set designer

Born in Minneapolis, he was a designer of stained glass windows, copper reliefs, and stage sets for the Minnesota Opera Company. Hoppin (2006): "During World War II, he even helped the US Army design new types of camouflage."

•

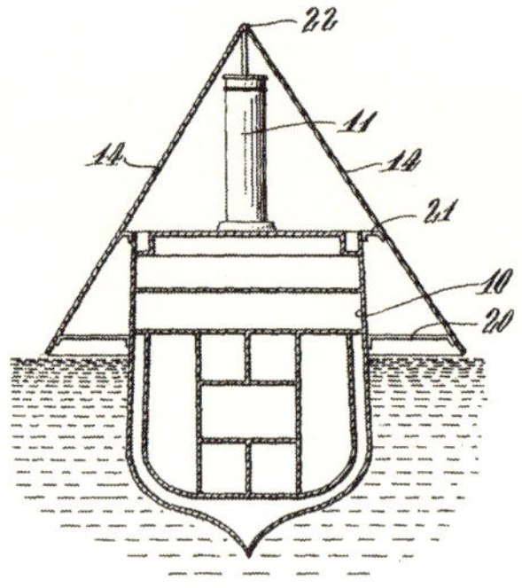

Above Drawing submitted for US Patent No. 1,310,533 titled "Means for Screening or Obscuring Ships," invented by Isidor Karton (1919). In this proposal, a ship would be camouflaged by an A-frame of plate glass mirrors, in which only the ocean would be reflected.

Ralph Samuels
(1907-December 20, 1972)
British-born American photographer, editor, administrator

Samuels was born in Manchester, England. He immigrated to the US, and studied at the University of New Mexico. He was an editor of *Popular Photography*, and a staff editor for *Modern Photography*. For thirteen years (1951-64), he was Dean of the New York Institute of Photography, during which he also edited the *Universal Photo Almanac*. *New York Times* ("Ralph Samuels" 1972): "He was noted for helping people in photography, ever since he trained pilots and bombardiers in the intricacies of camouflage detection and target identification in World War II, during which he flew more than 200 missions with the Fourth Air Force. A group of aerial photographs of American terrain by Mr. Samuels is in the permanent collection of the Museum of Modern Art."

•

Homer Saint-Gaudens (1941b:170-171): "About 1890, I remember, he [Abbott Thayer] went to live near [Mount] Monadnock. There he painted. There he engaged in wildlife discussions with Theodore Roosevelt. There he developed his theories as to the principles of protective coloration of animals and the possible application of these principles to what we now know as camouflage. His theories were correct enough. In practice, however, neither his naturalist nor his military contributions proved important. In reality such well-meant enthusiasms distracted him from his place of sound value."

John B. Sanger
(1892-August 23, 1955)
American architect

Educated at Columbia University and the Beaux Arts School of Design in Paris, he was head designer for Reginald Marsh and Starrett and Van Vleck, New York. He designed the Essex House, the Fox Theatre in Brooklyn, the RCA exhibits for the New York World's Fair, and various buildings at Northwestern University. *New York Times* ("John Sanger Dies"

1955): "During World War I he did camouflage design for the Government." He is also mentioned in *Literary Digest* ("Company A" 1917) as having been a camoufleur.

•

John Singer Sargent
(January 12, 1856-April 14, 1925)
American painter

While not a camoufleur himself, Sargent has several peripheral links to the subject. There are, for example, certain of his paintings that show the disruptions that come from the intersection of shadows on other figures, by which he anticipated the use of umbrella camouflage.

He was also well acqainted with Abbott H. Thayer (whom he had known in Paris) and agreed to use his social connections to arrange for a session in London at which Thayer might try to persuade the British Army to adopt a disruptively patterned field service uniform. The whole event was a tragic farce, and the source of great frustration for Sargent, because Thayer was unable to go through with it, and fled before the meeting began.

Two years later, in October 1917, Sargent was in Washington DC, painting a portrait of US President Woodrow Wilson, who invited him to ride along on a visit to the training camp of the American Camouflage Corps. Among the soldiers who demonstrated camouflage that day were Barry Faulkner (Thayer's cousin) and Sherry Edmundson Fry.

In the spring of 1918, Sargent was commissioned by British Prime Minister Lloyd George to spend four months on the battlefront, recording examples of the cooperation of British and American troops. He arrived with a huge, green-lined white beach umbrella, which he installed within view of the enemy lines. His position, as described in Harris (2005:161), "was an ideal spot for an artist to capture the realism of war. But Sargent had no idea what war was all about. He thought that on Sunday there was no fighting, that the soldiers on both sides actually rested. He failed to realize how much danger he was in, sitting out in the open, shielded only by his glaringly white umbrella. Obviously, the Americans feared for their visitor. It took some arguing, but they at least got permission from him to camouflage his obtrusive umbrella" [not to be confused with the quite different earlier meaning of umbrella camouflage].

•

Bruce W[ilder] Saville
(March 16, 1893-February 27, 1939)
American sculptor

Born in Quincy MA, Saville studied sculpture with T.A.R. Kitson at Boston Normal Art School. He became head of the sculpture program at Ohio State

University (1921-25), then taught at Columbus Art School. In the mid-thirties, he moved to Santa Fe NM. *New York Times* ("Bruce W. Saville" 1939:25): "During the war [WWI] he drove in the French transport service and after America's entrance was in charge of the modeling department, United States Army division of camouflage."

•

William Sayles
(dates undetermined)
American book illustrator

Sayles was educated at Pratt Institute. After serving with the Ghost Army in World War II, he shared a studio in New York with Arthur Shilstone, and worked as a book illustrator. His daughter is illustrator Elizabeth Sayles.

•

Robert (John) Schiffer
(September 4, 1916-April 26, 2005)
American makeup artist

Born in Seattle WA, Bob Schiffer worked initially on an ocean liner, where he made up passengers for costume parties. After moving to Hollywood in the early 1930s, he provided make-up for motion pictures, including *A Night at the Opera, The Wizard of Oz, An American in Paris, Gigi, Judgment at Nuremberg, What Ever Happened to Baby Jane, My Fair Lady,* and most famously (because of its difficulty) *Birdman of Alcatraz*, in which the main character ages from 18 to 80. For many years, he was the head of the make-up department at the Walt Disney Studios.

Fox (2005): "In addition to his Hollywood work, Mr. Schiffer helped create disguises for espionage and law enforcement agencies... He helped with camouflage work in World War II, and in 1961, during the Bay of Pigs invasion, he made up United States operatives to look more Cuban. In recent years, he created mock wounds for hundreds of Marines and sailors as part of battlefield training exercises."

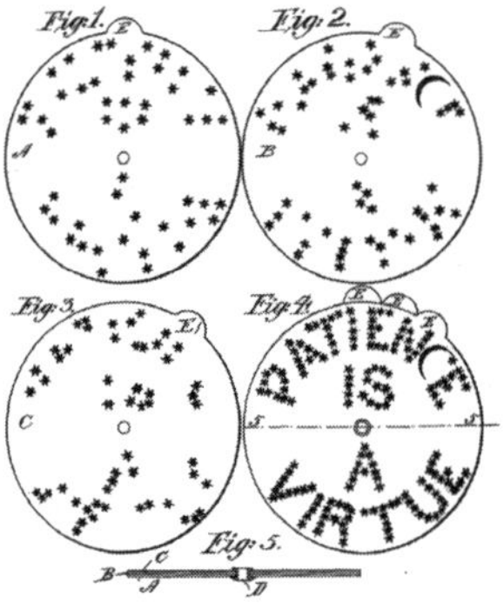

Above Drawing submitted for US Patent No. 602,175 titled "Puzzle," invented by J.T Smith (1898). The message is "camouflaged" until all three constellations merge.

Rubin (1943:13): "A popular camouflage game divides the campers into two teams which work on two sides of a hill. When the leader blows a whistle each team constructs a minature shelter of twigs, vines and leaves under which a single camper may be concealed. At another signal all the campers rush to the top of the hill after which they search the enemy's territory to discover the hidden camper."

...

Martin Esslin (1968:256), quoting a character in the play *Biedermann and the Incendiaries* by Max Frisch: "Jocularity is the third-best kind of camouflage; the second-best is sentimentaliy...But the best and safest camouflage is still the pure, naked truth. Funnily enough, no one believes it..."

Oskar Schlemmer
(September 8, 1888-April 13, 1943)
German sculptor, designer

Schlemmer was an influential teacher at the Bauhaus, the famous school of art and design in Germany. His most memorable achievements were crossdisciplinary, combining visual arts with dance, costume design, and theatre. Today he is often regarded as a pioneering "performance artist."

He was wounded during World War I, while serving in the German Army. Following his recovery, he was assigned to cartography. He was at the Bauhaus in Dessau from 1920-1929, then taught in Breslau briefly. With the rise of the Third Reich, he faced increasing restrictions and, in 1937, his work was included in the Nazi exhibition of "degenerate art" in Munich. From 1938 to 1940, he worked for a painting company in Stuttgart, for which, among other things, he designed building camouflage. There are references to this in his letters and diaries, which were published by his wife (Schlemmer 1972).

Oskar Schlemmer (Schlemmer 1972:377): "Well, tomorrow morning at seven I'll be off for Giebelstadt with a truck and two soldiers. Once there I shall have to experiment with camouflage on a grand scale. I feel that this empirical method is the only useful one."

Oskar Schlemmer (Schlemmer 1972:378): "The [Giebelstadt camouflage] job is very interesting. But there are many difficulties: the frost is a great nuisance; one cannot start painting before ten in the morning, and then only until three, otherwise the color peels off; an insufficient amount of scaffolding, with the result that the work progresses slowly, given the vast area we have to cover (two hundred seventy feet wide, fifty feet high, seventy-five feet deep), and there are seven barracks that size, not counting the many outbuildings, all of which have to be camouflaged, and as quickly as possible. I shall if I can make little models in Stuttgart which the painters could follow, so that I would only need to come now and then. The camouflage itself is turning out well. Gigantic paintings."

•

Schools for Camouflage Training

Harbord (1925:114): "We saw a wonderful school for 'camouflage,' an institution that has come up in this war [World War I], meaning the painting or otherwise preparing of ground or covering guns or trees, etc., so that in airplane photographs they do not show. It has a hundred different applications. It has also passed or is now passing into slang. Instead of saying 'He is a bluff,' we say 'He is nothing but camouflage.' We also saw many interesting things in the way of training."

Alphonse (Emile) Schuil
(October 10, 1911-1943)
British physicist

Schuil studied at Rutherford College and Armstrong College in Newcastle, England. In 1937, he joined the research staff of General Electric Company. According to Williams (1989:155), he "designed the haze attenuation and optical system for the Naval Camouflage Viewing Range at Leamington Spa," and later "engaged in camouflage research with Lieutenant Oliver Grahame Hall RNVR." There is more information about Schuil in Williams (2002), including the circumstances of his death, which occurred aboard a submarine that was attacked off Freetown, Sierra Leone.

Above The *USS Stockton*, in camouflage and heavily damaged, after colliding with another ship. Curiously, such collisions were sometimes regarded as proof of the effectiveness of dazzle painting. Courtesy US Naval Historical Center (NH 04236).

Peter Scott (1961:394): "If the enemy could not see the escorts there was an improved chance of surprising U-boats, but if the escorts could not see each other there was also an increased chance that they would collide. *Broke* and *Verity* were not the only escort ships to be involved in collisions for which good camouflage was accepted as a primary cause. The Admiralty called a conference to consider whether these disadvantages outweighed the advantages, and finally came down on the side of camouflage."

•

Remington Schuyler
(July 8, 1884-October 18, 1955)
American writer, artist

Born in St Louis MO, Schuyler was named after Frederic Remington, who was related to his mother. He grew up in the Midwest, where he studied engineering at Washington University in St. Louis. He dropped out of school, moved to the Rosebud Sioux Reservation in South Dakota (1902-03), and worked for a year as a cowhand. Returning to St Louis, he reentered Washington University as an art student. Later he studied with Howard Pyle, and (possibly) at the Académie Julian in Paris. When French Marshall Ferdinand Foch (who had been Allied commander during World War I) was given a grand tour of the US in 1921, Schuyler was the event's official American Legion artist.

Hamilton (1969:120-121): "In World War I Rem Schuyler was a First Camoufleur working with the Navy, and many widely varying camoufleur designs for ships are shown in a neat set of cards, designs which he prepared and which were used in that effort."

Smith (1979:175): "Remington Schuyler was a Western writer and artist. His poems, short stories, serials, and columns were widely read. Greater fame was achieved through his hundreds of magazine covers and illustrations, camouflage designs for ships, murals, book dust jacket sketches, and a variety of other illustrative materials."

•

Frank H[enry] Schwarz
(1894-September 5, 1951)
American mural painter

Born in New York, he studied at the Art Institute of Chicago, and at the American Academy in Rome (1921-24). After World War I, he worked as a designer and mural painter in New York. He was awarded a Guggenheim Fellowship in 1926-27, and taught briefly at Columbia University (1928-31). His commissioned murals are found in the US Senate and House of Representatives, the Oregon State Capitol, the Brooklyn Museum and the University of Nebraska. *New York Times* ("Frank H. Schwarz" 1951): "During the first World War, Mr. Schwarz was a sergeant in the Army's Ordnance Research Camouflage Corps."

•

Gabriel Scognamillo
(1906-1974)
Italian-French art director

Scognamillo worked with Jean Renoir on *La Chienne* (1931) before immigrating to the US in 1933 to work at the Metropolitan Opera. After moving to Hollywood, he was the art director for *The Merry Widow, The Great Caruso, The Seven Faces of Doctor Lao*, and other films. According to Naversen (1989:184-185), during World War II, he contributed to the March Air Force Base camouflage battalion, but then returned to Europe to serve with the French Air Force Combat Intelligence and the Italian Underground.

•

Peter (Markham) Scott
(September 14, 1909-August 29, 1989)
British artist, naturalist

The son of Artic explorer Admiral Robert Falcon Scott, Scott is credited with a ship camouflage method called the Western Approaches System. He was the author of *The Eye of the Wind* (Scott, 1961) and other books. See Hartcup (1980).

In a 1950 letter to Mary F. Boynton (the daughter of bird illustrator and Thayer student, Louis Agassiz Fuertes, Scott said this about Abbott H. Thayer: "As a boy of twelve I spent a good deal of time studying Thayer's great illustrated book on camouflage and was much influenced by it. Later I became a keen duck hunter and used a duck punt which was camouflaged in accordance with Thayer's principles of negative shading. Later yet, when serving in a destroyer in the North Atlantic I was confident that these same principles were applied in the tactics of night attack on convoys by German U-boats..." See White (1951:137).

Gerald H. Thayer (1923:165): "At the outset of the war [WWI], attempts were made in Europe...to *hide* ships at sea with the aid of contrasting patterns. My father [Abbott H. Thayer], meanwhile, supposed in some quarters to be the author of this *project*, did his uttermost to dissuade the governments concerned from any and all such attempts. He pointed out that the sky-lit ocean, surmounted by the blank brilliance of the sky, was almost the only realm on earth where strong patterns for concealment *would not work*, but, on the contrary, would *reveal* an object instead of hiding it, because of the entire lack, in the object's background, of anything which such patterns could 'cooperate' or combine [through coincident disruption]. He recommended *spotless white*..."

...

Everett L. Warner (in White 1951:137): "Persons without training in visibility always assume that a black ship is less visible than a white one, irrespective of the background against which it is seen—black is the color of night, and it is supposed to be intrinsically an *invisible color*...[But] In our observations...we found that the ships with the most white in their patterns were generally the least visible at the great ranges where there is any hope of concealment."

•

William A(lexander) Scott III
(January 15, 1923-March 7, 1992)
American journalist and civil rights advocate

Scott was born in Johnson City TN, but grew up in Atlanta GA, where his father was the founder of the *Atlanta Daily World*. He served in the US Army during World War II as (in his words) "a reconnaissance sergeant, photographer, camoufleur and part-time historian" in one of the first units to enter the Buchenwald concentration camp at the war's end.

•

Edward (Brian) Seago
(March 31, 1910-1974)
British painter

Seago was born in Norwich, England. He had little formal art training, but he did work closely with British poet John Masefield. As a painter, he was mainly known for equestrian, ballet and seascape subjects, as well as formal portraiture, including the royal family. Beginning November 1939, he was trained as a camouflage officer, then sent to France as a second lieutenant with the British Expeditionary Forces.

Goodman (1978.:154-155): "...alongside the technique of the artist he [Seago] brought the imagination of the showman to the business of camouflage as when he requisitioned a London film studio to produce hundreds of life-size models of Neville Chamberlain wearing battle dress and a tin hat. Every model had a rifle fixed at a different angle and was mounted on an individual turntable and used to man dummy anti-aircraft sites, strategically placed around the south coast to deter the invaders. The guns and the soldiers were remotely controlled from several central points so that they rotated and changed positions to give a most realistic effect when photographed from the air. Nothing delighted Seago's sense of humor more than to think of the model Neville Chamberlains he had created defending their country!"

Goodman (Ibid.:176-177): "[Seago's friend Harry] Colmer recalls the painter's emphatic dislike of the word 'camouflage' in favor of the term 'concealment.' To impress his theories about concealment on his audiences Seago enlisted the help of Sergeant Edwards, an amateur ventriloquist, and wrote an amusing script recapitulating the points of his lecture for the sergeant and his soldier-doll to enact."

•

Clem Seale
(dates undetermined)
Australian artist

Seale was a member of the Australian Camouflage Unit, which was part of the Department of Home Security during World War II, with headquarters on Goodenough Island. He was an associate of Max Dupain, William Dobell and Robert Emerson Curtis.

•

James M(ilton) Sessions
(September 30, 1882-November 14, 1962)
American painter, illustrator

Born in New York, Sessions studied painting at the Art Institute of Chicago (1903-06). For the next eleven years, he served in the US Navy, with 8 years on active duty and 3 years in the Naval Reserve. In 1933, at age 51, he returned to painting. For the rest of his life, he worked as an illustrator. He is described online ("James Milton Sessions" n.d.) has having been a "camouflage expert and brush reporter" during World War II.

•

Shadow Disruption

Comparable to Umbrella Camouflage

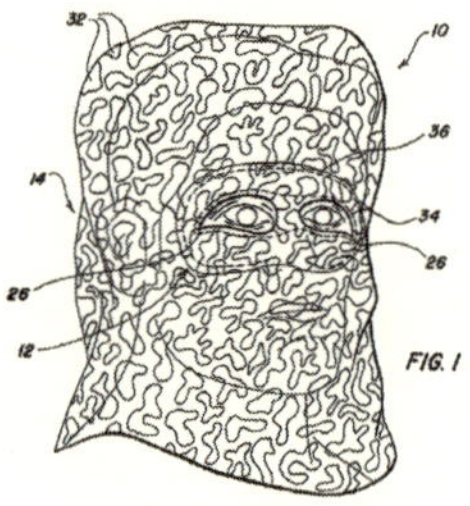

Above Drawing submitted for US Patent No. 4,285,068 titled "Camouflage Headware," invented by Daniel L. Ross (1981).

Perhaps the most common example of this is the incidental form disruption that comes from shadows being cast by venetian blinds or a chain link fence (as happens on the tennis court). World War I British camoufleur Solomon J. Solomon is usually credited with first making common use of this means of creating disruption, but as Elias notes below, the technique was "foreshadowed" prior to that, especially by Impressionists, in experiments in painting and photography.

Elias (2003): "Formally, there are strong connections between [Australian photographer Max] Dupain's work and [that of the] photographers of the Russian avant-garde and the Bauhaus, including Sergei Eisenstein and Laszlo Moholy-Nagy, who similarly used shadows as disruptive elements to form, in order to make the familiar look strange and forms difficult to read, with the intention of making the viewer look long and hard. The context in which this avant-garde photography was produced was the world of modern art and aesthetics, but similar practices were relevant to Dupain's wartime experiments. In the practice of camouflage the same principle of disorientation, of making familiar objects appear strange, and making unfamiliar objects appear familiar, is very important. In war it is greatly advantageous to prevent an easy reading of an object, and to capture the attention of the enemy for a dangerously long period of time while he tries to make sense of what he is looking at."

Of related interest is a very striking use of shadow disruption by Australian photographer Harold Cazneaux (1878-1953) titled *Pergola Pattern*, a portrait of Beryl Cazneaux (1931). It can be found by searching at the website of the National Library of Australia.

•

Julian Trevelyan (1957:130): "Occasionally I would be asked to give a demonstration of how to paint some piece of equipment so as to merge it with the broken country around. I would arrive on the barrack square with pots of paint and brushes, and set to work daubing the shield of some anti-tank gun with spots of different greens and brown, touching in the underside of the barrel itself with pure white on the principle of Dr. [Hugh B.] Cott's gazelles. Against the dreary barrack walls it looked an unholy mess, but when it was wheeled out into the country and placed against a hedge, there were cries of astonishment at my magic. This role, half-clown, half-magician, was one that I found camouflage officers were more or less expected to fill."

Arthur Shilstone
(dates undetermined)
American illustrator

Shilstone was a member of the Ghost Army. He illustrated the first public article on the subject (Park 1985). According to the website for an anticipated film about the unit (www.ghostarmy.org), he was an illustrator for NASA, *National Geographic, Life, Sports Illustrated* and other publications.

•

Charles Siclis
(June 10, 1889-May 8, 1942)
French architect

Siclis was born in Paris. According to his biographical entry at GroveArtOnline, "Siclis was held up in Lisbon at the time of the German invasion, and in 1940 he arrived in New York where he tried to sell his inventions in the field of camouflage technology on the American market."

•

Louis B. Siegriest
(1899-1989)
American painter

Siegriest was born in Oakland CA, where he studied art at the California College of Arts and Crafts, and in San Francisco at the California School of Fine Art. In the 1920s, he was associated with a group of California artists called the Society of Six. He worked in Milwaukee as a commercial artist, then returned to San Francisco, where he worked for the WPA under René d'Harnoncourt. Beginning in 1941, he served with the US Army as a camouflage artist. In an oral history interview in the Archives of American Art (Siegriest 1978), he recalls that "[during World War II] I went with the Camouflage. I was with the Camouflage up to almost the end of the war. I don't know how many guys worked in there–there were about eight or ten of [us]–what they called camoufleurs. I worked generally north, from Bolinas all the way to Eureka, and that was the best job I ever had in my life. I had a lot of fun."

•

Margaret Vandenburg (2001), in a fictional conversation between Pablo Picasso and Ernest Hemingway:

"'My ambition is to render reality, which is itself a form of autobiography, with unwavering precision,' Picasso said.

'How can you say that when your work is often so abstract?' Hemingway asked.

Picasso immediately took offense. 'Abstract? Never!'

'But what about your theory about camouflage?' Hemingway insisted. 'How the camouflage of each country is different, and how the subtle differences are the signatures of each nationality? And that camouflage itself proves the organic inevitability of modern abstraction?...'"

Above Dust Jacket of Ian Berry's *American Starts Here: Kate Ericson and Mel Ziegler* (Berry 2005). During 1985-1995, American artists Ericson and Ziegler collaborated on various installations and outdoor conceptual artworks. The best-known, as shown on this cover, was *Camouflaged History* (1991), in which they "camouflaged" a house in Charleston SC, using 72 paint colors that the municipal government had approved as the historic city's "authentic colors."

Oli Sihvonen
(January 31, 1921-1991)
American painter

Born in Brooklyn, he attended Black Mountain College (where Josef Albers was one of his teachers), and the Polytechnic Institute in Mexico City. According to Witt (2002:157), he "learned a great deal about abstraction from the military. He spent four years with a US Army unit during World War II engaged in camouflage and deception practices."

•

Hans Yohanan Simon
(November 3, 1905-January 31, 1976)
German-born Israeli painter and graphic designer

Born in Berlin, he initially studied medicine at Berlin University, then changed to the study of art. He enrolled at the art academy in Frankfort am Main, where among his teachers were Max Beckmann and Peter Rasmussen. He also studied in France under André Derain. In 1936, he immigrated to Israel, where he joined a kibbutz and worked as its graphic designer. In 1948, in conjunction with the outbreak of the Arab-Israeli War, he became a camouflage artist for the underground Zionist unit, the Palmach.

•

Arthur Singer
(?-1990)
American bird illustrator

Singer was a member of the Ghost Army, as mentioned in Park (1985). According to the website about the unit (www.ghostarmy.org), he illustrated more than twenty books, including the well-known *Birds of North America*. In addition, he and his son (Alan) were responsible for a series of US postage stamps on "the birds and flowers of the fifty states" (1982).

•

T. Fred Slack
(May 12, 1922-May 2, 2003)
American businessman

Born in Eugene OR, his online obituary (GoliathObits) states: "He served in the army in Italy during World War II, with the 369th Engineer Camouflage Company."

•

C[aleb] Arnold Slade
(August 2, 1882-1961)
American painter

Born in Acushnet MA, Slade graduated from Brown University in 1904, studied at the Art Students League in New York (1907), and then moved to Paris to study at the École des Beaux-Arts, and at the Académie Julian under Jean-Paul Laurens. Eldred (n.d.): "While many expatriate artists returned to the US at the onset of the First World War, Slade continued to paint in Paris and at his studio in Arras, France. His wife volunteered at hospitals, and Slade, like [John Singer] Sargent, painted emphatic narrative canvases of the human devastation of the war. He was issued a photo identification by the French government to remain in the country during the war...

Slade returned to the front in 1917, but this time it was to paint camouflage for extensive areas. As a Captain in the US Army of Engineers, he trained at Plattsburg (NY) and was then stationed in France for eighteen months... Slade took his role in the war effort very seriously and kept copious notes on camouflage production. He wrote, 'In no instance can camouflage be executed as an ordinary landscape or scenic painting. Effects must be studied at required estimated distances. As everything else at the front, it will be systematically destroyed, but if it succeeds in lengthening an object's existence, it is justified.' *The Philadelpha Public Ledger* reported in 1917 that the artist was 'studying how to hide retreats from the vulture airplanes of Germany and is enjoying his work tremendously.'"

•

Harold R[eeve] Sleeper
(1893-November 10, 1960)
American architect

Educated at Phillips Exeter Academy, he then studied architecture at Cornell University (1915) and New York University. An authority on architectural construction standards, he was the author of *Architectural Graphic Standards*, *Architectural Specifications*, and *Building Planning and Design Standards*. He taught at Columbia University, and served as Commissioner for the New York Board of Standards and Appeals. *New York Times* ("Harold Sleeper" 1960):

"During World War I, Mr Sleeper served as battalion adjutant of the Seventy-seventh Field Artillery, which was attached to the Fortieth Engineers, a camouflage unit."

•

Above WWI camouflage cover for two machine guns on a portable wagon.

Lieutenant F.G. Howkins, Royal Engineers, quoted in Barton (2005:45)."I also spent a day at the Special Works Park near Wimereux, on the coast. This was the place where the art of camouflage was being worked out. There were two of us and we were not really welcomed, for sufficient reasons. One was that they didn't want to be bothered showing us around, and secondly they didn't know we were coming. So we did the best we could 'on our own.' I started off by falling into a camouflaged gun-pit. This taught us a lesson, so we proceeded warily, as undoubtedly things were not what they seemed. At this time camouflage was in its experimental stage. The well-known Solomon J. Solomon was working there. I saw armored cars painted in various ways in order to find out the best method of deluding the enemy. It was explained that the object of camouflage was not necessarily to conceal, but to destroy identity. We saw parapets of sandbags painted on canvas. Steel trees indistinguishable even at close quarters from real ones. Models of heads mounted on poles from private's to staff officer's. There were 'lifelike' imitation corpses, and dummy guns. Nothing was real. All was trickery, fraud and delusion. A world of sham and make-believe."

Eric Sloane

(February 27, 1905-March 6, 1985)
American printmaker, painter and author

His birth name was Everard Jean Hinrichs. In 1919, his family moved to Long Island, where he became friends with his neighbor, the typographer Frederick Goudy, who taught him handlettering. He then studied briefly at the Art Students League (where he was influenced by John Sloan, whose name he would adopt c.1934, while adding an *e* at the end), and at the New York School of Fine and Applied Art, where he was enrolled as a "graphic advertising major." Due to poor attendance and failure to turn in his work, he never finished his coursework, yet he enrolled nevertheless at Yale University, where he lasted only several months. In the 1930s, through his interest in flying, he became friends with aviator Wiley Post (who would later die in a flight with humorist Will Rogers), as a result of which he decided to paint "cloudscapes." He studied meteorology at MIT, and, during World War II, served with the Army Air Force as an instructional interpreter of complex flight-related terms.

In 1942, he wrote and illustrated *Camouflage Simplified*, a 60-page overview of the subject, illustrated by pen and ink drawings. According to John D. Townsend, quoted in Mauch (2000:204): "He [Sloane] points out in the author's note that the genesis of the book came from a request from an architectural and engineering firm that wanted to compile a reference library on the topic of concealment."

•

Cyril H. Smith
(dates undetermined)
British military officer

Smith was a British Lieutenant-Colonel during World War II, presumably in the Royal Air Force. During that war, Pitman's published a series of "six penny" books, in which this or that subject was "simply explained" for the public, including such subjects as aerobatics, gliding, radio and the wireless. Smith's contribution was an illustrated text on *Camouflage Simply Explained* (Smith, c1940)

•

Elizabeth Bowser Smith
(1914-July 24, 1959)
American art museum administrator

She graduated from Wellesley College, with additional work at Radcliffe and the University of London. She served as assistant director when her husband, Gordon M. Smith (see below), was director of the Currier Gallery in Manchester NH, and the Albright-Knox Art Gallery in Buffalo. *New York Times* ("Mrs. Gordon Smith" 1959): "During World War II, Mrs. Smith worked in the plans and intelligence unit, camouflage branch, in Ft. Belvoir, Virginia."

•

Gordon M[ackintosh] Smith
(June 21,1906-May 20, 1979)
American art museum director

Born in Reading PA, Smith graduated from Williams College, then went on to advanced study at Princeton and Harvard. He was assistant regional director for New England for the WPA from 1936-41, Director of the Currier Gallery in Manchester NH, and of the Albright-Knox Art Gallery in Buffalo (1955-73). *New York Times* ("Gordon Smith" 1979): "After war service in a camouflage section in [Fort Belvoir] Virginia, he served from 1944 to 1946 in the Office of Strategic Services in Washington."

•

J(ules) André Smith
(1880-1959)
American architect, printmaker

He was born in Hong Kong. Initially, he studied architecture at Cornell University, then switched careers to visual art. According to Nibbe (n.d.), he was among the first eight American camoufleurs to arrive in France on March

15, 1918. He is mentioned in Embury (1917:138) as an officer in the Camouflage Corps, and described as someone "who is as well known for his etchings as for his architecture." For his own excellent report on World War I camouflage, see Smith (1917).

C.H.R. Chesney (1943:65): "The first British name that is connected with the subject is Solomon J. Solomon. Having conceived certain ideas at a very early date and having the necessary status, this famous artist was able to propound these ideas to the higher authorities in the army. They were, however, somewhat nebulous and even entrenched upon the field of strategy, so it is hardly to be wondered at that, at a time when no tangible accomplishment had yet been achieved, their author should be regarded rather as a visionary and impractical."

•

Sydney Ure Smith
(1887-1949)
Australian artist and publisher

He is listed as a member of the Sydney Camouflage Group in Elias (2003).

•

Eldredge Snyder
(1902-March 24, 1967)
American architect

Born in New York, Snyder graduated from Princeton University in 1922. He earned a Master of Architecture degree from the University of Pennsylvania, then studied at the American Academy in Rome (1927-28). He designed buildings at Oberlin, Princeton and Rutgers universities, as well as the Hong Kong Pavilion at the New York World's Fair in 1964-65. *New York Times* ("Eldredge Snyder" 1967): "Mr. Snyder was a lieutenant commander in the Navy in World War II. He headed the camouflage section of the Bureau of Yards and Docks in 1943-44 and served on the staff of Gen. Douglas MacArthur in the southwest Pacific in 1944-45."

Above The cover of British artist Solomon J. Solomon's early book on military camouflage, *Strategic Camouflage* (1920).

Mary Kelly (*Imaging Desire*): "The first tactic of camouflage is to become invisible, not unseen, but seen to be the same."

•

Solomon J(oseph) Solomon
(September 16, 1860-July 27, 1927)
British painter

Born in London, Solomon studied at the Royal Academy of Fine Arts, Munich Academy and the École des Beaux-Arts. He was particularly influenced by Alexandre Cabanel (his teacher), Frederic Leighton and Lawrence Alma-Tadema. During World

War I, he was a major contributor to British field camouflage and served as head of the British Camouflage Section within the Royal Engineers (1916-18). He also wrote a major book on the subject, titled *Strategic Camouflage* (1920). His son was Gilbert Solomon, who attended the Slade School of Art with Stanley Spencer, and Richard and Sydney Carline, who worked on camouflage during World War I with Canadian-born camoufleur Percyval Tudor-Hart.

Edmonds (1932:82-83): "The experience gained by this unit [the first French camouflage unit during World War I] was communicated to British G.H.Q., and in December 1915 the late Mr. Solomon J. Solomon, R.A., who was specially interested in the subject of camouflage, was invited to visit France and investigate it on the spot. It was on his initiative, under direction of the Engineer-in-Chief, that the first British detachment of officers and men was assembled." See also Alister Hardy.

•

Alan Sorrell
(February 11, 1904-1974))
British muralist

Sorrell studied at the Royal College of Art, and later taught there as well. He is mainly known for his archaeological murals and illustrations. During World War II, he served as a camouflage officer, "an experience that gave him useful insight into the value of the aerial perspective..."

•

Spanish landscapes
In relation to cubism and camouflage

Elliot Paul (1942:258): "For just before World War I got under way, Pablo Picasso, inspired by the sight of angular roof tops and chimneys of his native Catalonia, was experimenting with his first Cubistic landscapes."

Gertrude Stein (1933:89-90): "That summer they went again to Spain and he [Pablo Picasso] came back with some Spanish landscapes and one may say that these landscapes...were the beginning of Cubism... In these pictures he first emphasized the way of building Spanish villages, the line of the houses not following the landscape but cutting across and into the landscape, becoming indistinguishable in the landscape by cutting across the landscape. It was the principle of the camouflage of guns and the ships in the war."

•

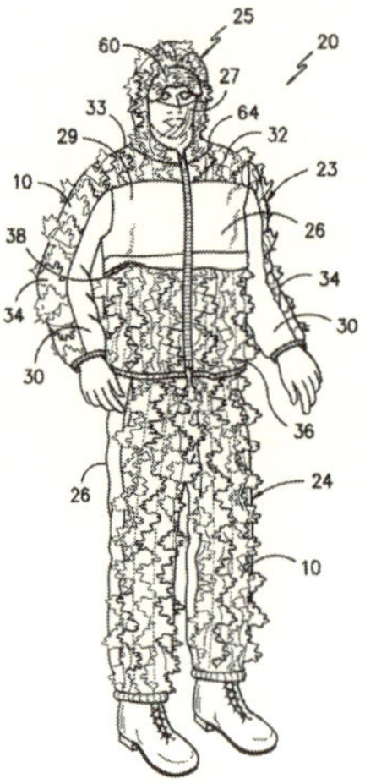

Above Drawing submitted for US Patent No. 6,499,141 titled "Mutlidimensional Camouflage Outer Wear Garment System," invented by James C. Egnew (2002). Protruding pieces of fabric "distort the smooth line silhouette of the wearer..."

Beston (1919:122-123): "I have an indistinct memory of a terrible mess [a WWI camouflaged ship] of milky-pink, lemon yellow and rusty black, which earned for the vessel displaying it the odious title of 'The Boil.' We saw the prize monstrosity in mid-ocean. Every school of camouflage had evidently had a chance at her. She was striped, she was blotched; she was painted in curves; she was slashed with jagged angles; she was bone gray; she was pink; she was purple; she was green; she was blue; she was egg yellow. To see her was to gasp and turn aside. We had quite a time picking a suitable name for her, but finally decided on the *Conscientious Objector*, though her full title was 'The State of Mind of a CO on Being Sent to the Front.'"

Austin (Osman) Spare
(December 30, 1886-May 15, 1956)
British artist and occultist

The son of a London policeman, he was apprenticed to a stained glass craftsman at age thirteen. He studied at the Lambeth School of Art, where his early drawings were sometimes regarded as similar to those of Aubrey Beardsley. In 1909 he became associated with the British poet and occultist Aleister Crowley. Online Wikipedia biography: "In 1917, during World War I, Spare was conscripted into the British army, serving as a medical orderly in the Royal Army Medical Corps in London hospitals. He did not see active service, and was commissioned as an Official War Artist in 1919."

Bowen (1956): "I was one of Spare's comparatively few pupils before the last war, and on the outbreak of hostilities he gave me some instruction in the art of camouflage which was to prove of considerable value in the years to come. Perhaps because he lived 'quietly and humbly,' he was prolific and many sided. Certainly his theories on camouflage were thoroughly sound. I heard some of them repeated as original and empirical long after 1939. No doubt his experience in the 1914-1918 war provided much of the background to those theories."

•

Basil (Urwin) Spence
(August 13, 1907-November 19, 1976)
British architect

Born in Bombay, India, his father worked for the Indian civil service as an analytical chemist. He entered the Edinburgh College of Art, intending to study sculpture, but turned instead to architecture. Throughout his career, he was especially known for public housing, church and school architecture, and exhibition design. Among his many accomplishments were the new Coventry Cathedral, the Scottish School of Art and Industry, and the British pavilion at Expo 67 in Montréal. During World War II, he served in the Camouflage Training and Development Unit at Farnham, during which he was responsible for the design of dummy tank landing craft and other diversionary equipment. See Morgan (1984) and Goodden (2007).

•

Georges Spitzmuller
(1867-1926)
French author

He was the author of a French language book of fiction (unseen, described as a "paperback booklet"), having to do with World War I, titled *Memoires d'un Camoufleur* (1918).

•

Steven Spurrier
(July 13, 1878-March 12, 1961)
British artist

Spurrier was a book and magazine illustrator, as well as a poster designer. During World War I, he worked with the British Admiralty on the development of dazzle ship camouflage.

•

Rudolf Staffel
(June 15, 1911-January 20, 2002)
American ceramic artist

Born in San Antonio TX, he studied at the Art Institute of Chicago. He moved to Philadelphia in 1940, and taught at the Tyler School of Art from 1940 to 1978. Sozanski (2002): "During World War II, he served in the Army Air Force, training as a glider pilot but also working as a camouflage instructor and radio operator and technician."

•

Rudolph Stanley-Brown
(1890-1944)
American architect

Originally from Mentor OH, he later moved to New York. He studied at Yale University and Columbia University School of Architecture, as well as at the École des Beaux-Arts in Paris. He was the grandson of US President James Garfield. He is cited in Nibbe (n.d.) as Brown (not Stanley-Brown) and as "probably a camoufleur" in World War I.

Clayton (Henri) Staples
(February 4, 1892-1978)
American painter

Born in Osceola WI, Staples studied at the Art Institute of Chicago, University of Illinois, and in Europe. He developed the art program at Wichita State University, c. 1930-1950. According to the Online Biographical Directory of Kansas Artists, "During World War I, he was an artist in a camouflage division."

•

Jacques Staples
(dates undetermined)
American advertising artist

Staples is interviewed in Naversen (1989:163-164). During World War II, he served as a camoufleur at the US Army Air Base at Fort Belvoir VA, and at Mitchell Field in Hempstead NY.

•

Jack Gage Stark
(1882-1950)
American painter

He is mentioned in Love (1999:318) as a World War I American camoufleur.

•

Eunice (Minnie) Stelzer
(December 13, 1880-June 4, 1962)
Australian musician

Born in Sydney, Australia, she joined the staff of radio station 2GB in 1927, and initiated an afternoon program for women. In 1929, she founded the Happiness Club for the same station. Under the motto of "Others First," she raised money for charitable organizations.

Above Photograph of US camoufleur Gordon Stevenson by Peter A. Juley & Son. Courtesy Photograph Archives, Smithsonian American Art Museum (J0101660).

Phil Silvers [Sergeant Bilko] (1974:20): "He [his father] wouldn't let me see it [a chemical laboratory he had set up in a vacant lot in Brooklyn]. I knew he was working on a bright green dye to camouflage ships and he hoped to get a contract from the government. The only trouble was it was highly explosive."

...

In recent years, Hannah Rose Shell (historian, filmmaker and media scholar) has been working with students and colleagues at the Rhode Island School of Design, Harvard and MIT on art-related camouflage experiments. Accounts of her work will be published in a forthcoming book, titled *Hide and Seek: Camouflage, Photography and the Media of Reconnaissance* (Zone Books, in press). Also see her website at www.people.fas.harvard.edu/~shell/.

Grahame (2002): "Hundreds of women were attracted to its [the Happiness Club's] functions... During World War II the club donated two mobile canteens, eight mobile food trolleys and two ambulances, made 158,230 dressings and 9500 camouflage nets, knitted hundreds of garments for the Australian Comforts Fund and baked innumerable cakes for the troops." In the 1950s, she founded Eurobodalla Home, described as "the forerunner of retirement villages."

•

Gordon Stevenson
(February 28, 1892-1982)
American painter, printmaker

Stevenson studied at the Art Institute of Chicago. An influential teacher was Joaquin Sorolla y Bastida. During World War I, he worked with Everett L. Warner in the design subsection of the US Navy's ship camouflage section in Washington DC.

•

Victor (Frank) Stiebel
(1907-1976)
South African-born British fashion designer

Stiebel moved to England in 1924, to study at Cambridge University. During World War II, he was assigned to the East Command Camouflage School, where he designed camouflage attire.

Trevelyan (1957:118): "Victor Stiebel was also on the [Farnham Castle camouflage] course, his connection with camouflage being presumably in the design of snipers' suits."

•

Jean-Alphonse Stival
(November 11, 1879-April 10, 1944)
French painter, muralist

Born in Paris, he studied with Fernand Cormon. He was a friend of Paul Signac. He is listed in Coutin (1988) and Ducasse (1962) as a World War I French camoufleur.

•

George L. Stout

(1898-July 1, 1978)
American art restorer

An authority on the application of scientific methods to art restoration, he was Director of the Isabella Stewart Gardner Museum in Boston until 1970. *New York Times* ("George L. Stout" 1978): "He was an officer in the Navy in World War II and helped develop camouflage techniques. He later was assigned to Gen. Dwight D. Eisenhower's command as a member of the General's staff on monuments, fine arts and archives."

Anon, in World War I trench newspaper, *Rigolboche* (1917): "There are two very distinct schools of camouflage: one clearly reminds us of the Pointillist Impressionists; the other is infinitely closer to the Cubist school. There are a few independent camouflage artists but they are not very numerous."

•

David (Edgar) Strachan

(June 25, 1919-November 23, 1970)
British-born Australian painter, printmaker

Born in Salisbury, Wiltshire, England, Strachan studied in London at the Slade School of Art and in Paris at the Académie de la Grande Chaumiere. Pearce (2002): "He 'spent an erratic war' painting camouflage at Bankstown aerodome with other artists, among them (Sir) William Dobell, and dancing minor roles with Helene Kirsova's ballet company."

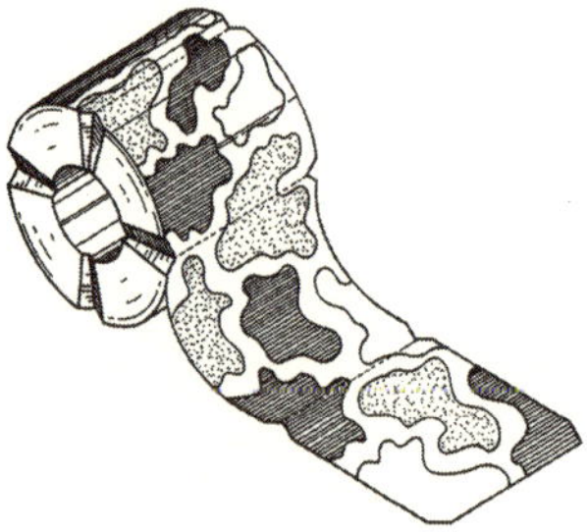

Above Drawing submitted for US Patent No. 6,499,141 titled "Camouflage Toilet Tissue," invented by James L. Marquart (1988).

•

Ian Strang

(April 11, 1886-March 23, 1952)
British printmaker

A painter of urban landscapes, he was the son of British artist William Strang. He studied at the Académie Julian in Paris. As a World War I camoufleur, his battlefield specialty was the construction and installation of artificial dead trees with observation posts inside, which were exchanged with the genuine tree during the night.

Gertrude Stein (1945:147-148): "It is funny why do the Germans wear camouflaged raincoats but not camouflaged uniforms now why do they. The first I saw was the other day, they went by on bicycles, and they reminded me of the chorus of the Tivoli Opera House in San Francisco, it used to cost twenty-five cents and the men in medieval costume looked so like these camouflaged coats, with sort of keys and crosses on them in contrasted colors. Oh dear. It would all be so funny if it were not so terrifying and so sad, this is January forty-four."

Chesney (1943:98): "There was Strang, an artist, who specialized in the OP Tree business. It was he who spent many hours by day in the trenches making pictures in color of the tree which was existing, and in reproducing a dummy at Wimereaux. Also many hours by night erecting the dummy when complete." Also see Hartcup (1980).

•

Styles of Camouflage
In relation to styles of art

Fry (1917:68): "In France, the [camouflage] work was begun by [Jean-Louis] Forain, [Lucien Victor] Guirand de Scevola and [Louis] Abel-Truchet. They began by disguising cannon... Other artists grouped themselves about the beginners and soon the schools of Classicists, Cubists and Impressionists presented a picture of perfect union."

Guirand de Scevola, quoted by Kahn (1984:37): "[In the ranks of the French camoufleurs] The Cubists rubbed elbows with the disciples of Ingres..."

Jones (2002): "[Henri] Matisse later told [Pablo] Picasso that during the first world war the rival French units of military camouflage painters had called themselves the Picasso and Matisse units."

Richardson (1996:414): "Above all he [Pablo Picasso] uses dots—black on white, white on black, multicolored ones—as a form of camouflage, just as he had told Apollinaire the artillery should do. (Ironically, the camouflage units of the French army took a reciprocal interest in modern art, to the extent that they named their regimental mascots 'Picasso' and 'Matisse')."

Richardson (1996:389-390): "On August 11 [1916], he [Jean Cocteau] wrote vaingloriously to Stravinsky, who was in Switzerland: 'In a week I shall return to my post on the Somme, where big guns are camouflaged as if by [Russian painter Léon] Bakst or the Cubists who are shaking all Picardy.'"

Stein (1933): "Another thing that interested us enormously was how different the camouflage of the french looked from the camouflage of the germans, and then once we came across some very neat camouflage and it was american. The idea was the same but as after all it was different nationalities who did it the difference was inevitable. The color schemes were different, the way of placing them was different, it made plain the whole theory of art and its inevitability."

Townsend [in Cornebise:180]: "[As the Germany Army was preparing to turn over 384 trucks to the Allies during World War I] German officers were inspecting them and making a list of their numbers. What a lot of them there were. Most of them were camouflaged. The covers on many of them were in unusual patterns that must have been decorated so, as much to please the fancy of some appreciative officer or man entrusted to do it, as for any real camouflage value it might have had."

•

Submarine warfare

Bond (1918-1919:503): "[World War I German submarines, called U-boats] used all sorts of camouflage. They were frequently equipped with sails, which they would raise to disguise themselves as peaceful sloops, and then steal up on a victim. They would seize a ship and hide behind it in order to get near their prey."

Bond (Ibid.:500-501): "In order to fire its torpedo with any certainty, the U-boat had to get within a thousand yards of its victim. A torpedo is really a motor boat that travels at from thirty to forty miles per hour. It takes time for it to reach its target, and if the target is moving at, say, fifteen knots, it will travel five hundred yards while a thirty-knot torpedo is making one thousand yards. And so, before he could discharge his torpedo, he had to know how fast the ship was traveling and how far away it was from him. He could not come to the surface and make deliberate observations, but had to stay under cover, not daring even to keep his eye out of water, for fear that the long wake of foam trailing behind the periscope would give him away. All he could do, then, was to throw his periscope up for a momentary glimpse and make his calculations very quickly; then he could move to the position he figured that he should occupy and shoot up his periscope for another glimpse, to check up his calculations. On the glass of this periscope there were a number of graduations running vertically and horizontally. If he knew his victim and happened to know the height of its smokestacks or the length of the boat, he noted how many graduations they covered, and then, by a set formula, he could tell how far he was from the boat. At the same time he had to work out its rate of travel and note carefully the course it was holding, before he could figure where his torpedo must be aimed.

Bond (1918-1919:502): "Early in the submarine campaign, one of our [American merchant] boats was given a coat of [dazzle] camouflage, and when the vessel sailed from its pier in the North River, New York, the owners sent a photographer two or three piers down the river to photograph the ship as she went by. He took the picture; but when the negative was developed, he found, much to his astonishment, that the boat was not all on the plate. In the finder of his camera, he had mistaken a heavy band of black paint as the stern of the ship, quite overlooking the real stern, which was painted a grayish white. The artist had fooled the photographer, and at a distance of not more than two or three hundred yards!"

...

Bement (1919:496): "The torpedoes [carried by a German U-boat] were resting in their shining cradles, ready to be inserted in their tubes. When the U-16 had left Bremen, there were fourteen of these death-dealing monsters on board; now there were but six. As each one had cost $8000, to waste one in an attack was a crime in the eyes of the German admiralty."

...

Charles DeKay (1918:105): "Our ports are full of strange monsters [dazzle-painted ships] such as the naval architects of the past never envisaged in their wildest dreams."

Above USS K5 (SS-36), a World War I American submarine painted in a striped camouflage scheme, proposed by William Andrew Mackay. At sufficient distances, the stripes of light and dark gray merged to an optical nondescript gray. Photo courtesy US National Archives (80-G-451710).

There was always more or less uncertainty about such observations, because they had to be taken hastily, and the camoufleurs were not slow to take advantage of this weakness. They increased his confusion by painting high bow-waves, which made the ship look as if it were traveling at high speed. They painted the bow to look like the stern, and the stern to look like the bow; and the stacks were painted so that they appeared to slant in the opposite direction, in order to make it look as if the vessel were headed the other way. U-boats came to have a very wholesome respect for destroyers, and would seldom attack a ship if one of these fast fighting-craft was about; and so destroyers were painted on the sides of ships as scarecrows to frighten off the enemy."

•

Henry Sutter
(1882-?)
American painter

He is mentioned by Faulkner (1957) as a World War I American camoufleur. Originally from Chicago, he studied with Charles Hawthorne in Provincetown MA (c1914). He is also mentioned in *Literary Digest* ("Company A" 1917) as a camoufleur.

•

Sydney Camouflage Group

Australian Citizens Unit, World War II

Elias (2003) has written extensively on this subject: "These 'civilians' were members of the Sydney Camouflage Group, which was formed in 1939 and was made up primarily of artists, photographers, architects, scientists, engineers and civil servants. [William] Dakin was the chairman, and on 26 April 1940 the group included Sydney Ure Smith, John D. Moore, Frank Hinder, Douglas Annand, Robert Emerson Curtis, Russell Roberts, Max Dupain and Adrian Feint. They were a group of concerned citizens who worried about the slowness of Australia to develop strategies with camouflage. They shared the concern expressed by local academic Eric Ashby [prominent British botanist] 'that camouflage is still regarded in some quarters as a hobby rather than as an instrument of war.' They were aware that in Britain 'film directors, artists, architects, commercial artists, picture restorers, sculptors and many others, including a magician, have been enlisted in a special corps of camouflage specialists'. The Sydney Camouflage Group undertook camouflage experiments, testing ideas with representatives of the Army, and Air Force. They collated their research into a book titled *The Art of Camouflage*, which was first printed in 1941 with approval by the Department of Defense Coordination and republished by the Government in 1942 for use by the military."

Another good source is Bogle (1998:97), who lists one other camoufleur, Leslie Wilkinson, as a member of the Sydney group.

Charles DeKay (1918:105): "The object of this camouflage [dazzle painting] is to get the enemy gunner groggy as to distance and direction, so that he will be unable to tell how far away the vessel is and what course it is holding, and so lead him to aim short or overshoot his mark."

...

Cuttwell (1936:386): "The escort kept the enemy [German U-boats] at a distance and forced him to discharge his torpedoes at long range with uncertain aim. Twenty or thirty ships elaborately camouflaged with streaks and blotches of violently contrasting colors, all zigzagging in formation, present an uncertain and bewildering target."

•

Stephen Sykes

(1914-January 22, 1999)
British artist

As described by Reit (1976), he worked with Geoffrey Barkas in WWII (p. 138). He provided his own account of the experience in Sykes (1990).

Above Dust jacket of Stephen Sykes' memoir about being a camouflage officer during World War II (Sykes 1990).

Left Pen and ink drawing of camouflaged US ships by Burnell Poole, c1918. Courtesy US Naval Historical Center (NH 50743).

Harrod (1999): "Sykes was an ingenious camouflage officer. He created a series of realistic dummy railheads in the Western Desert and was able to convince skeptical fellow officers of the value of his activities. During the D-Day landings of June 1944 Sykes worked tirelessly, camouflaging snipers and blocking enemy sightlines...[In later years] His most magnificent and idiosyncratic creation was his garden at his studio home, Hopkiln, near Midhurst in Susses, created out of a piece of rough ground he bought in 1967. It was a triumph of *bricolage* and improvisation, incorporating a maze, a grotto, a waterfall and small raised canal, statues and mosaic work. To meet him (naked) beside his swimming pool, which was embellished with a gold peacock, was to encounter a charming sun worshipper from some ancient lost culture who had taken up unexpected residence in a fold of the South Downs."

•

L.D. Symington
(dates undetermined)
British theatre designer

Symington remains somewhat of a mystery. He is most often said to have been a theatrical designer, yet he was also a prolific book illustrator. As a World War I camoufleur, he specialized in sniper suits, two drawings of which have been published in Blechman (2004:126). He also made clever frontline observation posts, including one known as a "beehive," consisting of a dome that looked like a pile of tin cans, old shoes and other rubbish. Two drawings of beehives (possibly Symington's own drawings) are reproduced in Barton (2005:43). Chesney (1943) goes on at length about the virtues of Symington's efforts.

•

Pied Beauty

Glory be to God for dappled things—
For skies of couple-color as a brinded cow;
For rose-moles all in stipple upon trout that swim;
Fresh-firecoal chestnut-falls; finches' wings;
Landscape plotted and pieced—fold, fallow, and plough;
And all trades, their gear and tackle and trim.

All things counter, original, spare, strange;
Whatever is fickle, freckled (who knows how?)
With swift, slow; sweet, sour; adazzle, dim;
He fathers-forth whose beauty is past change:
Praise him.

—Gerard Manley Hopkins (1918)

Above Plate from F. Nemos, *Europas bekannteste Schmetterlinge* (c1895). Public domain.

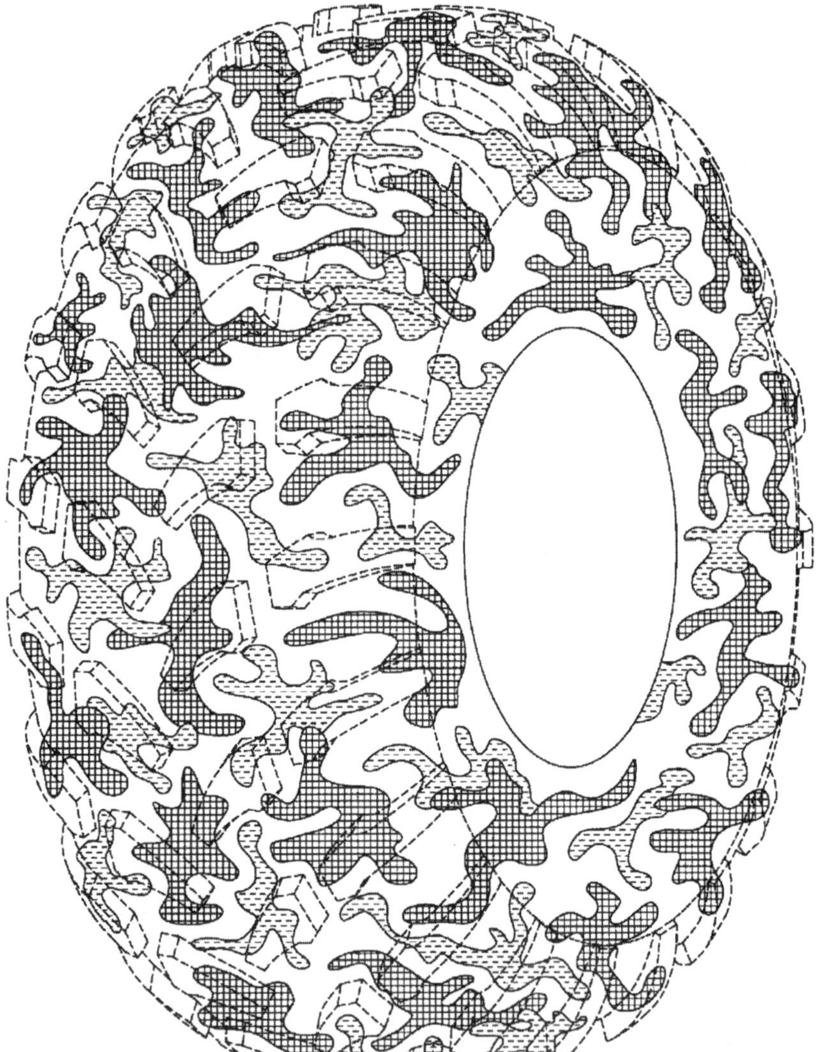

T

Richard B. Talbot Kelly
(1896-1971)
British artist, naturalist

Born in Birkenhead, England, Talbot Kelly (sometimes cited as Kelly) served in both World Wars, during the second as chief instructor at the War Office School of Camouflage. He was Art Master at Rugby School for many years. He wrote a memoir about his experiences in World War I, titled *A Subaltern's Odyssey* (Talbot Kelly, 1980).

Talbot Kelly (1980:86): "One day three camouflage experts arrived armed with a lorry-load of paint and brushes and repainted in jazz colors and patterns our beautiful green guns. They painted all over our polished leather work, ruining months of labor, and quite uselessly infuriated my Major and our limber gunners."

During World War I, he was wounded by a German shell. Following his recovery, he was reassigned to camouflage. In April 1918, he was sent to France "to study the latest concealment techniques, and returned to become the specialist instructor in camouflage at the school of Artillery, Larkhill."

Lewis (1994:93): "On my return to England, I was posted to the Camouflage Center at Farnham Castle as an assistant instructor under Talbot Kelly. He was a bird-like man and birds were his main interest. Both kinds. He was a dedicated ornithologist and the paper birds he made were brilliant creations."

Facing page Drawing submitted for US Patent No. D480,041 titled "Camouflage Tire," invented by David John Zandig, et al. (2003).

Loosmore in Talbot Kelly (1980:14), quoting a student: "What magic he made for us too—he would take an ordinary cork, and make a few deft cuts, add a scrap or two of paper and a flick or two of a brush; and there would be a small bird; a bird so alive that it seemed to breathe, so mortal that one scarcely dared to touch it."

Rockwell Kent (1955:100): "His [Abbott Thayer's] mutually conflicting dual interests—he was a distinguished naturalist as well as a painter—his declining prestige in the fickle world of art, the controversial nature of his biological tenets, and the family's too swift reversion to their earlier poverty, all combined with the abnormal nervous tension of his nature to undermine his creative faculties, his health and, near the end, his reason."

Above Photograph of Abbott H. Thayer, c1919, from Beatty (1921). *Facing page* The title spread from his famous book. Author's collection.

Richard Meryman, Jr. (1999:121): "Brilliant, isolated, ascetic, hyperintense, an almost pure example of late-19th-century romantic idealism, [Abbott] Thayer epitomized the popular image of a genius. His mind would race at full throttle in a rush of philosophies and certainties. His joy was exploring the imponderables of life, and he scrawled passionate, hardly readable letters, his second thoughts routinely continued in a series of postscripts."

•

L(eonard) Campbell Taylor
(December 12, 1874-1969)
British painter, illustrator

Born in Oxford, England, he studied art at the Ruskin School, St. John's Woods School, and the Royal Academy Schools of Art. He produced illustrations for magazines, and posters for London Midland Scottish Railways. He was also the director of the Whitechapel Art Gallery. He was a navy lieutenant in World War I, during which he made paintings of dazzle painted ships in dock. For reproductions of these, see Hurst (1919).

•

Boris Mikhailovich Teplov
(October 21, 1896-September 28, 1965)
Russian psychologist

Teplov was born in Tula, and was educated at Moscow University. He graduated from a military camouflage school in 1921, following which, according to Zusne (1975:441-442), he "did considerable work on camouflage, some of which was published in journals of military technology. He extended his work to problems of light and vision in architecture..." He was an associate head of the research department at the Institute of Psychology, Moscow, in 1929. He also lectured at Moscow State University, and was the editor of the research journal *Voprosy Psikhologii* (1958-1965).

•

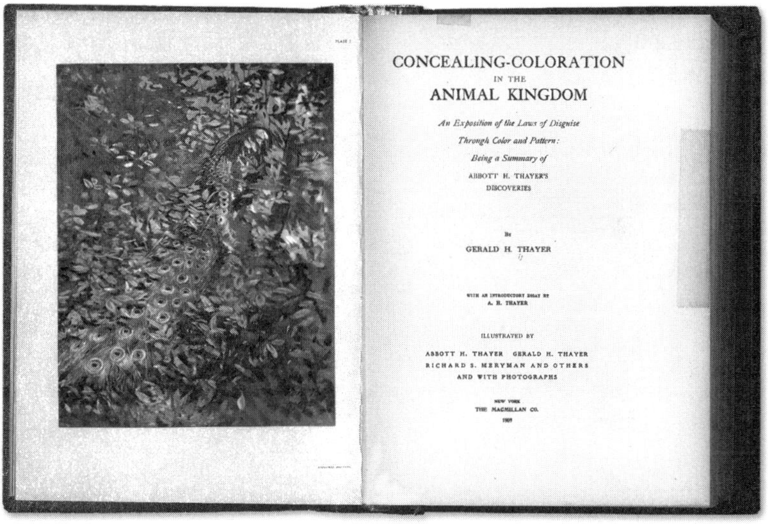

Abbott Handerson Thayer
(August 12, 1849-May 29, 1921)
American painter

Thayer was a prominent American painter at the end of the nineteenth century and the early years of the twentieth. Born and raised in rural New Hampshire, he studied painting in New York at the Brooklyn School of Art and the National Academy of Design. In the 1870s, he moved to Paris to study for four years under Jean-Léon Gerome at the École des Beaux-Arts. Returning to New York, he and his family eventually moved back to Dublin NH, where he continued to paint and to work with a small number of apprentices, and where he remained for the rest of his life. Having always been fascinated by nature (he described himself as "bird crazy"), he increasingly turned his attention to the appearance of animals. In 1896, in an article in a scientific journal, he disclosed his discovery of countershading (also known as "Thayer's Law"), which he frequently demonstrated using painted sweet potatoes and wooden duck decoys. Two years later, with the outbreak of the Spanish-American War, Thayer and his close friend George de Forest Brush proposed (unsuccessfully) that US warships should be countershaded, to reduce their visibility. In 1902, Thayer and Brush's son, Gerome Brush, jointly obtained a patent for this method of ship camouflage.

In subsequent years, prior to World War I, Thayer and his own son, artist and naturalist Gerald H. Thayer, continued to work intensively on speculative theories of biological camouflage. In 1909, they published a large, richly illustrated book on the subject titled *Concealing Coloration in the Animal*

Kingdom (Thayer 1909), in which they made new and largely original claims about the survival function of animal coloration. In addition to countershading, among the concepts they proposed were disruptive coloration and background picturing. The elder Thayer was persistent in his claim (disputed by zoologists and other naturalists) that there is no nuptial or advertising coloration, and that all animal coloration (however bright and conspicuous) contributes to its concealment.

William James (letter dated 1907): "Gerald [Thayer], who is staying with us now, is really a splendid fellow. I'm so glad he's taken to Aleck [Alexander James, his son], who now is aflame with plans for being an artist. I wish he might—it would certainly suit his temperament better than 'business'" (quoted in James (1920:275-276).

...

White (1951) referred to Thayer as the "father of camouflage," which to some degree he was. One consequence of Thayer's writings was the formation by Barry Faulkner (Thayer's cousin) and Sherry Fry of the New York Camouflage Society, through which American artists prepared to serve as camoufleurs during World War I.

Thayer (1923:150): "My father, Abbott H. Thayer (one of the founders and main promulgators of camouflage in the war), who now stands, both in this country and Europe, as the extreme believer, or *over*-believer, in protective coloration in nature, is, to my mind, at once an unequalled expert and an extremist."

Boynton (1956:128): "He [Abbott Thayer] was constantly devising new means of persuasion: placing woodpecker skins upon photos of trees against sky, hanging papier-maché models of patterned oryx heads in trees, talking people into the woods to look for themselves at a mounted peacock concealed in bright sunlight."

A number of interesting experiments in background matching, disruptive coloration, countershading, and other aspects of zoological camouflage have been conducted in recent years by biologists in the UK and other countries. Using computer-built models of moths and other innovative methods, they have made extensive studies of many of the same camouflage principles that were promoted and demonstrated by Abbott H. Thayer, but not tested empirically. See for example Merilaita (1998), Cuthill et al. (2005) and Stevens et al. (2007). More recently, Martin Stevens and Sami Merilaita have edited an entire theme issue on *Animal Camouflage: Current Issues and New Perspectives*, to be published by the Philosophical Transactions of the Royal Society B (2009).

Thayer's wife in a letter to Louis A. Fuertes dated February 28, 1917 (quoted in Boynton 1956:217-218): "As you probably know Abbott wore himself out trying to get the Allies to concealingly color their ships and men. At last they are to large extent doing it. A friend, who spent last spring and summer at the French front, tells us that everything there, motors, bridges, derricks, the canvas used to cover piles of ammunition back from the front, cannon, tents, and even white horses are painted with ruptive patterns, green, brown, etc., landscape colors. Snipers also paint their faces, and wear motley.

But all this came too late...to save Abbott from getting so worn out that, though he has been sick a year now, he seems to make very little progress toward health.

He is pronounced sound, but excessively fatigued. His nerves are in such a raw state that he exaggerates every feeling..."

Left Detail from a 1942 mural by Barry Faulkner in the New Hampshire State House in Concord NH. Abbott H.Thayer (right foreground) is shown demonstrating natural camouflage to followers, among them (left to right) Daniel Chester French, Thayer's wife Emma Thayer, George de Forest Brush, Barry Faulkner, Alexander James (son of William James), and Thayer's daughter, Mary Thayer. In the background is one of Thayer's angel paintings (a memorial to Robert Louis Stevenson).

For more on Thayer, see Anderson (1982), Boynton (1952), Chapman (1933), Meryman (1999), Faulkner (1973), White (1951), Behrens (2002) and numerous other sources. There is also a documentary film, titled *Invisible: Abbott Thayer and the Art of Camouflage* (2008).

•

Gerald Handerson Thayer
(September 5, 1883-June 5, 1939)
American painter, naturalist, writer

The son of Abbott H. Thayer, he was the author of record of *Concealing Coloration in the Animal Kingdom* (1909). Born in Cornwall-on-Hudson NY, he learned about painting and the natural sciences by working with his father. He married Alma Wollerman on Mount Monadnock near Dublin NH in October 1911. *New York Times* ("Gerald H. Thayer," 1939): "Survivors are his widow, Alma E. Thayer; two sons, David C. Thayer of New York and Bernard P. Thayer

New York Times ("Art Notes" 1917b): "[Harry] Thrasher is a Prix de Rome man, once a pupil of [Augustus] Saint-Gaudens, and considered one of the most brilliant of the younger sculptors. He is now in the Camouflage Corps, being so anxious to get to France that he refused to wait for the training that might land him a commission."

Right Gerald H. Thayer's remarkable watercolor painting of a Ruffled Grouse, which was published in full color in *Concealing Coloration* (Thayer 1909).

Maximilian Toch (1918:8): "A few years ago, at an exhibition of cubist paintings, there was a large picture which looked like pine shavings whirling about on the floor of a carpenter shop, and which was entitled 'Impressions of a Nude Woman Falling Downstairs,'* or something equally ridiculous. When the first camouflaged British steamer arrived here [in the US], someone called it 'Impressions of a Drunken Sailor Falling Down a Hatchway.' The colors used were exceedingly vivid, and the visibility of the ship seemed to have materially increased; but even so, it may have been difficult for an enemy to determine how far off the freakish looking vessel was and in what direction it was moving. This sort of camouflage, no matter how ridiculous it may appear, may often serve a useful purpose."

*Toch is referring to Marcel Duchamp's *Nude Descending a Staircase*, first shown in New York at the Armory Show in 1913.

of Dublin, and a sister, Mrs. Dawd Reasoner of Montecito, Calif."

Faulkner (1973:14): "Gerald, Thayer's only son, was his father's constant companion and from childhood knew wild nature. He became a well-trained ornithologist. Under his father's supervision he painted birds and fishes with exquisite skill, but he had little creative impulse. His late mother, Kate Bloede, had come from a distinguished German family of poets and scholars, and Gerald longed vaguely to emulate them but lacked the talent and drive to do so. He was a noble-looking boy with a fine muscular figure and with a nature generous, open-hearted and appealing."

G.H. Thayer (in a letter to L.A. Fuertes, quoted in Boynton, 1956:89): "About our book [*Concealing Coloration*] I don't like to talk. To me it has the smell of a shameful carcass and I doubt its ultimate triumph even. No fault of Papa's, but of mine—or, in part, the work of fate. But cheer up, the worst is yet to come! Things stand so: the text, left for finished when I went south last spring, and then re-read and deeply slashed into by Papa, has now at last been smoothed out again, from end to end. Work toward this result has been intersprinkled with all sorts of trafficking for pictures, with fruits of sad smallness in proportion to the time and labor spent..."

•

Thomas Elston Thorne
(dates undetermined)
American art educator

Thorne was a faculty member in the Department of Fine Arts at the College of William and Mary in 1943. He taught a camouflage course there, in which "Each student must, with aid of camouflage, start from the lake and reach school unobserved by sentries stationed in the woods" (Merriman, 1943).

•

(Charles) Leslie Thrasher
(September 5, 1889-December 2, 1936)
American sculptor

A student of Howard Pyle, he was known for his magazine illustrations, especially his covers for *Liberty* magazine (1924-30). *New York Times* ("Miss Jackson Weds" 1920): "During the war he served with the 40th Engineers, and was one of the first Americans to do camouflage work in France."

Reed (2001:199): "After his return [from Europe] in 1910, he studied briefly with Howard Pyle, and developed a close friendship with fellow-student Douglas Duer. They both served overseas during World War I, where they were assigned to camouflage with the Fortieth Engineers."

•

Harry D[ickinson] Thrasher
(1883-August 11, 1918)
American painter, sculptor

Harry Thrasher was born in Plainfield NH, where he was a student of Augustus Saint-Gaudens. He also studied in Rome at the American Academy. According to Barry Faulkner (1957), who was a good friend of his, when the US entered World War I, Thrasher enlisted as a private in the first squad of the American Camouflage Corps. He was killed in France in 1918, when he was struck by a shell fragment while inspecting a camouflage installation.

•

William D[enver] Timmons

(1898-May 30, 1960)
American industrialist, photographer

Timmons was the first American to be admitted to the Royal Society of Arts in London. He was President of the Pretty Products Company, which manufactured household wares. *New York Times* ("William D. Timmons" 1960): "He developed a treatment for canvas used in military camouflage to prevent detection by means of infrared photography and was the author of a book on textiles."

•

Shown here are photographs of four World War I ship camouflage treatments by American chemist and art expert Maximilian Toch (pictured *above* in a news photo), including *USS Kajeruna* (*above*), *USS Aeolus* and *USS Huron* (*below*, top and bottom), and *USS Warren J. Courtney* (*facing page*).

Maximilian Toch

(July 17, 1864-May 28, 1946)
American chemist

Born in New York, Toch earned his undergraduate degree at New York University (1886), then went on to graduate work at Columbia University. He taught chemical engineering at Cooper Union, and at the University of Beijing. He also lectured on organic chemistry at Columbia (1905-06) and the City College of New York (1909). He later taught industrial chemistry at Cooper Union (1919-1924), and painting chemistry at the National Academy of Design (1924-36). *New York Times* ("Dr. Toch" 1946:23): "The cement-filler treatment used in the construction of the Panama Canal was developed by him."

Toch's opinions regarding chemical evidence in the authentication of art became controversial in 1931, when he claimed that all but one of the Metropolitan Museum of Art's Rembrandt paintings (about thirty at the time) were not genuine, and that of four hundred paintings attributed to Rembrandt in collections worldwide, only forty-eight were genuine. He defended his claims in a book titled *Paint, Paintings and Restoration*. He was the uncle of art materials expert Ralph Mayer.

New York Times ("Dr. Toch" 1946:23): "He is credited with developing the original battleship gray formula used by the United States Navy, and during the

first World War was called America's first camoufleur. He had charge of camouflage of the East Coast defenses at that time also, and developed the Toch system of camouflage."

Toch (1919): "It was my good fortune the first week of the war [WWI] to be called down to the Navy Department to the Bureau of Yards and Docks, which department really took up the first camouflage of the United States, and I had the mission thrust upon me of going along the coast of the United States and designing and executing the camouflage of the yards and docks and a number of fortifications... When I was called to Washington during the first week of the war, it was because in 1915 I had camouflaged two forts at Panama. In those days the word 'camouflage' was totally unknown. The words used were 'military concealment,' and on the question of the visibility of color [blending] and juxtaposition of color [dazzle], these two fortresses which contained disappearing guns were given to me to distort and to lower their visibility...

The first workers on ship camouflage were Messrs. [William Andrew] Mackay, Gerome Brush, Louis Herzog, Lt. [Everett L.] Warner and myself, and of course in those days we didn't have the guidance of the brilliant scientific research of Messrs. [Harold] Van Buskirk, Warner and [Loyd A.] Jones."

Toch (1931:308): "In the beginning of the War [WWI] we all thought it was perfectly possible to lower visibility until you could not see the object. We learned that this is impossible, so I worked on the coloration concealment of Abbott Thayer, because I was one of the fortunate possessors of his book, which had long been out of print and which he personally had given me. I am frank to say that without his book, my slight knowledge of physics, chemistry, and optics would not have stood me in good stead."

Toch (Ibid.:308) "Early in May, I was hurriedly called down to the Shipping Board, and they read me a report from the Captain of the *Luckenbach*, in which he stated that the Toch system of camouflage was no good and did not lower visibility; that some other method ought to be tried, and he concluded by saying that one thing he did notice was that he could not tell which way the camouflaged ships were going.

I immediately saw that here we had at last the solution to our problem: that to lower visibility to the point of extinction was utterly impossible in physics and in optics, and I explained to the Shipping Board that every effort must be made to forget this but to try and distort direction."

Toch (Ibid.:309): "[Having been painted in disruptive camouflage]...a ship not only became foreshortened but its direction was so distorted that when I went out to sea in a submarine and fired some dummy shots at one of my own camouflaged ships, I missed the boat at as short a distance as 600 meters.

Early in January [1918], the Eastman Kodak Laboratories were enlisted in the studying of the lowering of visibility. I went to Rochester and consulted with their physicists; but I stood my ground that the lowering of visibility was not what we were after but distortion of direction, and in the end most of the ships were camouflaged by this method."

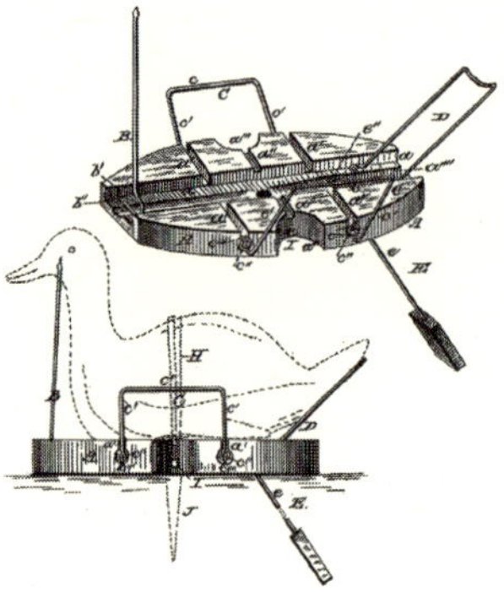

Above Drawing submitted for US Patent No. 230,600 titled "Frame for Supporting Dead Birds as Decoys," invented by F.A. Allen (1880).

Charles R. Knight (2005:62-63): "Like most geniuses, [Abbott H.] Thayer had his peculiarities, and a funny incident occurred soon after our return to Oak Lodge [in Forida]. Charlie Latham had a big white duck...which usually in duck fashion delighted in a bath in the creek just before retiring for the night. One evening I was in the house talking to Col. Dexter when we heard the roar of a shotgun coming from the direction of the creek. I rushed out only to meet Charlie Latham walking toward the house with a frown like a thunder cloud on his dark features. 'What's the matter?' I asked. 'Damn fool shot my tame duck,' said Charlie between his clenched teeth, and marched into the house. In a minute more came Abbott, hurrying after him, a bit sensitive, face lined with emotion. 'I'm so sorry, Knight—Gerald and I shot Latham's duck. It was against the sunset, and it looked black to us!' Here indeed was a wonderful example of Abbott's coloration theory, but in this case it didn't work out in just the way expected."

•

Ledyard H. Towle
(1890-November 8, 1973)
American painter, color consultant

Born in Brooklyn, Towle taught color theory at the College of William and Mary. He also served as a color consultant to General Motors, Pittsburgh Plate Glass Company, and E.I. duPont de Nemours and Company. *New York Times* ("Ledyard Towle" 1973): "[Towle] served in Army camouflage during World War I."

Harley Earl (1954): "About 1928, [when] Mr. [Alfred P.] Sloan saw that we... had taken hold of this General Motors Styling, he and Mr. Lamont duPont decided that it would be a good thing to put trimming selections and painting selections into our studios...into the Styling Section. So in 1928, Mr. Sloan called me from New York and said that Mr. Lamont duPont had offered the services of [World War I camoufleur] Capt. Ledyard Towle, who was head of duPont Duco service in New York. He's an artist. And, they realized that I had a lot to do, getting this other thing going, and he'd be very helpful. So he came out, and we moved... And we built a studio for Capt. Towle next to me, and he did a remarkable job of organizing it and getting the colors and the trim and all that, and about 1931 or 32, I've forgotten which, he resigned to take a job with Ditzler and Pittsburgh Plate Glass as their art director."

See also Harley Earl. David Jones, McClelland Barclay.

•

Evarts Tracy
(dates undetermined)
American architect

Tracy is mentioned in Embury (1917:137-138) as a World War I American camoufleur. In addition, Embury writes: "I think a great compliment was paid to the [architectural] profession when Major Evarts Tracy (of Tracy & Swartwout) was selected to begin the organization. He was assumed to have, and has, excellent judgment of men, and extensive acquaintance among the men fitted to be camoufleurs. He has certainly selected a wonderful lot of men." At the close of the war, Tracy published a "Memoranda on the Camouflage Service of the US Army" in the *Professional Memoirs of the US Corps of Engineers* (Tracy 1919).

•

Julian Trevelyan
(February 20, 1910-1988))
British painter and printmaker

Born in Dorking, England, he came from a prominent family, and was a nephew of the British historian George Trevelyan. He studied at Trinity College, Cambridge, then went to Paris, where he worked with British printmaker Stanley W. Hayter (1933-34). At the start of World War II, he collaborated with Hayter, Roland Penrose, John Buckland Wright, and (if not harmoniously) Erno Goldfinger in the formation of an industrial camouflage unit. He also served as a camouflage officer in the Royal Engineers (1940-43). He was the brother-in-law of (camoufleur) Robin Darwin. He recounted his wartime camouflage experiences in a highly readable autobiography (Trevelyan 1957).

Percival (Albert) Trompf
(May 30, 1902-July 17, 1964)
Australian graphic designer

Born in Beaufort, Victoria, Australia, Trompf began his career as a packaging designer for a confectionary firm. He is especially known for his posters for the Australian National Travel Association, Victorian Government Railways, and the Canadian Pacific Railway. Spearritt (1990): "On 4 May 1942 he enlisted in the Royal Australian Air Force and was commissioned as a pilot in June; he served mostly at Milne Bay, Papua, and also worked as a camouflage officer..."

Facing page A spread from Lancelot Law Whyte's influential book on *Aspects of Form* (Whyte 1951), which features Hugh B. Cott's pen-and-ink illustrations for his essay on protective coloration. Author's collection.

•

Casimir A. Truszkowski
(1895-?)
Polish-American artist

Born in Poland, he immgrated to the US at age two and grew up in Depue IL. He was drafted into the US Army in 1917, and assigned to Battery A, 68th Artillery, CAC. His role was that of an assistant camouflage technician. Following World War I, he moved back to Chicago and worked as an advertising artist.

Julian Trevelyan (1957:123): "Yet here I was [at the Farnham Castle camouflage school], an apostle of Camouflage, with the job of selling them the idea."

...

Annie Dillard (1974:18): "Specialists can find the most incredibly well-hidden things. A book I read when I was young recommended an easy way to find caterpillars to rear: you simply find some fresh caterpillar droppings, look up, and there's your caterpillar."

...

Salvador Dali (1993 [1942]): "Some young sardines are making their first outing under the supervision of their parents. A submarine passes by. The little fishes, alarmed, question their father:
'Papa, what's that?'
'That, my children, is our revenge which is passing: a great tin made of sheet-iron in which men, covered in oil, are held inside, pressed against each other.'"

•

[Ernest] Percyval Tudor-Hart
(December 27, 1873-June 8, 1954)
Canadian-born painter and teacher

Tudor-Hart was born in Canada, but studied art in Paris with Jean-Léon Gérome and Jules Lefebvre. He lived in England for much of his life. He was a color theorist, and the painting teacher of (camoufleurs) Wilford S. Conrow, and Sydney and Richard Carline. Hartcup (1980) talks about his World War I camouflage proposals, and they are also discussed and illustrated in Newark (2007:86). In 1917, while working on experimental ship camouflage for the British, he asked for and was granted the assistance

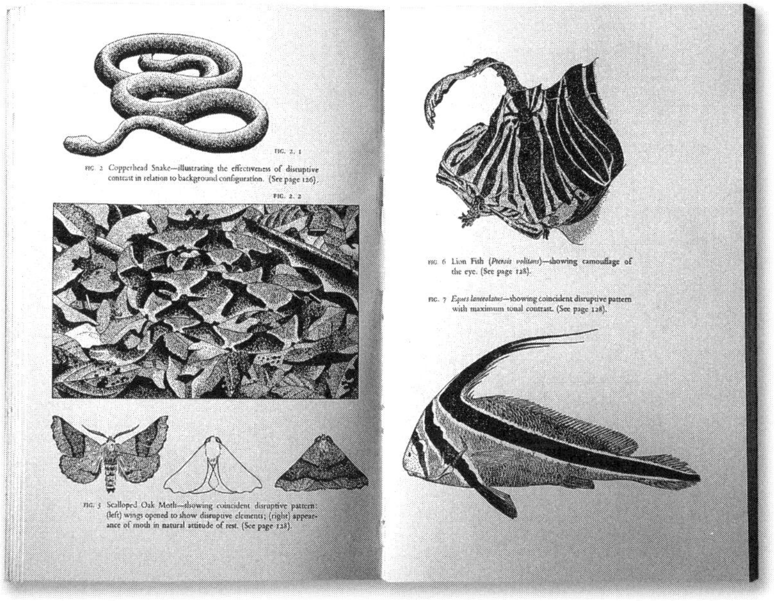

of the Carline brothers, Hugh de Poix and James Wood. See Hartcup (1980:41), and MacGregor (1961).

Mentioned in MacGregor (1961) is Tudor-Hart's student, US camoufleur Wilford S. Conrow. It is also noted there that Tudor-Hart was a friend of painter Owen Merton, the father of Thomas Merton.

•

Alfred James Tulk
(1899-1988)
British-born American muralist

Tulk was born in London, but studied art at Yale University and at the University of Guanajualo, Mexico. During World War II, he served as a US Army camoufleur, during which he also designed altars for military chaplains. During his career, he completed more than 300 large murals. See Tulk (n.d.).

•

William Twigg-Smith
(1883-1950)
New Zealand-born painter

Born in Nelson, New Zealand, Twigg-Smith came to the US around 1899 to study with Henry M. Walcott at the Art Institute of Chicago. He settled in Hawaii in 1916, where he collaborated on dioramas for the 1917 Pan-Pacific Carnival with Lionel Walden and David Howard Hitchcock. Later that same year, he joined the US Army and was assigned to the American Camouflage Corps in France. He returned to Hawaii in 1919 and resumed painting, while also working as an advertising illustrator for the Hawaiian Sugar Planters Association. An accomplished flutist, he held the second chair in the Honolulu Symphony. See also Forbes (1992).

Faulkner (1957) recalls that Twigg-Smith was among the first four American camoufleurs, along with Sherry Fry, Everett Herter, and himself. He also recalls (Faulkner 1973:89) that when they first arrived at their tent, "we found a minstrel [Twigg Smith], easing his solitude by playing Hawaiian airs on a ukelele. He came from the Islands and was pleasant and companionable."

Literary Digest (1917:33): "Twigg Smith, who has been painting the exquisite scenes of the Hawaiian Islands, has come back and enlisted."

Raphael Pumpelly (1918:659): "Speaking of [Abbott H.] Thayer's protective coloration, I remember an interesting evening [in Dublin NH]. It was at a time when, after the *Titanic* disaster, there was discussion as to the ability of lookouts to see an iceberg at night. Thayer took several of us, including an eminent physicist, into a field. The full moon stood about thirty degrees up in the east; to the west of where we stood the field was flat with a space of open sky. Thayer had erected nearly vertically a sheet of white cloth perhaps ten by twenty feet in size and about one hundred feet westward from us, and asked us to find it. It was absolutely invisible against the sky until he had it moved to where it hid the lower half of a tree."

...

Ward (2003:46): "[In December 1917, American painter Edwin] Dickinson decided that enlisting would give him some chance to decide how he would serve in the war. He and his friend, painter Herbert Groesbeck... attempted to be assigned together to a camouflage unit in some branch of the military. They traveled to the War Department in Washington in an effort to arrange this assignment, but without success. Dickinson commented in his journal upon his return to New York, 'camouflage *nihil faciens*.' Resigning himself to the failure of the plan, he went to the Brooklyn Navy Yard a week later and, on 12 December 1917, enlisted in the naval reserve."

•

Carroll (Sargent) Tyson
(November 23, 1878-1956)
American artist

Tyson is listed in Wolanin (1983:64) as having worked with Arthur B. Carles and other Philadelphia artists as a ship camouflage painter in the Philadelphia Navy Yard during World War I. Prior to that, he had studied with William Merrit Chase and Cecilia Beaux at the Philadelphia Academy of the Fine Arts, and in Munich, Germany, with Carl Marr.

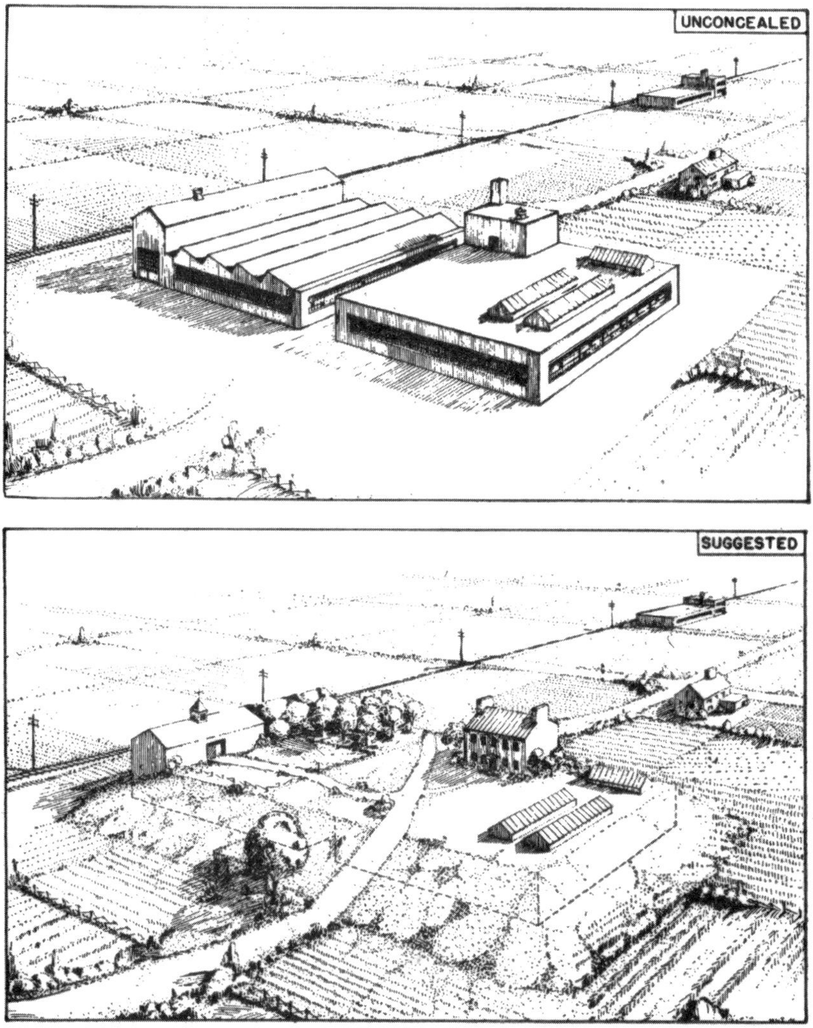

During World War II, American attitudes toward civilian defense uses of camouflage differed remarkably from that of the previous war, in part because of the attack on Pearl Harbor. In applications of industrial camouflage, vital landmarks were sometimes not merely disrupted or blended, but made to look like something else, such as the concealment of aircraft factories by the construction of dummy buildings, roads, trees and so on. See Breckenridge (1942) and Wittman (1942). Drawings from US War Department, *Civilian Defense: Protective Concealment* (1942:11).

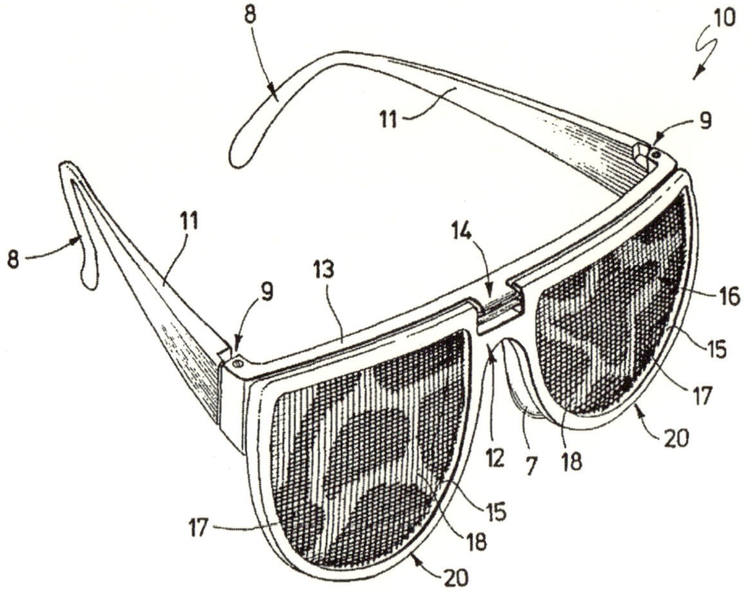
8
10
11
9
11
8
14
13
9
16
15
17
20
12
7
18
15
17
18
20

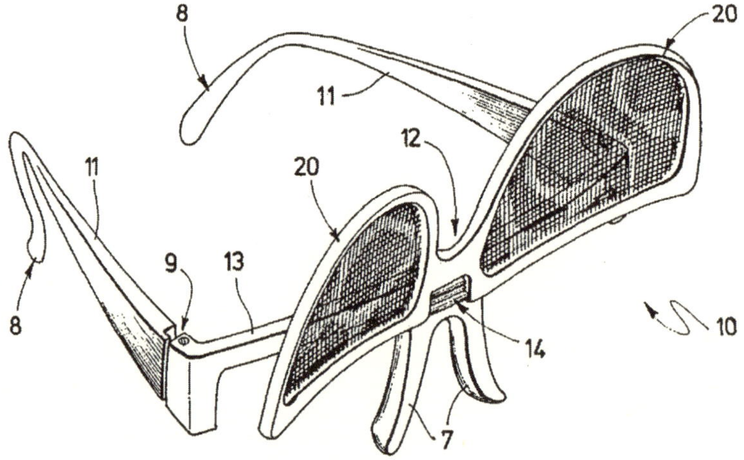
8
20
11
12
20
11
9
13
8
14
10
7
7

UV

Umbrella Camouflage
Overhanging garnished nets

Edmonds (1932:83-84): "Concealment from the air was first attempted by the use of canvas sheets painted to represent the ground as seen from an airplane. Experience proved these to be heavy to erect and difficult to maintain in position. On the suggestion of Mr. [Solomon J.] Solomon, they were replaced by a lighter and more manageable article [called "umbrella camouflage"], in the form of old fish nets or wire netting, garnished with tufts of painted or dyed raffia (gardener's bast). For the Somme thousands of rolls of wire netting were supplied to the divisional engineers who, when raffia was lacking, wove grass into it. In the course of time all batteries were equipped with these overhead covers. As the demand increased and the supply of raffia became inadequate, canvas strips were substituted for it, and were found to be less flammable."

•

Leon Underwood
(December 25, 1890-October 9, 1975)
British painter, printmaker and sculptor

Underwood was born in London to a family of antiquaries and numismatists. At age 20, he was awarded a scholarship to the Royal College of Art. An important influence that year was the first exhibition by Roger Fry of Post-Impressionist painting. He was an influential teacher, first at the Royal College of Art, then at his own school, and at St. Martin's School of Art. With Henry Moore (his student) and C.R.W. Nevinson, he founded a magazine called *The Island*.

Facing page Drawings submitted for US Patent No. 4,812,031 titled "Camouflage Eyeglasses," invented by Tony Evans (1989).

Redfern Gallery (n.d.): In 1916, having enlisted in the army at the start of World War I, he was "transferred to the Camouflage Section, Royal Engineers, at Wymereaux...

[Having been evacuated from the front, due to severe illness, he] Submitted trial drawings to the Ministry of Information, Propaganda Section, in June, with a view to painting camouflage pictures for the Imperial War Museum." Later, in April 1919, he was commissioned by that museum to complete two war-related paintings, one of which was *Erecting a Camouflage Tree*. "Both pictures, and many sketches made in France, [were included] in Camouflage Artists' Show at Burlington House in October [1919]." In 1940, he "joined the Civil Defense camouflage unit at Leamington Spa..."

•

Above Military photos of the use of overhanging garnished nets, or umbrella camouflage, in World Wars I and II, respectively.

Beston (1919:122): "Whether camouflage be valuable or not, it certainly is the fad of the hour. The good, old fashioned, one-color boat has practically disappeared from the seas, and the ships that cross the ocean in these perilous times have been docked to make a Cubist holiday; the Futurists are saving democracy."

Uniforms
In relation to camouflage

Horne (1964:68-69): "The French soldier of 1916 bore hardly a resemblance to the insouciant neophytes in their red pantaloons that had marched to war behind the regimental music in 1914. The *képi* [a flat circular French cap with a visor] had been replaced (despite Joffré's optimism) by the more practical steel helmet; in which, for once, the French were ahead of the Germans. Gone were the deadly pantaloons and gone was the pre-war-spit-and-polish, even among units normally as proud of their chic as the Chasseurs. A new uniform, the *horizon bleu* [horizon blue, or blue-gray], was slowly being issued (it was a compromise; though still not as well camouflaged as the British khaki [Urdu for "dust"] or the German *Feldgrau* ["field gray"], after a few days in the mud of the trenches it blended with surroundings as well as any other. And in its pristine state, as British troops relieved by French units frequently noted, the sight of the *horizon bleu* was a tremendous morale-booster to stained and jaded troops on the line.) But it was still not widespread; in the Bois des Caures the Chasseurs continued to wear a motley of tattered sheepskins and tunics so patched as to be no longer identifiable as anything military."

Among the best sources for detailed information on the history of military camouflage uniforms are the books of J.F. Bosarello, a leading collector, who was a major contributor to Newark (1996), as well as to the two-volume boxed edition of Blechman (2007), which includes a supplementary, largely pictorial volume about the camouflage uniforms of 107 nations, titled *DPM: Military Camouflage Patterns of the World.*

•

US Student Army Corps
College training program

Mordecai Gorelik, interviewed in Naverson (1989:36): "At the end of World War I, there was a new federal institution, the US Student Army Corps, based in universities, colleges, and even technical schools. I enrolled at Columbia University. We were in uniform, taking military drill and attending special courses. There was a course in camouflage. It attracted less than a dozen students, who were pleased to be so honored and were fond of the Army officer who was our instructor [probably H. Ledyard Towle]. In a few weeks the war [World War I] ended and we were each given an honorary discharge."

V

(Charles) Harold Van Buskirk
(February 20, 1894-October 25, 1980)
US Navy Officer, Olympic fencing champion

Usually referred to as Harold Van Buskirk, he was a three-time US Olympic fencer (1924, 1928 and 1932). He was also President of the Amateur Fencers' League of America, and his wife was the women's national fencing champion. Their son Charles Jr. died in 1936, at age 19, in a shooting accident. His father may have been a judge who played a role in the trial of Bruno Hauptmann, in connection with Lindberg kidnapping.

As explained by Buskirk (1919:227): On March 25, 1918, the US Navy (its Bureau of Construction and Repair) was placed in charge of all camouflage designs, while the civilian Emergency Fleet Corporation (US Shipping Board) was placed in charge of the actual application of the camouflage designs to ships' surfaces. "I had the honor of being placed in general charge of the Navy's organization and was fortunate indeed in being given the assistance of Lt. [Everett L.] Warner and Lt. [Loyd A.] Jones, who were placed in charge of the subsection of design and the subsection of research respectively.

The design sub-section originated and developed all designs on models, prepared type plans and transferred all approved designs to the type plans ready to issue.

The research sub-section had charge of all original investigations in connection with new systems of camouflage that were proposed, as well as a special investigation to determine the exact physcological [sic] value of the Dazzle system. This subsection also had charge of designing and building very specialized types of portable photometers and colorimeters for collecting the deserved quantitative data on the quality and distribution of light under various conditions at sea."

Above Camouflaged WWI American ship, the *USS Wolsey* (1918). Courtesy US Naval Historical Center (NH 100438).

•

George Vander Sluis
(dates undetermined)
American artist

He studied art in Cleveland and at the Colorado Springs Fine Art Center (under Boardman Robinson), where he also later taught. Crane (1944): "In 1942, he entered the [US] Army and is attached to the Engineers Camouflage Division." He served in the Ghost Army, where he became a friend of Bill Blass, in whose autobiography (Blass 2002) he is mentioned (as George Vandersluis [sic], "an army buddy...who later became a university professor") along with a wartime photograph.

Charles Burchfield (University of Arizona Art Gallery 1965:24): "[During World War I] All the arts seemed to have gravitated to the camouflage outfit—even musicians."

...

During World War II, someone came up with the clever idea of camouflaging the carrier pigeons that were used for sending messages on the battlefield. The method was not to paint them in disruptive camouflage—but to breed them in favor of dazzle patterns. For further information, see "World War Pigeons Are Bred with Camouflage for War Service" in *Popular Science Magazine* (January 1941), p. 81.

•

John Vassos
(1893-1985)
Greek-born American industrial designer

Vassos was born in Constantinople. While serving in World War I, he was rescued by US forces after his ship was torpedoed. He arrived in Boston in 1919, where he studied with John Singer Sargent. He then moved to New York in 1924, where he studied at the Art Students League. He opened an advertising design studio, and worked as an illustrator for major magazines, including *Harper's* and the *New Yorker*.

From 1933 to 1964, he was associated with the Radio Corporation of America (RCA) as its corporate designer. It was Vassos who designed the first RCA television sets, which premiered in 1939 at the New York World's Fair, for which he also designed the RCA pavilion. He worked with (camouflage expert) Alexander Kostellow at Pratt Institute in initiating programs in industrial design education in the 1940s.

During World War II, he served with the US Army Air Corps as a camouflage consultant. He also worked for the Office of Strategic Services (OSS), the precursor to the Central Intelligence Agency in connection with its efforts to support the efforts of underground forces in Greece. In the process, Vassos secretly parachuted into Greek territory on two occasions.

In the 1930s, Vassos published an impressive number of unusual books, including his poetry, nonfiction on such subjects as phobias and dog training, and his own illustrated fiction.

•

Charles Vildrac
(November 22, 1882-June 25, 1971)
French poet, playwright

Born Charles Messager, he served in the French camouflage section during World War I. He was associated with Othon Friesz and André Salmon. He is mentioned by Ducasse (1962), where he is referred to as a man of letters.

•

Jacques Villon
(né Gaston Duchamp)
(1875-1963)
French painter

Villon was the older brother of artists Marcel Duchamp and Raymond Duchamp-Villon. He is listed in Coutin (1988) as a World War I French camoufleur.

•

Esteban Vincente
(January 20, 1903-2001)
Spanish-born American painter

Born in Segovia, Spain, he grew up in Madrid. He began as a sculptor, but turned instead to painting, and moved to Paris, where he married an American. *New York Times* ("Estaban Vincente" 2001): "Civil war broke out in Spain just as the couple were planning to move to New York. Mr. Vicente immediately volunteered for the Loyalist army, for which he painted camouflage in the mountains outside Madrid for several months." In 1936, he moved to New York, and served as vice consul for the Spanish consulate in Philadelphia until 1939. Associated with the Abstract Expressionists, he was a founding member of the New York Studio School, and also taught at various times at Black Mountain College, Princeton University, University of California at Berkeley, New York University, and Yale University.

He once said: "New York is hell, and hell is the only place for people."

•

Above *Douglas Volk, 1917*. 1 photographic print: b&w; 25 x 18 cm. Courtesy of the Macbeth Gallery records. 1838-1968, Archives of American Art, Smithsonian Institution.

Right A camouflaged Ukrainian airplane (Su-25), photographed in 2002. Public domain.

Facing page Arrival at Liverpool, England, of camouflaged US transport ships, from *War of the Nations* (1919).

John Winsor (1996): "I asked [fashion designer Gianni] Versace, whose Autumn collection this year contains camouflage gear for both sexes, why camouflage was so popular. 'There'll be some philosophy behind it,' I was told, 'but fashion isn't very logical. It's dictated by designers. And what goes around comes around.'"

(Stephen Arnold) Douglas Volk
(1856-1935)
American painter

Born in Pittsfield MA, Volk was the son of the prominent sculptor Leonard Wells Volk, and was named after Abraham Lincoln's political rival, Stephen A. Douglas, who was his mother's cousin. He studied at the École des Beaux-Arts at the same time as Abbott H. Thayer and George de Forest Brush. Because he and Brush were look alikes and frequently seen together, their painting master Jean-Léon Gerome called them "the twins." In 1886, having returned to the US, he founded the

Minneapolis School of Fine Art (now the Minneapolis College of Art and Design), for which he was director until 1893.

Blechman (2004:166): "[Among proposals for ship camouflage during World War I was] Douglas Volk's concept of shrouding a ship in mist pumped out from angled pipes. In 1996, almost 80 years after Volk's idea, British shipbuilder Vosper Thornycroft experimented with a similar 'mist' concept for a stealth frigate that they called the *Sea Wraith*. This differed from the WWI idea in that it was designed to camouflage the ship's heat signature from infrared imaging missiles, rather than from human eyesight. However, due to the cost of implementation and the practical difficulties associated with the system, the VT *Sea Wraith*, like Volk's plan, was never put into production."

•

Will Irwin (1917): "This word [camouflage]—having none other for the process—they [the English] added to the vocabulary of the British Army; it was new, and it was susceptible to a great variety of metaphorical uses. At latest accounts the British soldiers were working it to death. They use it as a noun, verb and adjective. They use it for any variety of concealment—moral, spiritual, or intellectual."

Above All the photographs on these two pages are from the World War I design subsection of the US Navy camouflage section in Washington DC. Shown here are various stages in the process of devising ship camouflage schemes and the rooms in which they were designed, tested (as models), then converted into diagrams for use in actually painting the ships. In the top photo on this page is the model-making room in which the following five artists are working (left to right): Douglas D. Ellington, Kenneth MacIntire, Frederick C. Clayter, Richards, and Sullivan. In the lower photo, the camoufleurs painting the models are (left to right) Everett L. Warner, Frederick Waugh, John Gregory, Gordon Stevenson, M. Nash, and M. O'Connell. A sign on the wall above Warner reads "Keep It Simple." On the facing page, the top left photos are two views of the very same set-up, in which only the height of the camera

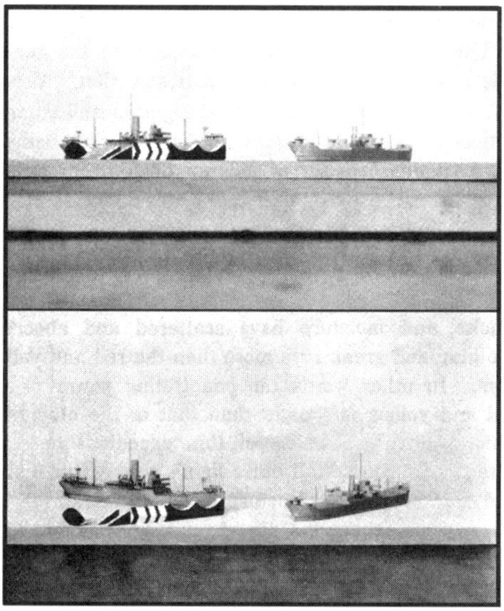

has changed. In this case, the camouflage pattern has not yet been applied to the ship model; and in fact it is only an arrangement of painted wooden blocks, a method that Warner invented as a faster, more expedient way of generating new designs. On the top right the US Secretary of the Navy (Josephus Daniels) is looking at models through a periscope in the testing theatre. The lower photo is a view of the drafting room, showing R.J. Richardson, the ensign in charge (front row center), Everett L. Warner (standing in the aisle on the left, looking down) and Frederick Waugh (in the center at the back of the room, with hand on chin). US Navy photographs, courtesy Everett L. Warner Archives.

tall-crowned top hat
leather boots with toe-caps
leather boots with dark toe-caps
fabrics

WXYZ

Edward (Alexander) Wadsworth
(October 29, 1889-June 21, 1949)
British painter

Wadsworth was born at Yorkshire. He studied in London at the Slade School of Art, where he became affiliated with Roger Fry and the Omega Workshops, and, following that, with the Vorticists, among them Wyndham Lewis, Ezra Pound, William Roberts and Alvin Langdon Coburn. While serving in the British Navy during World War I, by request of camoufleur Norman Wilkinson, he was assigned to help Wilkinson in the dazzle painting of British merchant ships.

Wadsworth (1989:5): " When Edward got back he was sent immediately on a training course to learn the intricacies of this new disguise system and its techniques. The course may well have been at the Royal Academy where, in Wilkinson's office, there stood a telescope [sic], a turntable and models, made by female Slade students, to which the camouflage designs were transferred. In the photographs Edward kept of his activities in this venture there are varying types of design, and it is interesting to compare these with the woodcuts he made at that period, for he seems to have been impressed not so much with the checkered or diamond shaped forms utilized on the ships but only by the stripes—stripes upright, horizontal and curved, calling to mind his earlier paintings in which stripes were a decorative feature."

Facing page Drawings by Russian designer Valeria Nekaeva (c2005) of the costumes for a theatrical interpretation (never actually performed) of *The Nose* by Nikolai Gogol. Courtesy the artist.

Wadsworth (ibid.:76): "In the early spring of that year Edward was in Bristol and Avonmouth, as his photographic records of camouflaged vessels show; the first is *New York City* [sic], taken in April 1918 at Bristol. By July he had been moved up to Liverpool with headquarters at 299 Royal Liver Building, and his work there seems to have connected him with various ships, some of them famous, such as the *Aquitania* and the *Mauretania*; but there were more modest vessels, too, *Hellenes*, *Tascalusa*, *Alsation*, *Virginia* and *Zealandia*—all dated 1918."

Above (top) According to his daughter, it was British Vorticist artist Edward Wadsworth, then serving as a British dock officer, who oversaw the camouflage painting (but probably not the design) of the *USS Zealandia*. (bottom) He may have also painted the *SS City of New York*, a passenger steamer that was commissioned by the US government for troop transport in 1917 and renamed the *USS Plattsburg*. Photos courtesy US Naval Historical Center.

•

Everett L[eslie] Waid
(1900-June 22, 1976)
American artist

Born in Denver CO, Waid served in the US Navy during World War I. After the war, he studied at the Art Students League, serving as vice president in 1937-39, treasurer in 1945, and later as president. The enrollment at the Art Students League quadrupled during his term. *New York Times* ("Everett L. Waid" 1976): "Mr. Waid continued his art studies here and abroad as well as with the league and in World War II served in the Army as a specialist in camouflage."

•

National Marine ("Baffling the Undersea Pirate" 1919:30) "Each nation seemed to have a characteristic type of [ship] camouflage and after a little practice you could usually spot a ship's nationality by her style of camouflage long before you could make out her ensign." [For examples of very evident stylistic differences, see photos of WWI ship camouflage by Germans, Japanese, French, Spanish and others in Williams (2002)].

Raymond S. Waldron, Jr.
(1913-2002)
American interior designer

Raymond Waldron was born in Kansas City MO. He was educated at the University of Missouri and Parsons School of Design. After graduating, he served in the US Army for four years (1942-1946), during which he designed maps and camouflage. He was also assigned to make drawings of the destruction at Nagasaki, Japan, in the wake of its bombing. Following the war, Waldron worked at Lord and Taylor as an interior designer until 1965, when he formed his own company. In 1963, he was commissioned by First Lady Jacqueline Kennedy to rennovate the Queen's Bedroom in the presidential guest quarters, Blair House. His papers are housed at the Kellen Archives Center at Parsons College. See Waldron (n.d.).

•

Ralph Walker
(1912-2003)
Australian artist

He attended East Sydney Technical College. During World War II, he served as camouflage instructor and an Official War Artist..

•

Ralph T. Walker
(1890-January 17, 1973)
American architect

Born in Waterbury CT, Walked was educated at Classical High School in Providence RI, and at MIT, where, in 196, he was awarded the MIT Rotch Traveling Scholarship. A principal in the firm of Voorhees, Walker, Smith and Smith, he designed the New York Telephone Company and Irving Trust Company buildings in New York, the Bell Telephone Laboratories in Murray Hill NJ, and the AFL-CIO Building in Washington DC. He designed various pavilions for the 1939-40 World's Fair, the Roger Williams Memorial in Providence RI, and the George Eastman Memorial in Rochester NY. He was described by the American Institute of Architects as "the architect of the century." He was listed in Embury (1917:138) as "of Goodhue's." *New York Times* ("Ralph T. Walker" 1973): "He was a lieutenant in the camouflage section of the Army Corps of Engineers in World War I."

•

John (Edwin) Walley
(1910-1974)
American designer, professor

Walley was born in Sheridan WY. He studied political cartooning, painting, mural painting and design at the Chicago Academy of Fine Arts. After graduating, he taught briefly at the University of Wyoming, then returned to Chicago to work as an advertising designer. He completed two murals for high school settings for the WPA, served as director of that agency's Illinois Art Project design workshop, and then as its assistant state director (1939-42). Gray (2001): "After a stint in the army in 1943, working as a draftsman, camouflage expert, and traveling lecturer-teacher," he joined the faculty of the Institute of Design, under the directorship of Laszlo Moholy-Nagy. He joined the architecture faculty of the University of Illinois at Chicago in

1953, and became head of its Art Department in 1971. During World War II, he served in the same camouflage unit as William Kienbusch (Selvig 1968). See also Mavigliano (1987).

•

As is well known, American artist Andy Warhol created an extensive series of paintings and prints that were based on camouflage fabric patterns (1986). For reproductions and a discussion of these, see Richardson (1998). It may or may not be significant that Warhol studied art at the Carnegie Institute of Technology, where two of the top US camoufleurs taught, Everett L. Warner and Homer Saint-Gaudens.

Above and facing page WWI-era photographs of Everett L. Warner (c1919), at his home and in the ship model painting room at the US Navy camouflage subsection.

John Lester Ward
(January 30, 1890-November 23, 1960)
American set designer

He served in the first American camouflage unit in France. Faulkner (1973:90) recalls that members of the American Camouflage Corps, prior to shipping off to France, wrote and produced a satirical play on army life, as a fundraiser. Using his theatrical expertise, "Wardie" was in charge of the scenery and lighting.

Faulkner (1973:110) also talks about a "riotous and fraternal" reunion of members of the Camouflage Corps, a year after the end of World War I, held in New York at 11 MacDougall Alley. Still an effective manager, it was organized by John Lester Ward.

•

Everett L(ongley) Warner
(July 16, 1877-October 20, 1963)
American artist

Born in Vinton IA, Warner studied at the Corcoran Museum, the Art Students League in New York, and the Académie Julian in Paris. He later taught for many years at the Carnegie Insitute of Technology, where (camoufleur) Homer Saint-Gaudens was head of the art department.

Today, he is all but unheard of in histories of American art, as well as in historic accounts of World Wars I and II. Yet, he was undoubtedly one of the chief contributors to the development of US ship camouflage in the twentieth century. Warner, who supervised the artists in US ship camouflage in both wars, was the American equivalent to British camoufleur Norman Wilkinson. During World War I, he was Wilkinson's official host when

the latter visited the US for the purpose of advising the Americans on British dazzle painting. In subsequent years, he and his team of artists went beyond the British method, and devised a systematic way of using solid geometry and sawed-up bits of wooden blocks to create ingenious course deception schemes. Nothing was more symptomatic of Modernism than Warner's startling dazzle designs, and yet, in his later life, his own artistic work was eclipsed by more fashionable gallery trends. See Warner (1919a, 1919b), Fusscas (1992), and Behrens (2002).

•

Alan Warren
(1919-1991)
Australian artist

He studied drawing and painting at the National Gallery of Victoria Art School in 1937-38. He later taught at the Melbourne Technical College, and, in 1951, became the art critic for the *Melbourne Sun*. ArtAngel.com: "Warren served in the camouflage corps during most of the war [WWII], and was responsible for disguising the perimeter defense system at Darwin which was subject to continual air raids by Japanese bombers."

•

Franklin (Chenault) Watkins
(December 30, 1894-1972)
American painter

Born in New York, Watkins (known as "Watty") grew up in Philadelphia and attended art school at the Pennsylvania Academy, where one of his teachers was Arthur B. Carles. During World War I, both he and Carles served as a US Navy camoufleurs, along with Adolphe Borie, Jean Knox, Waldo Pierce, and Carroll Tyson. See Wolanin (1983:64).

Philadelphia Museum of Art (*Arthur B. Carles* 1970): "Watkins explains the theory of camouflage in 1917-18 as follows: 'The idea was to make the boat appear to be going in a different direction than it was. Torpedoes went pretty slowly then and the U-boat captain had to calculate the speed and direction of the target vessel and fire his torpedo to meet the ship anywhere from one to five minutes later. Naturally, if the camouflage patterns made the U-boat captain plot a false course, the torpedo would miss the ship.'" See also Archives of American Art (1995:161).

•

Watson
(full name and dates undetermined)
American ship painter

During World War I, Watson was the master ship painter at the Norfolk Navy Yard, but no one seems to refer to him by his full name. His approach to ship camouflage, known as the Watson-Norfolk system, was one of five proposals to be formally adopted by the US Shipping Board.

•

Alec Waugh
(July 8, 1898-September 3, 1981)
British writer

Born in London. He was the older brother of British writer Evelyn Waugh. According to Hartcup (1980:78), he was briefly on the staff of a British camouflage factory during World War II.

•

Frederick J. Waugh
(September 13, 1861-September 10, 1940)
American marine painter

Born in Bordentown NJ, Waugh was the son of portrait painter S.B. Waugh. He studied with Thomas Eakins at the Pennsylvania Academy of

Above Two views of the *USS Nebraska* during World War I (c1918), with experimental camouflage plan known as Watson-Norfolk. Courtesy US Naval Historical Center (NH 101208).

Facing page (clockwise from top left) Photograph of US naval camoufleur Frederick J. Waugh (1929), as photographed by Peter A. Juley & Son (n.d.). Courtesy Photograph Archives, Smithsonian American Art Museum (J0002264). *top right* Dazzle scheme designed by Waugh (according to Everett L. Warner's lecture notes) for a US ship named *SS Congaree*. *bottom* Group photograph of Warner (far right) and other members of the design subsection of the WWI US Naval camouflage unit. Waugh is standing in the back row, just right of center, wearing a civilian hat. According to Warner's lecture notes, "The total number of persons employed in the Camouflage Section varied from time to time but never ran but slightly over fifty and this number included those who were assigned to the Research Section in Rochester. About a dozen women were employed, enlisted as yeowomen, as they were then called, now yeomen (F)."

Fine Arts, and with William-Adolphe Bouguereau at the Académie Julian in Paris. He remained in Europe afterwards, living in England until 1907, where he worked as a seascape painter. While serving under Everett L. Warner as a ship camoufleur during World War I, there may have been no one as innovative as Waugh. He produced a particularly memorable scheme for camouflaging the *USS Leviathan*, (sometimes credited to Norman Wilkinson) and a forced perspective plan for the *Proteus*. Oddly, although he was an ingenious practitioner of ship camoufleur, the work that he did as an artist—like that of Warner, Sherry Fry, and any number of other camoufleurs—is now largely forgotten.

Havens (1969:153-154): "On March 13, 1918, [Everett L.] Warner wrote to [Fred] Waugh from Washington.

'I have suggested your name to the Bureau as one who might be willing to take a hand with marine camouflage which is now being organized under the Navy Department.'

He then went on to propose a meeting in New York on the following Sunday or Monday, March 17 or 18, at the Arts Club or elsewhere, but with some uncertainty still as to time and place since, as he explained, 'I have Lt. Commander [Norman] Wilkinson in tow and am taking him around to show him our Navy Yards at Boston, New York, Philadelphia, and Norfolk. He is an English marine painter who is the father of marine camouflage in England and has been detailed by the Admiralty to cooperate with us in establishing the work here...'

Waugh threw himself into naval camouflage with all his energies, finding in this new work a ready outlet for this natural mechanical bent. Many large ships, including the *Leviathan*, were painted according to his designs. Through the enterprise was of course a team effort in which no man played a solo part, he had every reason to be proud of his record. Only one ship with his system of camouflage was lost during the war."

•

John Wells
(July 27, 1907-2000)
British painter

Wells was born in London. Trained to be a medical doctor, he turned to the study of painting in the late 1920s. From 1936 until after World War II, he lived on the Scilly Isles, where he was a general practitioner. According to the Tate Archives: "Wells was known to be interested in camouflage designs (including patterns, shapes and insignia) and he must have had direct contact with both naval and air force designs while on the Isles of Scilly. He was interested in the idea of concealing forms and places within a landscape."

•

Charles W(ilbert) White
(April 2, 1918-1979)
American painter

White was born in Chicago, where he studied at the Art Institute. He also studied at the Art Students League in New York, and in Mexico. As a muralist,

Warner (1919:106), contrary to the claims of Whiting (1928): "The Navy [WWI camouflage design subsection] worked so quietly and under such close censorship that few people were aware of the leading part that it played in the work. There exists even today a very widespread impression that the designs which the [US] Shipping Board camoufleurs applied to the ships originated with them. This belief is entirely without foundation. All designs were supplied by the Navy, and while it is true that at several of the ports the camoufleurs built small testing theatres copied after ours and did a certain amount of experimental model painting, this was wholly for their own education or relaxation, and none of the dazzle designs so made was ever authorized for application to any ship."

...

Whiting (1928:242-243) "[When WWI ended, William Andrew Mackay] was still abroad where an appreciative Navy had sent him to confer with the wizards of camouflage in England."

Above John Whiting's pen-and-ink drawing of US Shipping Board camoufleur Phil Duroc (which may have been a fictional name), viewing a dazzle painted ship model through a periscopic viewing device (Whiting 1928:70).

he worked with Diego Rivera and other Mexican painters. He himself painted murals for the WPA, including a mural for the auditorium at the Hampton Institute in Hampton VA, where Austrian art educator Viktor Lowenfeld was the Art Department Head.

White (1965:9): "After the fellowships, I had two consecutive Rosenwald Fellowships, then I was drafted into the Army in 1944. I spent about a year in a camouflage unit of the Engineers.

...I painted signs for quite a while as a camoufleur. Mostly on cans that said 'Edible' and 'Nonedible.' And I also had the opportunity of painting the Mess Hall once."

•

Victor (Gerald) White
(1891-1954)
Irish-born American painter

Born in Dublin, Ireland, he studied at Art Students League in New York, with George Bellows, William Merrit Chase, and Robert Henri. He created a WPA mural for the post office at Rockville Center NY. During World War I, as described by Rickard (1942:190), he was among the second group of officers assigned to the American Camouflage Corps, along with Laurance Hitt and Carroll T. Berry.

•

John D. Whiting
(dates undetermined)
American illustrator

In a 1921 issue of *Publishers Weekly*, it was stated that "The weird but efficacious designs which appeared on our transports during the war [WWI] were the results of experiments of John D. Whiting, of the Camouflage Department of the Navy, to find out just what lines and colors did in reducing the visibility of ships." *Convoy: A Story of the War at Sea* (Whiting, 1928), which he wrote and illustrated, is an apparent eyewitness account of the work of the World War I US Shipping Board camouflage team that was directed by William Andrew Mackay. Whiting was also the author of *Practical Illustration*, *Storm Fighters* and other books.

•

Ray Whiting
(October 7, 1898-?)
Australian cartoonist

Born in St. Leonards, Sydney, New South Wales. DAAO: "Whiting served in the Australian Imperial Forces during World War II." The same source refers to a photo caption that reads: "West Desert, Egypt. October 27, 1942. Ray Whiting of the 9th Australian Division Camouflage Training Unit (formerly a cartoonist of the *Bulletin Sydney*) at work on a sketch for the *AIF News*, a newspaper published in the Middle East for Australian Troops."

•

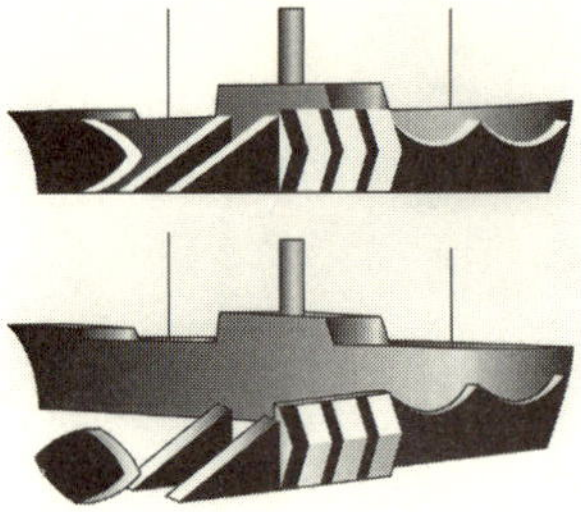

Above Illustrations by Ryan McAdam showing two views of the same set-up, as a way of demonstrating how Everett Warner and his WWI US ship camoufleurs produced new disruptive camouflage schemes by arranging painted wooden blocks (initially made by slicing up wooden ship models) in positions that contradicted the true course of the ship.

Boynton (1952:546) "Mr. [Abbott H.] Thayer placed a model of a zebra not many feet back from the bridle path in Central Park, New York City, and the story goes that no one reported seeing it, not even the mighty hunter Theodore Roosevelt, who rode daily on that path and maintained in print that the zebra was the most conspicuous of animals."

John (Codner) Whitlock
(1913-?)
British painter

He is listed as John Codner in Reit (1978:138) and in Barkas (1952) as a World War II camoufleur.

•

Paul Whitman
(April 23, 1897-December 13, 1950)
American printmaker

Born in Denver CO, he grew up in St. Louis. Whitman website: During World War I, he was a lieutenant in Europe with the 128th Field Artillery, Battalion A, while during World War II, he served in San Francisco for five years as a camouflage artist with the Army Corps of Engineers. Beginning in 1926, he lived in Carmel CA and was closely tied to the Carmel Art Association.

•

Frank C[lifford] Whitmore
(1888-June 25, 1947)
American organic chemist

Whitmore was born in North Attleboro MA. He completed his undergraduate and graduate studies at Harvard, then taught at Williams College, Rice Institute, University of Minnesota, Northwestern University, and Pennsylvania State College. He is known for his research of organic compounds of mercury, in connection with their medical use, for which he received the Willard Gibbs Medal in chemistry in 1945. *New York Times* ("Dr F.C. Whitmore" 1947): "During the recent war Dr. Whitmore served as a consultant to industrial groups and federal agencies. He headed here a staff of fifty men doing research work on aviation fuels, explosives, lubricants, synthetic rubber, penicillin and airplane camouflage."

•

George W. Wickstead
(1912-?)
American landscape architect

O'Connell (2000:35): "George Wickstead, FASLA, remembers being a 29-year-old, 'wet-behind-the-ears' landscape architect when he was sent to Puerto Rico in 1941 to help camouflage a major airstrip, a key part of the defense of the Panama Canal. He was there when the United States declared war. 'In a way it was the blind leading the blind,' he recalls. 'We were not trained camoufleurs, but we had the basic understanding of what was to be attempted.' Wickstead and his fellow designers were charged with diminishing the aerial view of runways and a large compound of buildings that were visible from the air forty miles away. With paint and vegetation, they carried the appearance of the surrounding landscape onto the runways. Unfortunately the camouflage worked too well, making it nearly impossible for aviators to land. They ultimately abandoned the effort."

•

Wiggly
Amoeboid design motif

British art historian Bevis Hillier (1983) has said that World War II camoufleurs should be partly credited with (or blamed for) the post-war ubiquity of an amorphous blob shape, a wiggly or squiggly. Hillier (ibid.:119): "These amoeba-like forms invaded design in the immediate post-war period—in the table tops of the French designer Jean Royere (1946), the cross-section of a glass vase by Alvar Aalto (1947), a Jacqueline Groag design for Hill Brown (1948) and a plywood table designed by Carl-Axel Acking for A.B. Svenska Mobelfabrikerna (1949)." They may also be a source for the so-called kidney-shaped swimming pool and the typical sandtrap on golf courses (especially in view of the fact that some golf course designers, such as Alister MacKenzie, were also camouflage experts). The adoption of the shape was probably also influenced by the abstract wood reliefs of Jean Arp. In another context,

Hillier (1975:38) recalls that when he met British camoufleur Julian Trevelyan, "he told me that they had used the term 'wigglies' for the amoeboid shapes painted to conceal targets."

•

John [Henry] Wilde
(December 12, 1919-March 9, 2006)
American painter

Geoffrey Barkas (1952:37):
You see these Greeny-Browny Blobs?
Well, that's a special kind of Paint
That makes things look like
What they Ain't.
No fooling, it's the lastest thing—
It's called Disruptured Patterning.

Sullivan Goss Gallery website: "Immediately after receiving his BS [degree] at [the University of Wisconsin at] Madison in 1942, Wilde was inducted into the army to fight in WWII. During his service in the US Infantry and Air Force, Wilde was assigned to the design of venereal disease prevention propaganda and camouflage patterns."

•

Above British artist and WWI ship camoufleur Norman Wilkinson near the end of his life, holding a dazzle-painted ship model. According to Crowell (1921:509), while Wilkinson was in the US in 1918, as a consultant to the navy, he "personally designed patterns for the troop transport *Siboney* and the navy ship *Prairie*. In England he designed the camouflage for the *Leviathan* and for thirty-five of our destroyers." This contradicts Havens (1949:154), who states that American painter Frederic Waugh was responsible for camouflaging the *Leviathan*.

Norman Wilkinson
(November 24, 1878-May 31, 1971)
British artist

Wilkinson was born in Cambridge, England. Although he studied figure painting in Paris, his primary achievement was in marine painting. During World War I, he served in the Royal Naval Volunteer Reserve. As described in his autobiography (Wilkinson 1969:79), in early 1917, he arrived at the idea of camouflaging ships by dazzle painting: "...that since it was impossible to paint a ship so that she could not be seen by a submarine, the extreme opposite was the answer—in other words, to paint her, not for low visibility, but in such a way as to break up her form and thus confuse a submarine officer as to the course on which she was heading." His proposal was adopted, and he was placed in charge of a Dazzle Section, in which he and his assistants made painted foot-long models of dazzle painted merchant ships, using studios at the Royal College of Art. The models were used for testing, in a laboratory-like setting, the degree to which a dazzle scheme might prevent an experienced naval officer from judging the course of the vessel.

Effective schemes were then applied to hand-colored construction plans, and used by various dock officers in supervising the painting of the actual ships. One of these dock officers, assigned to Liverpool, was Vorticist painter Edward Wadsworth.

In early 1918, Wilkinson was asked to spend a month in the US, to serve as an advisor to the US Navy in connection with establishing a unit for the dazzle painting of merchant ships, which eventually resulted in the painting of about 1,100 US merchant ships. Accompanied by his wife (functioning as his secretary), he sailed to the US on the *USS Leviathan*. Shortly after arriving, he met with Franklin D. Roosevelt, Assistant Secretary of the Navy. Wilkinson (1969:91): "In the course of developing the system I had to make a tour of all the major United States Dockyards, to deliver a lecture at each of them on my system, its objects, how it worked and was to be applied." During his US visit, the naval officer who functioned as his escort was Lieutenant Everett L. Warner, who was subsequently put in charge of the design division of the US ship camouflage section.

After World War I, Wilkinson applied for the Royal Commission of Awards to Inventors for formal acknowledgment of his having originated dazzle painting. Two other applicants claimed that they had invented the method, so a legal hearing became necessary, at the end of which Wilkinson was officially declared the inventor and given a monetary award. It was Wilkinson who wrote the article on Naval Camouflage for the 12th edition of the *Encyclopedia Britannica* (1922). In Goodden (2007:13), there is also a plan for a dazzle-painted World War I-era airplane, titled "Dazzle Design for Aeroplane Scout," signed by Wilkinson.

During World War II, he served the British Air Ministry as Inspector of Camouflage under the Directorate of Works, for which he was responsible not for ship camouflage, but for the concealment of airfields.

Warner (1919:216): "Lt. Commander Norman Wilkinson RNVR visited this country for a month [in 1918], and gave us the benefit of the British experience in ship camouflage. I was in close touch with him during that time, and he showed us all of the tricks and devices of pattern that they had found successful in creating course distortion. We adopted many of them and used them in our designs, but it was not then apparent to us or to Lt. Commander Wilkinson, that there was any general law by which all of these devices could be explained."

•

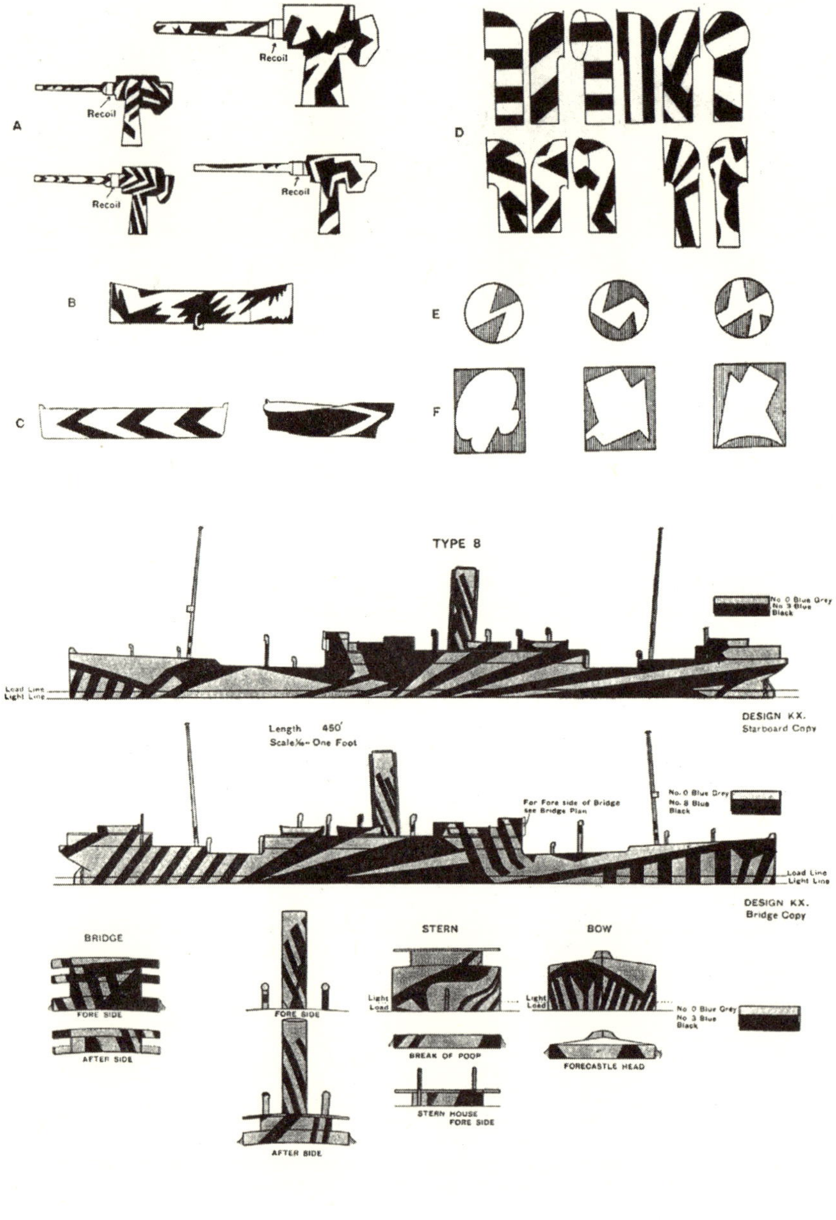
A
Recoil
Recoil
Recoil
Recoil
D
B
E
C
F
TYPE 8
No 0 Blue Grey
No 3 Blue
Black
Load Line
Light Line
DESIGN KX.
Starboard Copy
Length 450'
For Fore side of Bridge see Bridge Plan
No. 0 Blue Grey
No. 8 Blue
Black
Load Line
Light Line
DESIGN KX.
Bridge Copy
BRIDGE
FORE SIDE
AFTER SIDE
FORE SIDE
AFTER SIDE
STERN
Light
Load
BREAK OF POOP
STERN HOUSE
FORE SIDE
BOW
Light
Load
FORECASTLE HEAD
No 0 Blue Grey
No 3 Blue
Black

Louis Willaume
(May 31, 1874-1949)
French painter, printmaker

Willaume was a painter of urban landscapes, and a book illustrator. A student of William-Adophe Bouguereau, he is listed in Coutin (1988) as a World War I French camoufleur.

•

Chili Williams [Marian Sorenson Uhlman]
(1922-October 17, 2003)
American actress and model

Chilli Williams was a World War II model featured in *Life* magazine on September 27, 1943. Known as "The Polka Dot Girl," she posed for a series of pin-up photographs that were used as "visual aids" in lectures on the principles of camouflage at Fort Belvoir VA. *Life* received 100,000 requests for her photograph after she appeared in that issue. In 1944, she signed a Hollywood motion picture contract and was subsequently featured in 17 films, among them *Girl Rush* (1944), *Johnny Angel* (1945), *Wonder Man* (1945) and *Having a Wonderful Crime* (1945).

•

Keith S[haw] Williams
(1906-February 11, 1951)
American painter and printmaker

Born in Marquette MI, Williams studied with Ivan Olinsky, Charles Hawthorne and Karl Anderson at the National Academy School, then moved to Paris, where he continued his training at the Académie de La Grand Chaumiere and the Callorassie Académie. He was appointed head of the Fine Arts Department at the Brooklyn Central Branch of the YMCA, and in 1942, was designated a member of the National Academy of Art and Design. *New York Times* ("Keith S. Williams" 1951): "During the second World War, the artist was a sergeant in the camouflage division of the Army."

•

Facing page An example of the extensive diagrams that were provided to dock officers, who supervised the painters in applying ship camouflage. This World War I example is by British camoufleur Norman Wilkinson and was published by him in his article on camouflage in the 12th edition of the *Encyclopedia Britannica* (Wilkinson 1922).

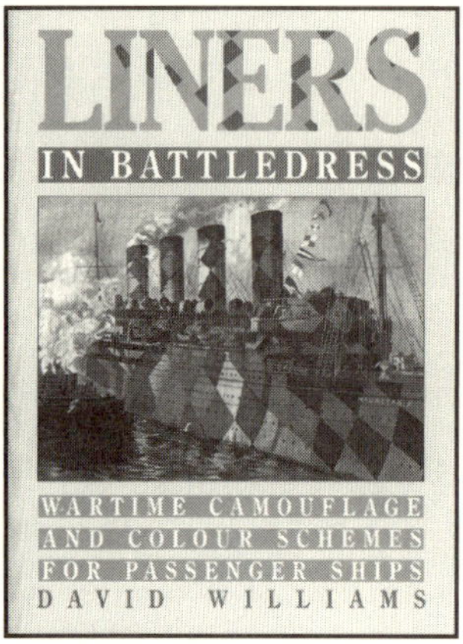

Above Dust jacket of David Williams' *Liners in Battledress: Wartime Camouflage and Color Schemes for Passenger Ships* (Williams 1989). In addition, see his more recent, excellent book *Naval Camouflage 1914-1945: A Complete Visual Reference* (Williams 2002).

J(ames Earnest) Elder Wills
(April 26, 1900-1970)
British filmmaker

Born in London. During World War II, he served in the British Army, in Special Operations, for which he was in charge of an undercover unit that developed camouflage, encoding equipment, and concealed bombs (in milk bottles, pens, cigarettes, and in simulated animal dung). The operation's headquarters were in a two-story mock-Tudor hotel in London called the Thatched Barn, which later became the Playboy Club. See Stevenson (1976).

•

Ezra Winter
(March 10, 1886-April 8, 1949)
American mural painter

Born in Manistee MI, Winter studied at the Art Institute of Chicago (1908-09), then in Italy, after winning an American Academy in Rome scholarship in 1911. He is particularly known for a large mural, titled *Fountain of Youth*, in the foyer of the Radio City Music Hall. Other works include aspects of the Cotton Exchange and the Cunard Building (with fellow camoufleurs Barry Faulkner and John Gregory) in New York, Eastman Theatre in Rochester NY (with Faulkner), George Rogers Clark Memorial in Vincennes IN, the Federal Reserve Building, and the Library of Congress in Washington DC. He committed suicide. *New York Times* ("Winter, Muralist, Takes Own Life" 1949): "During World War I he designed camouflage for the US Shipping Board."

•

Above Cover of the *Industrial Camouflage Manual* prepared during World War II by Konrad F. Wittman at Pratt Institute's camouflage laboratory in the Department of Architecture (Wittman 1942). The volume was dedicated to Homer Saint-Gaudens. Author's collection.

Konrad F. Wittman
(1889-April 16, 1951)
American interior designer

Born in Augsburg, Germany, he studied at the Technische Hochschule in Munich. After graduating from the University of Munich in 1917, he worked

Above Further drawings by Russian designer Valeria Nekaeva (c2005) of costumes for a theatrical interpretation of *The Nose* by Nikolai Gogol. Courtesy the artist.

as an architect in Hanover, Germany, from 1919-1938. He immigrated to the US in 1939, and joined the faculty of Pratt Institute in 1941. *New York Times* ("Prof K.F. Wittman" 1951): "In 1942, Professor Wittman wrote the industrial camouflage manual for the Office of Civilian Defense."

•

Van Horn D. Wolfe
(1891-1918)
American artist

Born in Chatham NY, he studied architecture at the University of Pennsylvania, and is listed in Embury (1917:138) as a member "of Walker & Gillette's office." During World War I, he was a sergeant in the American Camouflage Corps, in the course of which he died from wounds. *New York Times* ("Lieut. D.I. Reynolds Is Killed in Italy" 1918): "Sergeant Wolfe, who was an art student when war was declared, was a member of the camouflage company commanded by Lieutenant Homer Saint-Gaudens."

•

Grant Wood
(1891-1918)
American artist

Julian Trevelyan (ibid.:130): "[During World War II, while lecturing to soldiers on camouflage] Now and then I would pop in slides of nude girls under camouflage nets to wake the men up when they had dropped off."

Above Iowa artist and WWI camoufleur Grant Wood, as photographed by Peter A. Juley & Son (n.d.). Courtesy Photograph Archives, Smithsonian American Art Museum (J0002298).

Facing page Selected examples of the use of stripes for camouflage (clockwise from top): Zebra-like camouflage as applied to the French cruiser *Gloire*. Courtesy US Naval Historical Center (NH 60543). *right* A photograph from Thayer (1909), in which a stuffed zebra (Is this the one he placed near the bridle path in Central Park?) has been photographed in a thicket, with a donkey on the right, all this supposedly from the viewpoint of a crouching predator. *left* Two views of zebra-like stripe applications to US Navy Patrol Torpedo boats, called "PT boats" (c1944). US Navy photographs.

Born in Anamosa IA, Wood served briefly in the US Army during World War I. Garwood (1944:59): "Grant's stay at Camp Dodge promised to be pleasant. He was excused from the more onerous military duties, given a sidecar motorcycle, and told to go about making historic records. Then he contracted anthrax, and nearly died. When he recovered, he was sent to Washington to do camouflage work. He served as a private in Company B, Regiment 97, United States Engineers, and was in charge of the paint tent. He made clay models of field-gun positions and helped camouflage artillery pieces. Before-and-after photographs of a cannon he helped camouflage were on display for several years at the Smithsonian Institution."

•

Allan Wood-Thomas
(January 12, 1920-c1976)
American artist

He was born in Paris, where his parents were students at the École des Beaux Arts. His father was an architect, and his French mother was a sculptress. Later, when the family moved to the US, he completed two years of architectural study at Princeton University, at which time he withdrew and moved to New York. During World War II, Wood-Thomas served in the Ghost Army, and is mentioned and quoted in Goossen (1973:13). Following the war, he continued to live in Europe, but, unable to make a living as an artist, he moved back to the US. Shortly after age 55, he died of Lou Gerhig's disease or ALS.

•

William (Henry Kemble) Yarrow
(September 24, 1891-1941)
American painter

Born in Glenside PA, Yarrow studied art at the Pennsylvania Academy of Fine Arts, and in Paris at the Académie Grand Chaumiere, Colorossi.

Charles Burchfield (University of Arizona Art Gallery 1965:24): "[During his service in World War I] The camouflage corps [at Camp Jackson] was under the command of 1st Lt. William Yarrow, in private life a portrait painter from

Philadelphia. After the war's end he did murals for Princeton University. Lt. Yarrow can only be described as an 'officer and a gentleman.' In addition, he was sensitive artist...

Shortly after I joined the corps, Lt. Yarrow interviewed all of us in order to determine the sort of work for which we were fitted. When I told him that I admired the work of Cézanne very much, he turned to his aide and said, 'See! Here's another artist who likes Cézanne'."

•

John (Erskine Joseph) Yunge-Bateman
(1897-September 16, 1971)
British book illustrator

He joined the Royal Navy in 1910, and retired as a Lieutenant-Commander in 1926, then retired again with the rank of Commander in 1937. He apparently returned to service a third time, because Hartcup (1980) notes that, during World War II, he was in charge of a naval camouflage section, and that, previously, he had worked with L.M. Glasson. As a book illustrator, he was active after World War II until the 1960s. Two of the books he illustrated were published by the Golden Cockerel Press.

•

Zébrage
French slang for disruptive patterning

Kahn (1984:18): "This technique was called *zébrage*, the multicolor painting of patterns." While promoted by camoufleur Louis Abel-Truchet, it was largely opposed by French non-artists.

Warner does not use the French term, but he does mention the increased use of zebra-like patterns (high contrast black and white stripes) for the camouflage on Allied ships near the end of World War I.

•

Rudyard Kipling (*How the Leopard Got His Spots*): "I can smell Zebra and I can hear Zebra, but I can't *see* Zebra."

...

Norman Wilkinson ((1919): "Some time before the end of the war [WWI] we had arrived at the striped type of design [for ship camouflage] which was the most successful type. These striped designs were commented on by a great number of seamen as being by far the best for upsetting the calculation of a ship's course."

...

Annie Dillard (1974:18): "Nature is like one of those line drawings of a tree that are puzzles for children: Can you find hidden in the leaves a duck, a house, a boy, a bucket, a zebra, and a boot?"

...

Rudolf Arnheim (1989): "Nothing is more humbling than to look with a strong magnifying glass at an insect so tiny that the naked eye sees only the barest speck and to discover that nevertheless it is sculpted and articulated and striped with the same care and imagination as a zebra. Apparently it does not matter to nature whether or not a creature is within our range of vision, and the suspicion arises that even the zebra was not designed for our benefit."

Camouflage

They tell us tales of camouflage,
The art of hiding things;
Of painted forts and bowered guns
Invisible to wings.
    Well, it's nothing new to us,
    To us, the rank and file;
    We understand this camouflage
    —We left home with a smile.

We saw the painted battleships
And earthen-colored trains,
And planes the hue of leaden skies,
And canvas-hidden lanes.
    Well, we used the magic art
    That day of anxious fears;
    We understand this camouflage
    —We laughed away your tears.

They say that scientific men
And artists of renown
Debated long on camouflage
Before they got it down.
    Well, it came right off to us,
    We didn't have to learn;
    We understand this camouflage
    —We said we'd soon return.

We understand this camouflage,
This art of hiding things;
It's what's behind a soldier's jokes;
And all the songs he sings.
    Yes, it's nothing new to us,
    To us, the rank and file;
    We understand this camouflage
    —We left home with a smile.

—Anonymous (c1918)*

*Written by a World War I American soldier and originally published in *Stars and Stripes*. On the *facing page* is an illustrated feature from *War of the Nations* (1919).

Camouflage of Ships to Deceive the Submarines

THE ZEBRA STRIPED BRITISH TRANSPORT OSTERLEY AS SHE APPEARED IN NEW YORK HARBOR NOV. 11, 1918. SHE WAS DECKED WITH FLAGS IN CELEBRATION OF THE ARMISTICE SIGNED ON THAT DAY.
(© Underwood & Underwood.)

A MYSTIFYING CAMOUFLAGE WHICH WOULD MAKE IT VERY DIFFICULT FOR AN UNDERSEA PIRATE TO TORPEDO THIS SHIP. (© Underwood & Underwood.)

A BRITISH SHIP WHOSE BEWILDERING MARKINGS HAVE ENABLED IT ON MANY TRIPS TO BAFFLE THE SUBMARINES. (© Underwood & Underwood.)

other camoufleurs

In the mid-1960s, when I began to research art and camouflage, one of the questions I commonly asked was "Who was involved in camouflage?" Now, more than forty years later, perhaps a more sensible question would be "Who was *not* involved in camouflage?"

As I struggled to complete this book, it became apparent that, as long as I allowed it, I would continue to find new sources and new camoufleurs, in which case this would never end. In the final weeks of working on it, I repeatedly redesigned sections in order to insert new material in sequence.

I finally agreed to a "cease fire," in the process of which I established this new category of belated findings. From that point on, this is where any new entries were placed. Are these seeds for a second edition? That was certainly not my intention. Instead,these are provided as potential research opportunities for younger scholars, in the hope that those researchers might discover more about these people, along with dozens of others in the book about whom complete information is lacking.—RB.

Michel Acard
(dates undetermined) French sculptor. Listed in Coutin (1988) as a World War I French camoufleur.

Eleanor Arnett
(?-1985) American artist. Studied at the Pennsylvania Academy of Fine Arts, and with Howard Pyle. During World War I, she served in the Camouflage Corps.

G.F. Axt
(dates undetermined) American architect. Listed in Embury (1917).

Georges Barriere
(1881-1944) French painter. Born in Chablis, France. Listed in Coutin (1988).

Franklin A. Bates
(?-1974) American painter. A native of Lodi

Facing page American artist Maurice Freedman was a ship camoufleur for the US Shipping Board during World War I. Following the war, he gave his alma mater (Rhode Island School of Design) 455 lithographic plans for camouflaging US merchant ships, and twenty black-and-white ship camouflage photographs. Shown here are two photographs of the *SS Everglades* (c1918) at the Oscar Daniels Shipyard in Tampa FL. Courtesy the Fleet Library at RISD.

OH, Bates served as a camoufleur with the US Navy during World War II.

Frank C. Bates
(dates undetermined) Unknown. In Roskam (1987:49) someone of this name is cited as having contributed to US Navy camouflage during World War I. His solution is said to have come from an old magic trick in which two mirrors are arranged in a V-shape, resulting only in a reflection of the water. Oddly, it may have been all but identical to US Patent No. 1,310,533 titled "Means for Screening or Obscuring Ships," invented by Isidor Karton (1919), reproduced on page 320.

Paul Béat
(1874-1945) French painter. Born in Lille, France. Listed in Coutin (1988).

René Bentz *(dates undetermined) French painter.* Listed in Coutin (1988).

Jane Berlandina
(dates undetermined) French-born American painter, muralist and scenic designer. Born in Nice, France. Studied with Gustave Moreau and Raoul Dufy. Came to the US in 1928, and married Henry Howard. Lived in New York and San Francisco. Designed sets and costumes for San Francisco Opera Association. Designed camouflage during World War II (1942-1945).

André Berthon
(dates undetermined) French painter. Born in Paris, student of Gustave Moreau and Jules Lefebvre. Listed in Coutin (1988).

Victor Boucher
(August 24, 1877-February 21, 1942) French film actor. Served in a World War I French camouflage unit. Source unknown.

Léon-Émile Bouchet
(1880-1940) French furniture and interior designer. Listed in Coutin (1988).

Henri Boudrie
(dates undetermined) French painter. Listed in Coutin (1988).

Louis Bouillot
(dates undetermined) French painter. Listed in Listed in Coutin (1988).

Paul Bourgeois
(dates undetermined) French draughtsman. Listed in Coutin (1988).

Henri-Georges Bréard
(1873-?) French painter. Born in Paris, his paintings were often concerned with the legal profession. Listed in Coutin (1988).

Jacques Brissaud
(1880-1960) French illustrator. Born in Paris, he was the brother of Pierre Brissaud. Listed in Coutin (1988).

Jean-Louis Brown
(?-1930) French painter, sculptor. Listed in Coutin (1988), he preferred equestrian subjects.

Pierre Caen
(dates undetermined) French painter. Born in St. Denis, France. Listed in Coutin (1988).

Jacques Cancaret
(?-1932) French painter. Born in Clessy, he was a student of William-Adophe Bouguereau at the Académie Julian. Listed in Coutin (1988).

Baptiste Carlotti
(dates undetermined) French painter. Listed in Coutin (1988).

H.S. Chandler
(dates undetermined) American scientist(?) During World War I, he worked for the New York District of the US Shipping Board, where he was an assistant marine camoufleur. May have been an MIT graduate.

Jean Chaperon
(dates undetermined) French painter, illustrator.

Born in Paris, Chaperon was a draughtsman and humorist. Listed in Coutin (1988).

Raymond (Louis) Charmaison
(1879-1955) French landscape painter. Primarily known for his urban landscapes. Listed in Coutin (1988).

Georges Chepfer
(dates undetermined) French painter. Listed in Coutin (1988).

H. Cillard
(full name and dates undetermined) French artist. Listed in Coutin (1988).

Marius-Léon Cladel
(April 15, 1883-January 1948) French painter, sculptor. A student of Émile Antoine Bourdelle. Listed in Coutin (1988).

Walter C. Clifford
(dates undetermined) American architect. Embury (1917).

Kenneth J. Conant
(dates undetermined) American soldier. A Harvard graduate (1915), he was wounded while serving in France in the Camouflage Section of the 40th Engineers in 1918.

P. Damilot
(dates undetermined) French painter. Listed in Coutin (1988).

Georges-Édouard Darcy
(1883-?) French painter. Listed in Coutin (1988).

Fernand Daviot
(dates undetermined) French draughtsman. Listed in Coutin (1988).

Richard Day
(1896-1972) Canadian-born American film set designer. Day planned to be an artist, but after moving to Hollywood in 1920, he became a leading art director for motion pictures. Forty of his films were nominated for Oscars, among them *A Street Car Named Desire*, *On the Waterfront* and *Dodsworth*. In the 1930s, he continued his earlier interest in lithographic printmaking, and was featured in a book by Merle Armitage titled *The Lithographs of Richard Day* (1932). Day had served in the Canadian army during World War I, but in the papers of (Thayer student) Rockwell Kent in the Archives of American Art, there is correspondence dated 1942-1943, talking about "Day's work as a camouflage instructor for the US Navy."

Lockwood de Forest, Jr.
(1896-1949) American landscape designer. In Karson (2007), he is described as having served as a volunteer in the Camouflage Divsion of the US Air Corps ifrom 1942 to 1944. See Wright Ludington.

Paul-Léon Delacroix
(dates undetermined) French painter. Listed in Coutin (1988).

Marcel Delaunay
(1867-1959) French painter. A landscape painter born in Rouen. Listed in Coutin (1988).

Alexis Demarle
(1872-?) French painter, printmaker. Listed in Coutin (1988).

Auguste-Théodore Desch
(February 28, 1877-1924) French painter. Listed in Coutin (1988).

Frédéric-Léon Deshayes
(December 30, 1883-1970) French painter. Listed in Coutin (1988).

Dewer
(Full name and dates undetermined) American artist? This may be Douglas Duer. Mentioned as Dewer in *Literary Digest* ("Company A" 1917).

G. Dexter
(dates undetermined) American architect. Embury (1917).

Serge Dorizon
(dates undetermined) French painter. Listed in Coutin (1988).

Fernand Durozé
(dates undetermined) French painter. Listed in Coutin (1988).

John H. Eastman
(dates undetermined) American architect. Embury (1917).

Michael M. Engle, Sr.
(1896-1969) American painter, collector. Director of publicity for M. Grumbacher. Married to artist and author Mary Black Diller. His papers in the Archives of American Art include files on camouflage.

Richmond Knapp Fletcher
(1885-1965) American architect. A Harvard graduate (1908), he designed four gates on that campus. He was both a practicing architect and a portrait painter. During World War I, he painted ships for the US Shipping Board.

Bruno Funaro
(dates undetermined) American architect. His widow was Viola Peters Funaro (1900-November 30, 1999), a vocal coach, pianist and faculty member at the Julliard School of Music. Source unknown.

(Pierre) Arthur Gaillard
(dates undetermined) French painter. Born in Chaumont, France, he studied with Jean-Léon Gérome. Listed in Coutin (1988).

Émile Gallois
(October 16, 1882-February 28, 1965) French painter, illustrator. Described as a Neo-Impressionist. Listed in Coutin (1988).

Robert-Pierre Gaudefroy
(dates undetermined) French painter. He appears to have had some connection with theatre designer and camoufleur Joseph Pinchon. Listed in Coutin (1988).

G. Gautier
(full name and dates undetermined) French artist. Listed in Coutin (1988).

Paul-Jules Gérard
(dates undetermined) French painter. Listed in Coutin (1988).

André Germain
(1881-1971) French writer and editor. Listed in Coutin (1988).

Ludovic Gignoux
(dates undetermined) French painter. Listed in Coutin (1988).

Anne Furman Goldsmith
(October 16, 1890-?) American artist. Originally from Salem MA, she was the commander of the first Camouflage Corps of the National League for Woman's Service during World War I. James (1921).

Frederick Gore
(November 8, 1913-) British painter. A prominent artist, he served as a camouflage officer in World War II.

Arthur-Edmond Guillez
(February 8, 1885-1916) French printmaker. Born in Valenciennes, France, he died during World War I. Listed in Coutin (1988).

Raoul-Marie-Jean-Frédéric Guinet
(dates undetermined) French draughtsman. Listed in Coutin (1988).

José Hamman
(dates undetermined) French illustrator. He illustrated the works of Honoré Balzac. Listed in Coutin (1988).

William J(oseph) Hammer
(February 26, 1858-March 24, 1934) American inventor. He was a consulting electrical engineer who became a laboratory assistant to Thomas A. Edison in December 1879. He worked primarily on the incandescent electric light. During World War I, he was

chairman of the camouflage committee of the Aeronautical Society of America.

Myrta M. Hanford
(dates undetermined). American government official. A New York State government worker, she was the commander of the second Camouflage Corps of the National League for Woman's Service during World War I. See James (1921).

Perkins Harnly
(1901-1986) American artist. He taught camouflage at Fort Belvoir during World War II, as mentioned in an oral history interview at the Archives of American Art.

I.B. Hazelton
(dates undetermined) American scientist(?) During World War I, he worked for the New York District of the US Shipping Board, where he supervised ship camouflage painting, and engaged in camouflage research. May have been an MIT graduate.

Oscar Howe
(1915-1963) Native American painter. During WWII, he served for three years in North Africa where"he painted camouflage or 'dazzle painting' on military equipment." For many years, he was on the faculty at the University of South Dakota at Vermillion. See Anthes (2006).

Arthur Howland
(dates undetermined) American color expert. A graduate of MIT (1891), he was associated with Wadsworth, Howland and Company, during which he studied color and color relationships. During World War I, he worked on ship camouflage.

Clifford C. Jones
(dates undetermined) American architect. See Embury (1917).

Henri Joubert
(1873-?) French painter. Listed in Coutin (1988).

Louis Jourdan
(March 7, 1872-May 3, 1948) French painter. Born in Bourg-en-Bresse, France. Listed in Coutin (1988).

Oliver Larson
(dates undetermined) American architect. See Embury (1917).

G. Lavignac
(dates undetermined) French set designer. Listed in Coutin (1988).

Léon Leclerc
(dates undetermined) French painter. Listed in Coutin (1988).

Paul Émile Leconte
(1877-1950) French landscape painter. Listed in Coutin (1988).

Le Gourvil-Gautier
(dates undetermined) French artist. Listed in Coutin (1988).

Henri Pierre Lejeune
(June 27, 1881-?) French painter. Listed in Coutin (1988).

Alfred Marie Le Petit
(1876-1953) French painter, printmaker. Listed in Coutin (1988).

Raoul Lespagne
(dates undetermined) French painter. Listed in Coutin (1988).

Ernest-Louis Lessieux
(1848-1925) French painter. He was 66 years old when the war began. Listed in Coutin (1988).

Alfred Lévy
(dates undetermined) French painter. Listed in Coutin (1988).

Lucien Lievre
(May 27, 1878-1936) French painter, designer. Listed in Coutin (1988).

Pierre Llorens
(dates undetermined) French painter. Listed in Coutin (1988).

Wright S. Ludington
(1901-1992) American painter, art collector. According to Karson (2007), he "worked at Warner Brothers Studio during World War II preparing camouflage plans for airplane plants, and from 1942 to 1946 taught camouflage..." He was the owner of Val Verde in Santa Barbara CA. See Seldis in Lipman (1971).

Alexander MacLean
(dates undetermined) American architect. Embury (1917).

Alfred de Maghellen
(July 28, 1871-?) French painter. Born in Paris, he was a student of Jean-Paul Laurens and Benjamin-Constant. Listed in Coutin (1988).

Henri-Marcel Magne
(November 11, 1877-1946) French architectural scholar, painter and sculptor. Born in Paris, Magne was a professor at the Conservatoire des Arts et Métiers. Listed in Coutin (1988).

Charles Magnier
(December 7, 1883-January 23, 1950) French painter. Listed in Coutin (1988).

Fernand Maissen
(May 8, 1873-?) French painter. Listed in Coutin (1988).

Louis-Ferdinand Malespina
(June 21, 1874-1940) French painter. As an artist, he was primarly known for his paintings of horses and horse racing. Listed in Coutin (1988).

Henri Manset
(dates undetermined) French painter. Listed in Coutin (1988).

Marsier
(full name and dates undetermined) French artist. Listed in Coutin (1988).

Edmond Mazelly
(dates undetermined) French painter. Listed in Coutin (1988).

Charles-Louis Menneret
(1876-?) French painter. Listed in Coutin (1988).

Morainvillieurs
(full name and dates undetermined) French camoufleur. He apparently served during World War I. Source unknown.

Paul-Bernard Morchain
(1876-1939) French painter. Born in Rochefort sur Mer. Listed in Coutin (1988).

Rowley (Walter) Murphy
(May 28, 1891-1975) Canadian artist. Camoufleur in the Royal Canadian Navy during World War II.

Edwin Fairfax Naulty
(March 16, 1869-?) American engineer, inventor. According to Roskam (1987:48), he offered prosposals for World War I ship camouflage.

Robert Noël
(dates undetermined) French painter, printmaker. Listed in Coutin (1988).

Henry Allen Nord
(1904-1981) American mosaic artist. Archives of American Art (1995:116): In a 1965 interview, he "speaks of working in the Army camouflage unit designing three-dimensional maps."

Georges Offner
(dates undetermined) French painter. Listed in Coutin (1988).

Frederick Law Omsted, Jr.
(July 24, 1870-December 25, 1957) American landscape architect. He served as a camoufleur during World War I under the Quartermaster Corps.

Léon Pastereau
(dates undetermined) French painter. Listed in Coutin (1988) .

Paul Pauchot
(dates undetermined) French painter. Listed in Coutin (1988).

Frederick (Alexander) Pawla
(September 6, 1876-December 18, 1964) British-born American painter. In Crowell (1921:506), he is described as "head of the camouflage section of the Embarkation Service" and "one of the most valuable men in the Government for this sort of work. He camouflaged many of the army transports, particularly cargo carriers."

Auguste Peltier
(dates undetermined) French designer. Listed in Coutin (1988).

Pérille
(full name and dates undetermined) French painter. Listed in Coutin (1988).

Albert Petitclerc
(dates undetermined) French bronze caster. Listed in Coutin (1988).

André Petit-Jean
(dates undetermined) French painter. Listed in Coutin (1988).

John C. (Jack) Phillips, Jr.
(December 24, 1908-?) American artist and architect. Descended from Wendell Phillips and the founders of the Phillips Academy. His father knew Frank Benson and Louis Agassiz Fuertes, and used their illustrations in *A Natural History of Ducks*. Phillips, Jr., painted ship camouflage at the Brooklyn Navy Yard during WWII. See www.ccmht.org/phillips.html, as accessed on December 20, 2008.

Eugéne Pomilly
(dates undetermined) French painter. Listed in Coutin (1988).

Albert Pommier
(January 11, 1880-1944) French sculptor. Coutin (1988) and Ducasse (1962).

Loys Joseph Pratt
(1879-?) French painter. Listed in Coutin (1988).

Henri Privat
(dates undetermined) French painter, medalist. Listed in Coutin (1988).

Pierre Prunier
(January 29, 1875-?) French painter. Coutin (1988) and Ducasse (1962).

David Pye
(1914-1993) British designer and teacher. Author of *The Nature and Aesthetics of Design*, he served as a camoufleur for the Royal Navy during World War II.

Antoine Racadot
(dates undetermined) French painter. Listed in Coutin (1988).

Frédéric Émile Jean-Baptiste Ragot
(December 7, 1872-?) French painter, sculptor. A student of Jules Joseph Lefebvre at the Académie Julian. Listed in Coutin (1988).

Thomas I. Raguere
(dates undetermined) American architect. Embury (1917.

René Henri Ravaux
(dates undetermined) French painter. Born in Paris, he studied with Jean-Paul Laurens at the École des Beaux Arts. Listed in Coutin (1988).

Jesse Reichek
(August 16, 1916-July 18, 2005) American artist, designer. Professor of Design at UC-Berkeley. Student of Moholy-Nagy at the New Bauhaus in Chicago. Taught there and at Armor Institute and IIT. During WWII, served as a camouflage instructor at air bases in France with the US Army Air Corps. See obituary in *Point Reyes Light* (July 28, 2005).

Eugène Renain
(dates undetermined) French painter. He was a designer for the Paris Opera. Listed in Coutin (1988).

Maurice Renders
(November 13, 1877-March 16, 1951) French painter, miniaturist. A student of Édouard Vimont. Listed in Coutin (1988).

F. Reuflet
(full name and dates undetermined) French painter. A painter of Parisian landscapes and still-lifes. Listed in Coutin (1988).

Guillaume Rianni
(dates undetermined) French painter. Listed in Coutin (1988).

Reah deBourg Robinson
(October 29, 1889-?) American architect. Listed in Embury (1917:138) as "from Wilmington."

René Rochette
(?-1920) French painter. He died from war-related injuries. Listed in Coutin (1988).

Charlotte Rothstein Ross
(dates undetermined) American artist. In the abstract of a 1965 interview (Archives of American Art 1995:132), she is described as having contributed to "the Index of American Design, where she researched the history of camouflage in nature..."

Jean-Baptiste Rousselet
(dates undetermined) French theatre designer. Listed in Coutin (1988).

Pierre-Laurent Roustan
(October 11, 1877-?) French sculptor, designer. Born in Toulouse. Listed in Coutin (1988).

Joseph-Alexandre Ruelland
(1864-?) French painter. A student of Jean-Paul Laurens, he was fifty years old when World War I began. Listed in Coutin (1988).

Henri Paul Royer
(January 22, 1869-1938) French painter. Born in Nancy, he studied at the École des Beaux-Arts and the Académie Julian. He was awarded the Legion of Honor. Listed in Coutin (1988).

Prentice Sanger
(dates undetermined) American architect. Embury (1917).

Albert-Camille Seraphin
(dates undetermined) French sculptor. Listed in Coutin (1988).

Marie-Georges Seveau
(dates undetermined) French painter. An urban landscape painter who was born in Poitiers. Listed in Coutin (1988).

Severin
(dates undetermined) French mime. An important, well-known mime, he served as a World War I French camoufleur. Source unknown.

Thomas E. Seyster
(dates undetermined) American architect. He is listed in Embury (1917:138) as "a Princeton man who has come from Chicago."

André Sieffert
(dates undetermined) French painter. Listed in Coutin (1988).

V.P. Spalding
(full name and dates undetermined) American architect. He is listed in Embury (1917:138) as a World War I American camoufleur "who gave up his office to enlist."

Harold Sterner
(dates undetermined) American scientist(?) During World War I, he worked for the 3rd Naval District, where he made drawings of the camouflage of ships entering New York harbor, then forwarded these to Washington DC. May have been an MIT graduate.

Maurice Taquoy
(1878-1952) French painter, illustrator. He was primarily known as a painter of horses and sporting scenes. Listed in Coutin (1988).

Julien-Louis Tavernier
(August 22, 1879-?) French painter. A student of Léon Bonnat at the École des Beaux-Arts. Listed in Coutin (1988).

Auguste-Henri Thomas
(dates undetermined) French painter. Listed in Coutin (1988).

Georges Henri Tribout
(1884-1962) French artist, draughtsman. His primary interest was in religious subjects. Listed in Coutin (1988).

Antony Troncet
(May 23, 1879-1939) French painter. Troncet was a student of Jean-Paul Laurens and Jean Joseph Benjamin-Constant at the Académie Julian in Paris. He was awarded the Legion of Honor. Listed in Coutin (1988).

Tubesing
(full name and dates undetermined) American camoufleur. He is mentioned in *Literary Digest* ("Company A" 1917).

Turgy
(full name and dates undetermined) French painter. Listed in Coutin (1988).

Vallée
(dates undetermined) French illustrator. Listed in Coutin (1988).

Verhoeven
(full name and dates undetermined) Belgian camoufleur. In the autobiography of American painter and writer Hilaire Hiler, in the Archives of American Art, there is mention of Hiler having "worked briefly with the Belgian camouflage expert Verhoeven in Providence [RI]."

Paul Verlet
(dates undetermined) French sculptor. Coutin (1988) and Ducasse (1962).

Sheldon Viele
(dates undetermined) American architect. Embury (1917).

Derek Von Berg
(dates undetermined) South African architect. Listed in Reit as working with Barkas in WWII (p. 138).

Louis F. Voorhees
(dates undetermined) American architect. Embury (1917) also mentions a Nelson Voorhees, described as "a University of Michigan man" (p. 138).

R. de Walstrom
(full name and dates undetermined) French painter. Listed in Coutin (1988).

Jacob Weidemann
(1923-2001) Norwegian painter. He produced dazzle-like surface designs for Norwegian merchant ships in the 1950s.

Austin Whittlesey
(dates undetermined) American architect. Listed in Embury (1917:138) as "also from [B.G.] Goodhue's, who was Le Brun scholar last year [1916]."

Wielhorski
(full name and dates undetermined) French illustrator. Listed in Coutin (1988).

James R. Wilson
(dates undetermined) American architect. Embury (1917).

Beckford Young
(1905-1979) American art administrator. In the abstract of a 1965 interview (Archives of American Art 1995:173), he is described as having overseen "the camouflage for the Navy Air Base at [the] Oakland airport" [during WWII].

(James) Milford Zornes
(January 25, 1908-February 24, 2008) American artist. In a 1965 interview (see Archives of American Art 1995:175), Zornes mentioned having served with a US Army camouflage battalion, before being reassigned, during World War II.

afterthoughts

I first published an article on art and camouflage in 1970. I had written it the year before while teaching art at a high school. It described a problem I had given to my students in which I asked them to "paint out" a small object attached to a background, to make the object disappear.

Ironically, by the time the article came out, I was no longer teaching. It was the Vietnam War, and, unwittingly, I was a soldier instead. I had been drafted out of the classroom one month before the school year's end, and, at the induction center, I and a handful of others were diverted to the Marine Corps, which needed to strengthen its numbers that month. After the initial shock of boot camp, I was assigned to be a legal clerk and court reporter, and was eventually promoted to the rank of Sergeant. But I never grew accustomed to the strangeness of military life, and the days that I served were a challenge.

How did I become interested in camouflage, if not through the military? I don't remember exactly, but I do know that it started while I was still an undergraduate art student. I also recall that it came from my involvements in writing and painting, and that it partly began with a need to find out what qualities, if any, are indispensible in art. I concluded that excellence doesn't depend on ones choice of subject matter, style or medium. So I turned my attention to vision instead. How do human beings see? What aspects of vision are global? How does vision differ among cultural groups, or between individuals? Does the vision of artists differ substantially from that of scientists? And so on.

Facing page Daniel Weiss, *Visitation Nighthawk* (2008). 15.5h x 9.5w x 2d in. Wood, paint, cardboard, nails. Collection of Roy R. and Mary Snyder Behrens. One wonders if Abbott H. Thayer would have admired this artwork. Inspired by the coloration of a nighthawk, it is oddly reminiscent of Thayer's concept of "background picturing." For full-color views of this and other assemblages by the same artist, go online to <www.bobolinkbooks.com/Iowa/dwhome.html>.

In the process of asking these questions, I somehow inadvertently struck a rich vein of provocative writings, the single most important of which may have been a collection of essays by scholars in such diverse disciplines as biology, biochemistry, astronomy, physics, botany, art history, embryology, brain physiology, ethology, zoology, and perceptual psychology. Edited by Lancelot Law Whyte, a Scottish industrial engineer, this modest yet amazing book was called *Aspects of Form: A Symposium on Form in Nature and Art*.

Rodney Dangerfield: "Comedy is a camouflage for depression."

...

Edgar Varése: "There is no avant garde. There are only people who are a little late."

Whyte's anthology included essays by (among others) Gestalt psychologist Rudolf Arnheim (whose book on *Art and Visual Perception: A Psychology of the Creative Eye* would become an invaluable resource), art historian E.H. Gombrich (*Art and Illusion: A Psychology of Pictorial Representation*), ethologist Konrad Lorenz (*King Solomon's Ring*), and zoologist Hugh B. Cott (*Adaptive Coloration in Animals*).

Equally thought-provoking in Whyte's book were two appendices: a Chronological Survey on Form, and an annotated Selected Bibliography on Form. It was especially the latter that prompted me to search out other publications by other scholars whose writings continued to influence me for years to come, among them György Kepes (*Language of Vision*; and *The New Landscape in Art and Science*), Morse Peckham (*Man's Rage for Chaos*), D'Arcy Wentworth Thompson (*On Growth and Form*), Christopher Alexander (*Notes on the Synthesis of Form*), Adolf Portmann (*Animal Forms and Patterns*), Claude Lévi-Strauss (*The Savage Mind*), Edmund Leach (*Culture and Communication*), Michel Foucault (*The Order of Things*), Gregory Bateson (*Steps to an Ecology of Mind*), and Arthur Koestler (*The Act of Creation*).

Les Coleman: "When we get old, we lie through our dentures."

...

Alan Bennett (*Writing Home*): "I am reading a book on Kafka. It is a library book, and someone has marked a passage in the margin with a long, wavering line. I pay the passage special attention without finding it particularly rewarding. As I turn the page the line moves. It is a long, dark hair."

...

Aldous Huxley (*Time Must Have a Stop*): "Facts are ventriloquist's dummies. Sitting on a wise man's knee they may be made to utter words of wisdom; elsewhere they say nothing or talk nonsense."

At the same time, I continued reading other books, especially autobiographies by Modern-era scholars and artists. In doing that, I was increasingly struck by the frequency of art-related references to camouflage. One day, for example, I found an interview with Georges Braque in which he recalled when he realized the affinity between Cubism and camouflage (Liberman 1969:143). On another occasion, while reading the diaries and letters of Bauhaus artist Oskar Schlemmer, I came upon his account of being assigned to camouflage buildings in World War II Germany (Schlemmer 1972). And, on another, I discovered that the painter Arshile Gorky had taught a civilian course in camouflage at an art school in New York (Schwabacher 1957).

Gradually, the number and frequency of these findings increased, and of course it wasn't long before I realized that I was onto something. By the early 1970s, when my military service ended and I entered graduate school at the Rhode Island School of Design, the search had greatly intensified. It was while living in Providence that I wrote to Rudolf Arnheim (who was then teaching at Harvard), saying that I was beginning to write a book on art and camouflage, and asking for recommendations from him of further sources. At about the same time, I also sent inquiries to Alister Hardy, Hugh B. Cott, György Kepes, E.H. Gombrich, Arthur Koestler and others.

In those days, there were no personal computers, no internet, and no search engines. If a printed book lacked an index, one had to painstakingly visually search the entire volume, page by page, and line by line. In doing that, I somehow grew proficient at spotting the word "camouflage" (as if looking for arrowheads) while rapidly browsing through books, regardless of whether I consciously searched. As a result, I continued to find new references to camouflage almost daily, and as the years passed and the findings accumulated, I began to write essays and, later, a book.

In 1981, I published *Art and Camouflage: Concealment and Deception in Nature, Art and War* (Behrens 1981), a book that I see now as glaringly flawed. While working on it, I had no idea that an important and far more reliable view of a comparable subject range with a surprisingly similar title, *Camouflage: A History of Concealment and Deception in War* (Hartcup 1980), had been published in England the year before. And then, a few years later, I luckily discovered *The Neglected Majority: "Les Camoufleurs," Art History, and World War I* (Kahn 1984). In the years since, as I have explored other aspects of art and camouflage, those two pioneering books have continued to be both helpful and inspiring.

In 2002, having amassed a substantial amount of new research material, I decided to revisit the subject, in the hope that I might make amends for some of the inaccuracies, sins of omission, and other mistakes in my earlier book. The result (which has its own shortcomings) was *False Colors: Art, Design and Modern Camouflage* (Behrens 2002).

In the meantime, six more years have quickly passed. With increasing access to online archives and instantaneous search options, I suspect I've found more new information on art and camouflage in the past twelve months than in all the years before.

To publish yet another book on the subject is a way to distribute the findings I've made since 2002 (many of which are unresolved, and in some cases unverified), so that other, younger scholars can research aspects of the same subject and move ahead. It is also an opportunity for me to rethink my priorities, in part because, in recent years, writings on art and camouflage have grown so rapidly that the subject can easily start to consume nearly every moment in ones life. This book is a perfect example of that.

In order to take on a project of this magnitude, it is often necessary to forego others. Several years ago, I reluctantly ended production of *Ballast Quarterly Review,* which I had personally, lovingly published since 1985. Subtitled Books-Art-Language-Logic-Ambiguity-Science-and-Teaching, I thought of it as a periodical "commonplace book." Now, looking at this book, I have to wonder if I haven't abandoned *Ballast*, since this book looks and reads as if it were an extended issue of the same magazine.

•

For help in putting this book together, I am especially grateful to Thomas Warner, whose father was the American Impressionist painter Everett L. Warner, a largely unheralded leader in US naval camouflage in both World Wars, to whom this book is dedicated. Since locating Tom Warner in the late 1990s, he has graciously supplied me with photographic scans, lecture notes, and US Navy documents from his father's few remaining files (the rest having perished in a studio fire many years ago), which Tom has preserved and maintained as the Everett L. Warner Archives.

I am also greatly appreciative of the efforts of staff members at the Rod Library at the University of Northern Iowa (on the campus where I teach), and especially the people responsible for interlibrary loan services, including Rosemary Meany, Linda Berneking and Robin Darland.

It was at the same library, as an undergraduate in the 1960s, that I first found such important books as the

Rudolf Arnheim [as told to the author]: "The physicist George Gamow was also an entertaining popularizer. He once told the story of how with his wife and their baby daughter he visited the Leaning Tower of Pisa. As they climbed the steps, they noticed an increasingly musty smell, which they first attributed to the ancient walls of the building. Then, however, they began to suspect their little girl, and by the time they reached the top it was clear that she needed immediate attention. 'And from the very place,' explained Gamow, raising his arm and his voice dramatically, 'where Galileo launched his experimental objects we also propelled…"

…

Franklin P. Jones: "Experience enables you to recognize a mistake when you make it again."

…

Edward Rothstein: "There are only three things that are sure in life: Marshmallow, plum pudding, orangutans, and chairs."

…

Les Coleman: "Fat chance of losing weight."

…

Walter Hamady: "Like all artifacts, books are evidence of the process that made them. And the process is as complex and as interesting as the human beings perceiving it. One can see it in the world as much as there is to see in oneself. Oneself. One's Elf."

Thayers' *Concealing Coloration in the Animal Kingdom*, and Hugh B. Cott's *Adaptive Coloration in Animals*. After moving back to Iowa in 1990, I once again had access to the Rod Library, where I continue to make discoveries whenever I go there. For this massive project, it was particularly helpful to have direct access to that library's compact periodical shelves, consisting of archival copies of old periodicals that date back to the 19th century. It was while combing those issues (in tandem with internet searches) that I found the most pivotal articles on protective coloration, camouflage and related subjects.

At the same time, I could never have completed this project without the internet. It truly has revolutionized the expediency with which any of us can search for long forgotten "facts" on the most arcane subjects. The internet has its problems, as we all know, in that, as is commonly noted, much of what appears online is absurdly untrue or misleading. However, as is easily confirmed by a search of back issues of nearly any magazine or newspaper, much of the printed information from the past—and no doubt as much from the present—is equally spurious.

Of particular help in this project were Google, Wikipedia, Library of Congress, Archives of American Art, Smithsonian Museum of American Art, and other treasure-troves. I am also grateful to my students, from whom I have gained so much in 37 years of teaching. I am appreciative of all the support I've received from Reinhold Bubser, Dean of the College of Humanities and Fine Arts; from Jeffery Byrd, Head of the UNI Department of Art; and from various colleagues and friends.

•

In recent years, our much-loved animal companions Ruskin, Fiona and Keeko have died, and all are resting now beneath the tree in the backyard. But Zooey is still very present (she's laying eggs like crazy now), and Saskia is as beautiful as ever. All of us who live and work here will always be indebted to the angelic artist who cares for us all: Zooey and Sakia's mother—who is of course also my wonderful wife, Mary Snyder Behrens.

—Roy R. Behrens
February 20, 2009

CAMOUFLAGE

art, science and popular culture

www.cs.uni.edu/~wallingf/camouflage-conference/

Presentations by worldwide scholars, plus exhibits, films, and panels. $20 pre-registration fee. Free to all speakers, UNI students, faculty and staff.

PLEASE POST

Poster / Sparky Pullman (go, Sparky!)

an international conference at the University of Northern Iowa Cedar Falls

THIS IS a one-day interdisciplinary mini-conference on any and all dimensions of camouflage. It includes but is not limited to artistic, scientific, natural, military, psychological, magical, and fashion-related aspects of concealment and deception. Running from **8:30 am to 8:30 pm** in the **Kamerick Art Building**, it will feature half hour slide talks and other sessions by 37 artists, designers, scientists and scholars from Spain, Australia, UK and throughout the US. In addition, on continuous display will be exhibitions, films, book sales, and a camouflage gift shop. **See schedule of events at website.**

All presentations are open to the public, **for a registration fee of $20** per person. To pre-register, send a check for $20 (made out to the UNI Department of Art) to: **Camouflage Conference, UNI Art Department, Cedar Falls, Iowa 50614-0362.** Clearly indicate the name of the person registering, and include appropriate contact information (mailing and e-mail addresses). In answer, we will send you a complete updated schedule of all conference events. On-site registration will also occur. **Registration is free for all conference presenters, and all UNI students, faculty and staff.** However, those people are also asked to pre-register, by writing to the same address, stating their intention to attend, and providing contact information.

If there are questions, please address them to the conference organizer, **Prof. Roy R. Behrens at <ballast@netins.net>.**

all day Saturday 22 April 2006

• Participants will not be discouraged from wearing camouflage-related attire.
•• All participants are responsible for their own meals, transportation and lodging.

Speakers include—

- **Marvin Bell** (poet)
- **Maria Teresa Mendez-Baiges** (art historian)
- **Eugene Wallingford** (computer scientist)
- **Richard Koenig** (photographer)
- **Ron Sandvik** (writer)
- **Marcia Cohen** (artist)
- **Mark Rawlinson** (literary scholar)
- **Herbert J. Rieth III** (artist)
- **Jenny Roberts** (artist)
- **Harry Brod** (philosopher)
- **Samuel Gladden** (literary scholar)
- **Steve Rose** (writer)

- **Taryn Packheiser** (dancer)
- **Luke Kautz** (architect)
- **David Bower** (artist)
- **Harvey Opgenorth** (artist)

- **Ronald Naversen** (theatre designer)
- **Ann Elias** (art historian)
- **Henrietta Goodden** (art historian)
- **Monica Duncan** (video artist)
- **Lara Odell** (video artist)
- **Robert Ladislas Derr** (artist)
- **Artur Golczewski** (art historian)
- **Stephen Samerjan** (artist)
- **Deirdre Bucher Heistad** (literary scholar)
- **Catherine MacGillivray** (literary scholar)
- **Carrie Paterson** (artist/architect)

- **Arshia Mahmoodi** (architect)
- **Dennis Dake** (art educator)
- **Sue Hettmansperger** (artist)
- **Charles Adelman** (art historian)
- **Craig Ede** (artist)
- **Mary Frisbee Johnson** (artist)
- **Chad Thurman** (historian)
- **Roy R. Behrens** (designer)

bibliography

A List of Sources and a Camouflage Timeline

Abrioux, Yves (1992). *Ian Hamilton Finlay: A Visual Primer.* Cambridge MA: MIT Press.
Ackerman, Carl W. (1930). *George Eastman.* Boston: Houghton Mifflin.
Adams, Henry (1989). *Thomas Hart Benton: An American Original.* New York: Knopf.
"Air Raid Protection" (1940) in *Architectural Forum* (November), pp. 429-436.
"A.J. Kostellow, 56, Designer, Is Dead" (1954) in *New York Times* (September 1), p. 27.
"Alan Beeton" (n.d.) Online biography at <www.alanbeeton.com/index.php?option=com_content&task=view&id=13&Itemid=29> as accessed on March 15, 2008.
"Alan Warren" (n.d.) Online biography at <www.artangel.com.au/alanwarren/alan_warren.html> as accessed on March 13, 2008.
"Albertus De Friest, Rail Bridge Designer" (1953) in *New York Times* (July 27), p. 19.
"Eric Aldwinkle" (n.d.), excerpts from online articles pertaining to him at <http://nscadesign.ca/aldwinckle> as accessed on May 23, 2008.
Alexander, Christopher (1964). *Notes on the Synthesis of Form.* Cambridge MA: Harvard University Press.
"Alister MacKenzie" (n.d.). Wikipedia online biography as accessed on April 16, 2008.
"Alister MacKenzie, Links Designer, Dies" (1934) in *New York Times* (January 7) p. 31.
"Alister MacKenzie: World Golf Hall of Fame Profile" (n.d.) online at <www.wgv.com/hof/member.php?member=1078> as accessed on February 1, 2008.
Allen, Francis H. (1912), "Remarks on the Case of Roosevelt vs. Thayer" in *Auk* Vol 29 (October), pp. 489-507.
Allen, J.A. (1910), "Recent Literature: Thayer on Concealing Coloration in Animals" in *Auk* Vol 27 (April), pp. 222-225.
___(1911), "Roosevelt's 'Revealing and Concealing Coloration in Birds and Mammals" in *Auk* Vol 28 (October), pp. 472-480.
Allen, William Rodney, ed. (1988). *Conversations with Kurt Vonnegut.* Jackson MS: University of Mississippi Press.
"Allied War Salon Opens" (1918) in *New York Times* (December 10).
Alterman, James M. (2005). *New Hope for American Art.* Lambertville PA: Jim's of Lambertville.

Facing page Camouflage conference poster designed by the author (2006).

camouflage timeline

1858

British naturalists Charles Darwin and Alfred Russel Wallace arrive independently, at nearly the same time, at the theory of *natural selection*. Wallace defers to Darwin, and in the following year, the latter publishes a detailed account of his findings in *Origin of the Species*.

•

c1880

British Army adopts mono-colored field service uniforms (*khaki* is Urdu for "dust").

•

1896

American painter Abbott H. Thayer begins to share his discovery of the natural phenomenon *countershading*, by which a solid rounded form (such as the body of an animal) may appear flat and insubstantial when colored darker on its upper parts and lighter on its undersides. He demonstrates this publicly on numerous occasions, using hand-carved wooden ducks or painted sweet potatoes. His first published description of this appears in an issue of *Auk*, along with initial suggestions of other,

Amato, Ivan (1998). *Pushing the Horizon: Seventy-Five Years of High Stakes Science and Technology at the Naval Research Laboratory*. Washington DC: Government Printing Office.

"American 'Camouflage'" (1917) in *Literary Digest* (May 12), pp. 1411-1412.

"American Snipers in Green" (1918) in *New York Times* (October 20), p. 11.

Anderson, Ross (1982). *Abbott Handerson Thayer*. Exhibition catalog. Syracuse NY: Everson Museum.

"A New Art" (1917) in *Rochester* (NY) *Post Express* (July 18).

Annual Report of the Secretary of the Navy (1919). Washington DC: US Government Printing Office.

Anthes, Bill (2006). *Native Moderns: American Indian Painting, 1940-1960*. Durham NC: Duke University Press.

Antupit, Samuel N. (2000), "Tribute: Gene Federico, 1918-1999" in *Print* Vol 54 No 1 (January-February), pp. 22-24.

Aranda, Benjamin, and Chris Lasch (2005). *Tooling*. New York: Princeton Architectural Press.

Archives of American Art (1995). *Government and Art: A Guide to Sources in the Archives of American Art*. Washington DC: Smithsonian Institution.

Arnheim, Rudolf (1954). *Art and Visual Perception: A Psychology of the Creative Eye*. Berkeley: University of California Press.

___(1964). *Entropy and Art*. Berkeley: University of California Press.

___(1971). *Visual Thinking*. Berkeley: University of California Press.

___(1984), "My Life in the Art World." Printed transcript of a talk delivered at the School of Art, University of Michigan, Ann Arbor, on February 8.

___(1989). *Parables of Sun Light*. Berkeley: University of California Press.

"Art and Life of Paul Whitman" (n.d.) Online biography at <http://paulwhitman.org/Bio1.htm> as accessed on March 22, 2008.

Arthur B. Carles (1970). Exhibition catalog. Philadelphia PA: Philadelphia Museum of Art.

"Arthur Lismer" (n.d.) Wikipedia online biography as accessed on December 12, 2008.

"Art Notes" (1917a) in *New York Times* (November 17).

"Art Notes: An Exhibition by the Sculptors Who Are Now In Uniform" (1917b) in *New York Times* (December 17).

"Art Notes: Paintings Showing Nature's Models for Camouflage Students" (1918), in *New York Times* (May 5).

Atterbury, Paul (1975), "Dazzle Painting the the First World War" in *Antique Collector* Vol 46 No 4, pp. 25-29.

Aubrey, Michel (2001). *Michel Aubrey*. Exhibition catalog.

France: Le Quartier, Centre d'Art Contemporain.
"Auguste Herbin" (n.d.), online biography at <ArtFact.com> as accessed on March 22, 2008.
"Auguste Herbin" (n.d.), online biography at Fundación Telefónica, as accessed on March 22, 2008.
"Austin Osman Spare" (n.d.), Wikipedia online biography as accessed on March 23, 2008.
"Aymar Embury, Architect, Dead" (1966) in *New York Times* (November 15), p. 47.
"Aymar Embury II" (n.d.). Wikipedia online biography, as accessed on April 3, 2008.
"A Young Sculptor Wins E.H. Harriman's Daughter" (1910) in *New York Times* (May 8).

Baden-Powell, Robert (1915). *My Adventures as a Spy*. London: C. Arthur Pearson Ltd.
"Baffle Painting" (1918) in *Baltimore Sun* (August 25).
"Baffling the Undersea Pirate" (1919) in *National Marine* (March),pp. 30-32.
Bairnsfather, Bruce (1919). *From Mud to Mufti*. New York: G.P. Putnam.
Ballantine, Betty, ed. (1990). *The Art of Bev Doolittle*. New York: Bantam Books.
Banta, Martha (1987). *Imaging American Women: Idea and Ideals in Cultural History*. New York: Columbia University Press.
Barclay, McClelland (1940). "Camouflaging." US Patent No. 2190691.
"Barclay Draws in Jail" (1933) in *New York Times* (October 31), p. 18.
Barkas, Geoffrey (1952). *The Camouflage Story: From Aintree to Alamein*. London: Cassell.
Barnett, H.G. (1953). *Innovation: The Basis of Cultural Change*. New York: McGraw-Hill.
Barron, Clarence W. (1919). *War Finance: As Viewed from the Roof of the World of Switzerland*. Boston: Houghton Mifflin.
Barry, Ursuła M. (1943), "Mother Nature and Camouflage" in *School Arts* (September), pp. 32-33.
Barton, Peter (2005). *The Battlefields of the First World War: The Unseen Panoramas of the Western Front*. London: Constable.
"Basil Spence" (1984) in Morgan, Ann Lee, ed. *Contemporary Designers*. Detroit MI: Gale Research Company.
"Basil Urwin Spence" (n.d.). Dictionary of Scottish Architects, as accessed online on March 22, 2008.
Baur, John I.H. (1982). *The Inlander: Life and Work of Charles Burchfield, 1893-1967*. Newark: University of Delaware Press.
Beaux, Cecilia (1930). *Background with Figures*. Boston: Houghton Mifflin.
Bedford, Faith Andrews (2000). *The Sporting Art of Frank W. Benson*. Boston: David R. Godine.
Beechey, James (2006), "Guy Roddon" in *The Guardian* (September 7).
Behrens, Roy R. (1978), "On Creativity and Camouflage: The Dissolution of Boundaries" *Journal of Creative Behavior* 2nd quarter, pp. 91-97.
___(1978), "On Visual Art and Camouflage" in *Leonardo*, pp. 203-204.
___(1981). *Art and Camouflage: Concealment and Deception in Nature, Art and War*. Cedar Falls IA: North American Review.
___(1987a), "The Art of Dazzle Camouflage" in *Defense Analysis* [London] (Autumn), pp. 233-243.
___(1987b), "The Life and Unusual Ideas of Adelbert Ames, Jr." in *Leonardo*, pp. 271-279.
___(1988), "The Theories of Abbott H. Thayer: Father of Camouflage" in *Leonardo* Vol 21 No 3, pp. 291-295.

more controversial concepts such as *background picturing* and [dis]*ruptive coloring*.

While teachng Military Science and Tactics at Cumberland University in Lebanon TN, Charles Gerhardt develops a low visibility uniform for the US Military Wheelmen.

•

1899
Abbott H. Thayer and artist George de Forest Brush (his friend and neighbor) propose to the US Navy that American fighting ships should be countershaded, so that surfaces that usually fall into shadow should be painted white, while sunlit surfaces should be darkened, as in the coloration of animals. Their appeal is unsuccessful.

•

1902
Abbott H. Thayer and Gerome de Forest Brush (the 14-year-old son of George de Forest Brush) jointly obtain a US Patent for the use of countershading in ship camouflage.

•

1907
Spanish artist Pablo Picasso, while living and working in Paris, produces a highly unorthodox painting titled *Les Demoiselles d'Avignon*. It is now widely contended to have been the first Cubist painting.

___(1991), "Blend and Dazzle: The Art of Camouflage" in *Print* (January / February), pp. 92-99ff.
___(1992), "Drawing in the Dark: Rembrandt, Pearl Harbor, and the Flash Lab" in *Print* (September / October), pp. 96-101.
___(1994), "Adelbert Ames and the Cockeyed Room," in *Print* (March/April), pp. 92-97.
___(1996a), "Ames in Relation to Historic Uses of Anamorphic Distortion: From the Renaissance to the Present Day" in David Bisno, ed., *Proceedings of the Dartmouth Eye Institute Commemorative Symposium* [held at Hanover, New Hampshire: Dartmouth College, 1995].
___(1996b), "Among the Dazzle Painters: Sherry Fry and the Invention of American Camouflage" in *Tractor: Iowa Arts and Culture* (Fall), pp. 26-28.
___(1996c), "Camouflage" in Jane Turner, ed., *The Dictionary of Art* (32 vol. encyclopedia) London / New York: Macmillan and Grove Dictionaries.
___(1997a), "Art and Camouflage: An Annotated Bibliography," published on the internet at the Leonardo Online website (August; updated 2008).
___(1997b), "Iowa's Contribution to Camouflage" in *Iowa Heritage Illustrated* (Fall), pp. 98-109.
___(1998a), "On Max Wertheimer and Pablo Picasso: Gestalt Theory, Cubism and Camouflage" in *Gestalt Theory* Vol 20 No 2 (June), pp. 109-118.
___(1998b), "Art, Design and Gestalt Theory" in *Leonardo* Vol 31 No 4, pp. 299-303.
___(1998c), "Rudolf Arnheim: The Little Owl on the Shoulder of Athene" in *Leonardo* Vol 31 No 3, pp. 231-233.
___(1998d), "The Artistic and Scientific Collaboration of Blanche Ames Ames and Adelbert Ames II" in *Leonardo* Vol 31 No 1, pp. 47-54.
___(1998e), "Introspectives: Encountering Koestler" in *Print* (September / October), p. 38ff.
___(1999), "The Role of Artists in Ship Camouflage During World War I," in *Leonardo* Vol 32 No 1, pp. 53-59.
___(1999), "Adelbert Ames, Fritz Heider, and the Chair Demonstration" in *Gestalt Theory* Vol 21 No 3 (November), pp. 184-190.
___(2000), "Revisiting Gottschaldt: Embedded Figures in Art, Architecture and Design" in *Gestalt Theory* Vol 22 No 2, pp. 97-106.
___(2001), "Abbott H. Thayer's Anticipation of a Computer-Based Method of Working" in *Leonardo* Vol 34 No 1, pp. 19-20.
___(2002a), "How Form Functions: On Esthetics and

Gestalt Theory" in *Gestalt Theory* Vol 24 No 4, pp. 317-325.
___(2002b) *False Colors: Art, Design and Modern Camouflage*. Dysart IA: Bobolink Books.
___(2004) Various text sections in Hardy Blechman, ed., *DPM: Disruptive Pattern Material: An Encylopedia of Camouflage*. London: DPM Limited.
___(2008) "Ocultar lo que existe mostrando lo que no existe. El camuflaje como fragmentación coincidente" [Concealing What Is, While Revealing What Is Not. Camouflage as Coincident Disruption] in Jorge Lozano, ed., *Camuflaje una metáfora contemporánea* (theme issue) in *Revista de Occidente* No 330 (November).
___(in press), "Revisiting Abbott H. Thayer: Non-scientific Relections about Camouflage in Art, War and Zoology" in Martin Stevens and Sami Merilaita, eds., *Animal Camouflage: Current Issues and Perspectives* (theme issue) in *Philosophical Transactions of the Royal Society B*.
___(in press) "Camouflage" in Bruce Goldstein, ed., *Encyclopedia of Sensation and Perception*. Thousand Oaks CA: Sage Publications.
Bel Geddes, Norman (1960). *Miracle in the Evening: An Autobiography*. Garden City NY: Doubleday.
Bement, Alon (1917). "Camouflage," in *Teachers College Record* 18 No 5, pp. 458-462.
___(1919a). "The Report of the U-16," in *St Nicholas* XLVI (November 1918-April 1919), pp. 495-498.
___(1919b). "Tricks by Which You Can Camouflage," in *American Magazine* 87 (May), pp. 44-46.
Benezit (2006). *Dictionary of Artists*. 14 vols. Paris: Éditions Gründ.
Bennett, Chloe (1996). *Colin Moss: Life Observed*. Ipswich UK: Malthouse Press.
"Benny Hubbard Is Dead; Floral Designer Was 54" (1964) in *New York Times* (March 30), p. 29.
Berry, Henry (1978). *Make the Kaiser Dance*. New York: Doubleday.
___(1996). *Semper Fi, Mac: Living Memories of the US Marines in World War II*. New York: Harper.
Berry, Ian, and Bill Arning, eds. (2005). *America Starts Here: Kate Ericson and Mel Ziegler*. Cambridge MA: MIT Press.
Bess, Piet (1993), "Infrared: Surveillance and Concealment" in *Military Illustrated Past and Present* Nos 62 and 63 (July and August).
Beston, Henry B. (1919). *Full Speed Ahead: Tales from the Log of a Correspondent with Our Navy*. Garden City NY: Doubleday, Page and Co.
Bhownagary, Jehangir (1972), "Creativity and the Magician" in *Leonardo* 5, pp. 31-35.
"Birds and Other Animals" (1910), in *New York Times* (July 2), p. BR4.
Bittinger, Charles (1940). "Naval Camouflage" in *US Naval Institute Proceedings* (October), pp. 1394-1398."
"B.J.O. Nordfeldt" (n.d.), biographical data on Luther College Fine Arts Collection website at <http://finearts.luther.edu/artists/nordfeldt.html> as accessed on March 1, 2008.
"B.J.O. Nordfeldt" (n.d.), online biography at <http://www.sullivangoss.com/bjo_Nordfeldt/> as accessed on March 11, 2008.
Blackburn, Morris (n.d.). Morris Blackburn Papers at Archives of American Art.
Blackford, Stan (2002). *One Hell of a Life: An Anglo-Indian Wallah's Memoir from the Last Decades of the Raj*. Fulham Gardens, Australia: Self-published.
Blass, Bill (2002). *Bare Blass: Bill Blass*. Cathy Horyn, ed. New York: HarperCollins.
Blechman, Hardy, ed. (2004). *DPM: An Encyclopedia of Camouflage*. London: DPM.

1909
After years of work and revision, Abbott H. Thayer and his son Gerald (the author of record) produce a large, richly illustrated volume titled *Concealing Coloration in the Animal Kingdom: An Exposition of the Laws of Disguise Through Color and Pattern: Being a Summary of Abbott H. Thayer's Discoveries*. It is received with a clamorous mixture of acclaim and controversy, which leads to its later reissue in 1918 in a second edition.

•

1910
Gestalt theory begins in Germany under the leadership of psychologists Max Wertheimer, Wolfgang Köhler and Kurt Koffka. They conduct experimental studies and begin to publish the results. Two years later, their work is interrupted by World War I.

•

1911
Russian-born painter and designer Sonia Delaunay begins her experimentation with clothing design. Using discarded fabric scraps, she makes harlequin-like garments, not unlike a "crazy quilt."

•

1913
The opening (in a regimental armory in New York) of the Armory Show, an international exhibit of Modern Art, provides Americans with their

Bogle, Michael (1998). *Design in Australia 1880-1970*. Sydney, AU: Craftsman House.

Bolles, Edmund Blair (1991). *A Second Way of Knowing: The Riddle of Human Perception*. New York: Prentice Hall.

Bond, A. Russell (1918-1919), "Warriors of the Paint-Brush" in *St. Nicholas* Vol XLVI (November 1918-April 1919), pp. 499-505.

Bouché, Louis (n.d.). Louis Bouché Papers at Archives of American Art.

Bowditch, Nancy Douglas (1970). *George de Forest Brush: Recollections of a Joyous Painter.* Peterborough NH: Noone House.

Bowen, Donald (1956) "Austin Osman Spare: An addendum to The Times Obituary" in *London Times* (May 18).

Bower, David (1980). *David Bower.* Exhibition brochure. Muncie IN: Ball State University Art Gallery.

___(1986). *David Bower: Painted Wood Constructions*. Exhibition brochure. Maitland FL: Maitland Art Center.

Boxer, Adam J., comp. (1993) *The New Bauhaus School of Design in Chicago: Photographs 1937-1944*. Exhibition catalog. Chicago: Banning and Associates.

Boyd, Rutherford (1918), "Camouflage—In the Year One and 1918" in *Everybody's Magazine* (March), pp. 45-46.

Boynton, Mary Fuertes (1952), "Abbott Thayer and Natural History" in *Osiris* Vol 10, pp. 542-555.

___(1956). *Louis Agassiz Fuertes: His Life Briefly Told and His Correspondence*. New York: Oxford University Press.

Bracke-Desrousseaux, Alexandre (1916), "Le Rire de la Semaine" in *Le Rire* (August 3).

Brainerd, Eveline [unsigned] (1942), "Our Own Camoufleurs" in the *New York Times* (August 23), p. E7.

Breckenridge, Robert P. (1942). *Modern Camouflage: The New Science of Protective Concealment*. New York: Little and Ives.

Brimmer, F.E. (1918), "The Trapper's Tricks of Camouflage" in *Fur News* Vol XXVII (October), pp. 48-49.

British War Artists (1989) Exhibition catalog. New York: Guillaume Gallozzi Gallery.

Brown, D.R.E. (1963) *Ship Concealment Camouflage Instructions*. Washington DC: Bureau of Ships, US Navy.

Browne, Malcolm W. (1987), "Air Force Sees Beauty in Ugly Ducklings" in *New York Times* (August 18), Science Section.

Browse, Lillian (1978). *Forain: The Painter 1852-1931*. London: Paul Elek.

"Bruce W. Saville, 46, Noted as Sculptor" (1939) in *New York Times* (February 28), p. 25.

Brush, Gerome (1902) [with Abbott H. Thayer]. "Process of Treating the Outside of Ships, etc., For Making Them

Less Visible," US Patent No. 715013.
___(1919). "System for Reducing the Visibility of Objects," US Patent No. 1296753.
___(1943). "Method for Preventing Wake Formation," US Patent No. 2414632.
Brush, Mittie Taylor (1919a). "Aircraft," US Patent No. 1293688.
___(1919b). "Reenforced Fabric for Aircraft," US Patent No. 1442323.
___(1921). "Aircraft," US Patent No. 1367294.
___(1922). "Aircraft," US Patent No. 1619100.
___(1927). "Reenforced Transparent Sheet and Process of Making the Same," US Patent No. 1686649.
Buckman, David (1998) "Obituary: A.H. Gerrard" in *The Independent* (June 30).
Bullard, Robert Lee (1925). *Personalities and Reminiscences of the War.* Garden City NY: Doubleday.
Burchfield, Charles (1965). *His Golden Year: A Retrospective Exhibition of Watercolors, Oils and Graphics by Charles E. Burchfield.* Exhibition catalog. Tucson: University of Arizona Press.
Burford, Byron (1998), Letter to the author dated August 26.
"Burnham Hoyt, 73, Dies" (1960) in *New York Times* (April 8), p. 31.
"Burnham F. Hoyt" (n.d.), online biographical article at Colorado Architects website, as accessed on March 8, 2008.
Burston, David (2002), "The Art of Combat" in London Times Sunday Magazine (March 3), pp. 44-49.
Busby, John (2004). *Drawing Birds*. Portland OR: Timber Press, p. 139.
Butler, Roger (1996), "Feint, Adrian George (1894-1971)" in *Australian Dictionary of Biography* Vol 14. Melbourne AU: Melbourne University Press, pp. 146-147.
"By Any Other Light" (1935) in *New York Times* (March 3), p. X18.
Bygott, Ursula, and K. J. Cable (1981) "Dakin, William John (1883 - 1950)" in *Australian Dictionary of Biography* Vol 8. Melbourne AU: Melbourne University Press, pp. 190-191.

Cabanne, Pierre (1977). *Pablo Picasso: His Life and Times*. New York: William Morrow.
Caillois, Roger (1984), translated by John Shepley, "Mimicry and Lengendary Psychasthenia" in *October* No 31, pp. 16-32.
"Call for 'Fakers' to Fool Germans" (1917) in *New York Times* (September 4), p. 7.
Callaway, Anita (n.d.). "Phyllis Margaret Cilento" in Dictionary of Australian Artists Online, as accessed on March 7, 2008.
"Cameron Clark, Architect, Dead" (1957) in *New York Times* (March 24), p. 86.
"Camouflage" (1922) in *Encyclopedia Britannica*. 12th edition.
"Camouflage" (1939), in *Time* (September 25), p. 42.
"Camouflage" (1942) in *Architectural Forum* (January), pp. 14-25.
Camouflage (1988). Exhibition catalog. Edinburgh, Scotland UK: Scottish Arts Council.
"Camouflage Against U-Boats: The Art of Dazzle Painting" (1919) in *Illustrated London News* (January 4), pp. 20-22.
"Camouflage: Art's Aid in Modern Warfare" (1917) in *Current Opinion* Vol 63 (July), pp. 50-51.
"Camouflage Course for N.Y. Soldiers" (1918) in *New York Times* (May 19), p. 50.
"Camouflage in Modern Warfare" (1940) in *Nature* Vol 145 No 3686 (June 22), pp. 949-951).
"Camouflage in Nature" (1927) in *Science* (August 12), p. xii.

first official look at Post-Impressionism, Fauvism, Cubism and Futurism. The public is scandalized ("This is not art," proclaims Theodore Roosevelt), and one hundred thousand people attend. From New York, the exhibition travels to Boston and Chicago.

•

1914

Attention is brought to the urgent need for battlefield camouflage, largely because of the confluence of three wartime technologies: aerial reconnaissance, aerial photography, and long-range bombardment. This prompts not only concealment from aerial observation but also the construction of false or inaccurate landmarks, dummy guns and so on.

The term *camouflage* is coined by French infantry, and its practitioners are called *camoufleurs*. Within months, the terms have been adopted by English speakers as well.

In wartime France, artillery gunners begin wearing paint-splattered hooded cloaks, to prevent their being spotted by aerial observers. These camouflage cloaks, called *cagoules*, are disputably invented by Lucien-Victor Guirand de Scevola, Louis Guingot or Eugene Corbin (all of whom are French camoufleurs).

The first *section de camouflage* in military history is officially established by the French

"Camouflage Now a Regimental Essential" (1918) in *New York Times* (May 26).

"Camouflage School" (1940) in *Time* (October 28), p. 32.

"Camouflaged Bat Bewilders the Pitcher, But Gets Banned" (1932) in *Modern Mechanics* (July).

"Camouflaged Roads of the War" (1918) in *Scientific American* 119 (August 17), p. 121.

"Camoufleur First in the Plattsdale" (1919), in *New York Times* (June 11), p. 19.

"Camoufleurs in the Making" (1942), in *House and Garden* (August), p. 28.

"Caruso Buys Painting" (1914) in *New York Times* (November 25).

Casey, Dennis (n.d.) An excerpt from "The 'Disguise' of California," online at <www.flatrock.org.nz/topics/flying/the_disguise_of_california.htm> as accessed on February 26, 2008.

Castle, Charles (1986). *Oliver Messel: A Biography*. London: Thames and Hudson.

"Catching an Angel in a Net: A Marine Biologist with a Spiritual Bent Wins a Top Award" (1985) in *Time* (March 11), p. 59.

"Chaplin as Soldier Drops Old Disguise: Camouflaged Star in Shoulder Arms at Strand Has Fun in Capturing the Kaiser (1918) in *New York Times* (October 21).

Chapman, Frank M. (1933). *Autobiography of a Bird Lover.* New York: Appleton Century.

"Charles Bittinger, 91, Dies" (1970) in *Washington Post* (December 20), p. B12.

"Charles Burchfield Dies at 73; Artist Known for Water-colors" (1967) in *New York Times* (January 11), p. 25.

"Charles C. Rumsey Dies in Auto Crash on Jericho Turnpike" (1922) in New York Times (September 22).

"Charles Gerhardt" (n.d.) in *National Cyclopedia of American Biography*. New York: J.T. White.

"Charles Loloma" (n.d.), online biographical article at <www.eindianjewelry.com/?page=info.php&view=loloma> as accessed on April 4, 2008.

"Charles Siclis" (n.d.). Biographical article at <GroveArt.com>.

Chatton, Mary (1993). *Using Poetry Across the Curriculum.* Phoenix AZ: Oryx Press.

Cherkasoff, Baron Nicholas (1941). *Basic Principles of Camouflage*. New York: Self-published.

Chesney, C.H.R. (1941). *The Art of Camouflage*. London: Robert Hale.

"Childe Hassam Arrested" (1918) in *New York Times* (April 17), p. 24.

"Chili Williams—The Polka-Dot Girl" (n.d.), online biographical data at Skylighters website at www.

skylighters.org/girlsinfocus, as accessed on March 23, 2008.
Cirker, Hayward and Blanche, eds. (1967). *Dictionary of American Portraits*. New York: Dover Publications.
"Clarence H. Hegeman" (1950) in *New York Times* (January 6), p. 21.
Clark, Walter (1939). *Photography by Infra-Red: Its Principles and Application*. New York: John Wiley.
Claudy, C.H. (1918) "Gun Camouflage" in *Scientific American* (December 7), pp. 460ff.
"Clayton Henri Staples" (n.d.), article at Online Biographical Directory of Kansas artists, accessed on April 15, 2008.
Clement, Russell T. (1994). *Les Fauves: A Sourcebook*. Westport CT: Greenwood Press.
Coleman, Lee (2008), "WWI Battle Mural Finds New Home" in *Saratoga Springs Daily Gazette* (April 1).
Collins, Ian (2006), "Obituary: Colin Moss" in *The Guardian* (January 14).
___ (2007), "Obituary: Henry Holzer" in *The Guardian* (October 11).
"Color-Blind Flyers: In Demand Now; They See Through Camouflage" (1941) in *Scientific American* (February), pp. 103-104.
"Color-Blindness and Camouflage" (1940) in *Nature* Vol 146 (August 17), p. 226.
"Color Photographs at Night Penetrate Camouflage" (1940) in *Science News Letter* (October 26), p. 270.
"Company A: American Camoufleurs" (1917) in *Literary Digest* (October 13), pp. 32-33.
"Completing Military Camouflage with Camouflage Robes" (1917) in *Scientific American* (September 29), p. 225.
Cone, Michele (1975). *The Roots and Routes of Art in the 20th Century*. New York: Horizon Press.
Connelly, Marc (1972), foreword to John Held, Jr., *The Most of John Held, Jr.* Brattleboro VT: Stephen Greene Press.
Conot, Robert (1979). *A Streak of Luck: The Life and Legend of Thomas Alva Edison*. New York: Seaview Books.
Cordier, Marcel (1980), "Louis Guingot (1864-1948)" in *Le Pays Lorrain* Vol 61 No 3, pp. 173-174.
Cork, Richard (1976). *Vorticism and Abstract Art*. Berkeley: University of California Press.
___(1994). *A Bitter Truth: Avant-Garde Art and the Great War*. New Haven: Yale University Press.
Cornebise, Alfred E., ed. (1991). *War Diary of a Combat Artist: Captain Harry Everett Townsend*. Niwat CO: University Press of Colorado.
Cortissoz, Royal (1923). *American Artists*. New York: Charles Scribner's Sons.
Cott, Hugh B. (1938). "Camouflage in Nature and War" in *Royal Engineers Journal* (December), pp. 501-517.
___(1940). *Adaptive Coloration in Animals*. London: Methuen.
___(1951). "Animal Form in Relation to Appearance" in Lancelot Law Whyte, ed., *Aspects of Form*. London: Percy, Lund Humphries.
Coulter, Cornelia C. (1918), "Ancient Camouflage: Letter to the Editor" in *The Nation* (June 22), p. 738.
Coutin, Cecile (1988/1989), "Les Artistes de La Guerre: Le Camouflage Pendant La Premiere Guerre Mondiale." Parts 1-3 in *Historiens-Geographes*. Nos 321 (December) and 322 (March-April).
Coveney, Peter, and Roger Highfield (1990). *The Arrow of Time*. New York: Fawcett Columbine.

government. The first six camouflage artists are Lucien-Victor Guirand de Scevola, Eugene Ronsin, Jean-Baptiste Corbin, Louis Roger, Joseph Pinchon, and Georges Mouveau.

One winter evening, as Gertrude Stein, Pablo Picasso and their friends are walking down the Boulevard Raspail in Paris, they see the first examples of camouflaged vehicles (apparently trucks and cannons). They had heard about it, but had not seen any examples. Picasso, Stein remembers, was awestruck by its resemblance to Cubist style painting. "We originated that," he said, "it's Cubism."

In September, British embryologist John Graham Kerr recommends that British ships be camouflaged, using such methods as countershading and *parti-coloring* (his term for disruptive coloration). A supporter of the findings of Abbott H. Thayer, and the teacher of Hugh B. Cott, Kerr has advocated ship camouflage since the 1890s, and will later claim to have invented "dazzle painting" years in advance of Norman Wilkinson.

As a 10-year-old child in Spain, Salvador Dali sees and is astonished by camouflaged ships for the first time. He likens them to a leaf mimic insect that he encountered in earlier childhood.

Covert, Claudia T. (2007). "Art at War: Dazzle Camouflage," in *Art Documentation* 26 No 2, pp. 50-56.

Crane, Amiee, ed. (1944). *Art in the Armed Forces*. New York: Hyperion Press.

Crowell, Benedict (1919). *America's Munitions 1917-1918.* Washington DC: Government Printing Office.

___(1921). *The Road to France: The Transportation of Troops and Military Supplies 1917-1918*. New Haven CT: Yale University Press.

Cruickshank, Charles (1979). *Deception in World War II*. UK: Oxford University Press.

Cruttwell, C.R.M.F. (1936). *A History of the Great War 1914-1918*. UK: Oxford University Press.

Culkin, John (1980), "A Rattner Portfolio" in *Archives of American Art Journal* Vol 20 No 4, pp. 2-9.

Cuthill, I.C., M. Stevens, J. et al. (2005), "Disruptive Coloration and Background Pattern Matching" in *Nature* Vol 434, pp. 72-74.

Cutright, Paul Russell (1956). *Theodore Roosevelt, the Naturalist*. New York: Harper and Brothers.

Dali, Salvador (1942), "Total Camouflage for Total War" in *Esquire* Vol 18 No 2 (August), pp. 64-66 and 129-130.

___(1993). *The Secret Life of Salvador Dali*. New York: Dover Publications (originally published 1942).

Danchev, Alex (2005). *Georges Braque: A Life*. New York: Arcade Publishing.

"Daring Q-Boats Upheld Royal Navy's Tradition" (1919) in *Popular Mechanics*, Vol 31 No 3 (March), pp. 423-426.

Daugherty, James (n.d.). James Henry Daugherty Papers. The Children's Literature Research Collections website at <http://special.lib.umn.edu/findaid/xml/CLRC-122.xml> as accessed on April 4, 2008.

Deak, JoAnn (2002). *Girls Will Be Girls: Raising Confident and Courageous Daughters*. New York: Hyperion.

Dean, Roger and Martyn Dean (1984). *Magnetic Storm*. New York: Harmony Books.

Dearinger, David B. (2004). *Paintings and Sculpture in the Collection of the National Academy of Design*. NY: Hudson Hills Press.

"Decoy Blackouts" (1942) in *Scientific American* (January), p. 8.

DeKay, Charles (1918), "Ships That Fade Away" in *The Nation* Vol 107 No 2769 (July 27), pp. 105-107.

Delouche, Danielle (1993), "Cubisme et Camouflage" in *Guerrres Mondiales et Conflits Contemporains* 43, p. 171.

DeMeis, Richard (1987), "Visual Stealth" in *Aerospace America* Vol 25 No 7 (July), pp. 12-15.

___(1988), "Now You See It Now You Don't" in *Wings* Vol 18 No 4 (August), pp. 18-33.

"Description of Theatre at Rochester, Designed and Built Under the Direction of Lt. L.A. Jones" (1919) in *Transactions of the Illuminating Engineering Society* Vol 14 (July 21), pp. 232ff.
"Detect Camouflage" (1942) in *Science News Letter* (November 28), p. 342.
Dewey, John (1958). *Art as Experience*. New York: Capricorn Books.
Dixon, Jeremy (2000), "David Jones" in *The Guardian* (April 26).
"Donald M. Oenslager, 73, Dies; Designed Stage Sets 50 Years" (1975) in *New York Times* (June 23), p. 29.
Dooley, William G. (1969). *Great Weapons of World War I*. New York: Army Times Publishing.
"Douglas D. Ellington" (n.d.). Wikipedia online biography accessed on November 23, 2008.
Dow, Arthur Wesley (1997). *Composition*. Berkeley: University of California Press.
"Dr. F.C. Whitmore, Noted Scientist" (1947) in *New York Times* (June 25), p. 25.
"Dr. Matthew Luckiesh Is Dead; GE Lighting Expert Was 84" (1967) in *New York Times* (November 3), p. 43.
"Dr. Royal B. Farnum Is Dead; Art Educator and Executive, 83" (1967) in *New York Times* (August 30), p. 43.
"Dr. Toch, Chemist and Art Expert, 81" (1946) in *New York Times* (May 31), p. 23.
Dube, Ilene (2001), "Urban Beautification grew from the soil of Princeton Nurseries" in *New York Times* (October 5).
Ducasse, André, Jacques Meyer, and Gabriel Perreux (1962). *Vie de Mort des Francais 1914-1918*. Paris: Hachette.
Duer, Douglas (1916). *The Vanished World*. Boston: Sherman, French and Co.
Dunbier, Lonnie (2004), "Augustus Dunbier: Paint for the Love of Color" in *Resource Library Magazine* (March 18).
Dunoyer de Segonzac, André (1970). *Dessins 1900-1970*. Geneve: Pierre Cailler.
Dykhuis, Peter, curator (1999), *Dazzle Ships and Figure Paintings. September 18-November 21, 1999*. Exhibition catalog. Halifax, Nova Scotia: Art Gallery of Nova Scotia.

Eagle, Mary (1996), "Dobell, Sir William (1899 - 1970)" in *Australian Dictionary of Biography* Vol 14. Melbourne AU: Melbourne University Press, pp 10-13.
Earl, Harley (1954), interviewed by Stanley Brams at General Motors Styling Section (January) online at <www.carofthecentury.com/detroit_interview_with_harley_earl.htm> accessed on March 28, 2008.
Eby, Kerr (1936/1971). *War*. New York: Garland Publishing.
Edmonds, James E., ed. (1932). *History of the Great War: Military Operations in France and Belgium 1916*. Vol 1. London: Macmillan.
Edward Wadsworth 1889-1949 (1974). Exhibition catalog. London: P&D Colnaghi and Co. Ltd.
"Edwin Blashfield, Noted Artist, Dies" (1936) in *New York Times* (October 13), p. 27.
"Edwin W. Deming, Painter, Sculptor" (1942) in *New York Times* (October 16), p. 19.
"E.F. Molyneux, 62, Advertising Aide" (1958) in *New York Times* (March 26), p. 37.
Ehrenzweig, Anton (1953). *The Psychoanalysis of Artistic Vision and Hearing*. London: Routledge and Kegan Paul.
___(1971). *The Hidden Order of Art*. Berkeley: University of California Press.
Eldred, Julia Carlson (n.d.), "True Visions: The Paintings of C. Arnold Slade (1882-1961)." Online article at <www.tfaoi.org/aa/3aa/3aa139.htm> as accessed on July 14, 2008.

1916

French painter André Dunoyer de Segonzac is appointed head of the *Section de Camouflage du Nord*, a position he continues in until 1918.

French painter Pierre-Louis Gatier serves as a ship camoufleur for the French Navy at Rochefort.

British artist Solomon Solomon becomes head of the Camouflage Section in the Royal Engineers. He advocates the use of garnished overhanging nets, called *umbrella camouflage*.

Sixteen French artists begin collaborating on a huge circular diorama (the world's largest at the time) that pays tribute to Allied soldiers in World War I. Completed in 1918 and titled *Pantheon de la Guerre*, it portrays five thousand full-length figures, including a handful of French camoufleurs. It remains on exhibit in France until 1927, when it is purchased by US businessmen.

•

1917

In mid-April, German submarines (called U-boats) sink twenty-three British ships in one week, for an average of eight ships per day.

British seascape painter and Naval Reserve officer Norman Wilkinson recommends the adoption of disruptive ship camouflage, which he calls

"Eldredge Snyder, Architect, Dead" (1967) in *New York Times* (March 28), p. 39.

Elias, Ann (2003). "The Organization of Camouflage in Australia in the Second World War" in *Journal of the Australian War Memorial* (April), pp. 23-30.

___(2008), "William Dakin on Camouflage in Nature and War" in *Journal of Australian Studies* Vol 23 No 2 (June 2008), pp. 251-263.

___(2009), "Campaigners for Camouflage: Abbott H. Thayer and William J. Dakin" in *Leonardo* Vol 42 No 1, pp. 36-41.

Elkins, James (1996). *The Object Stares Back: On the Nature of Seeing*. New York: Simon and Schuster.

Elliot, Fran (n.d.), "Kate Cory" in *Plateau Journal: Magazine of the Museum of Northern Arizona*, as accessed online on March 18, 2008 at <www.bluecoyotegallery.com/KateCorypaintingsandart.htm>.

"Elliot Paul" (n.d.) Wikipedia online biography, as accessed on December 24, 2007.

"Elliot Paul" (n.d.) on SUNYSB Special Collections website at <www.sunysb.edu/libspecial/collections/manuscripts/payne> as accessed on December 24, 2007.

Elman, Kimberly, ed. (2001). *Percival Goodman: Artist, Planner, Teacher, Painter*. New York: Princeton Architectural Press.

Embury II, Aymar (1917). "Architects and the Camouflage Service" in *Architectural Forum* 27 (November), pp. 137-138.

"Emile Mollard, Camouflage Specialist, 96" (1991) in *New York Times* (October 19).

"Emilius R. Ciampa" online biography at <http://en.wikipedia.org/wiki/Emilius_R._Ciampa> as accessed on March 23, 2008.

"Eric Aldwinckle" online biography at <http://en.wikipedia.org/wiki/Eric_Aldwinckle> as accessed on May 23, 2008.

"Eric Maunsbach" (1969) in *New York Times* (May 5), p. 47.

"Erno Goldfinger" (n.d.) at Design Museum website at <www.designmuseum.org/design/erno-goldfinger> as accessed on March 20, 2008.

Esslin, Martin (1968). *The Theatre of the Absurd*. London: Pelican Books.

Etchells, Frederick (1919) "A Note" in *Exhibition of Original Woodcuts and Drawings by Edward Wadsworth*. London: Adelphi Gallery.

"Eveline Brainerd, Feminist, 78, Dead" (1948) in *New York Times* (August 1), p. 56.

"Everett L. Waid, 76, Art League Head" (1976) in *New York Times* (June 29), p. 34.

"Ezra Jack Keats" (n.d.), online biographical essay at <www.ezra-jack-keats.org/about/index.html> as accessed on April 4, 2008.

"**Faking** As An Art In Conducting War" (1917) in *New York Times* (June 24), p. 35.
Falk, Peter Hastings, ed. ((1999). *Who Was Who in American Art*. Madison CT: Sound View Press.
Fallico, Arthur B. (1962). *Art and Existentialism*. Englewood Cliffs NJ: Prentice Hall.
Farrer, Reginald John (1918). *The Void of War: Letters from the Three Fronts*. Boston: Houghton Mifflin.
Faulkner, Barry (c1957a). Unpublished notes for a series of radio talks on his WWI camouflage service, broadcast on WKNE radio, Keene NH. [See Rumrill (2007).]
___(c1957b). "Harry Dickenson Thrasher, 2nd Lieutenant, Engineers," unpublished draft of an essay.
___(1973). *Sketches from an Artist's Life*. Dublin NH: William Bauhan.
Faxon, Susan (1989), "Frank Weston Benson, Abbott Thayer, and Dublin, New Hampshire" in *Frank W. Benson: A Retrospective*. Exhibition catalog. New York: Berry-Hill Galleries, pp. 134-143.
Ferris, Keith (1976), "The Keith Ferris Deceptive Aircraft Paint System: Another Nickel on the Grass?" in *Topgun Journal* Vol 1 No 2 (Spring), pp. 11-17.
___(1978). *The Aviation Art of Keith Ferris*. New York: Peacock Press.
Feuer, A.B. (1999). *The US Navy in World War I: Combat at Sea and in the Air*. Westport CT: Praeger.
"First 'Invisible' Murals in Franklin Institute" (1934) in *New York Times* (June 8), p. 14.
Fisher, David (1983). *The War Magician*. New York: Coward and McCann.
Fleischmann, Gerd (2001). "A Poster by Max Bill or the Love of Geometry" in *Design Issues* Vol 17 No 2 (Spring), pp. 53-63.
Flower, Cedric (1986), "Moore, John Drummond Macpherson (1888 - 1958)" in *Australian Dictionary of Biography* Vol 10. Melbourne AU: Melbourne University Press, pp. 566-567.
"Fooling the Enemy's Eye" (1917) in *Literary Digest* (December 22), pp. 21-22.
Forbes, David W. (1992). *Encounters with Paradise*. Honolulu HI: Honolulu Academy of the Arts.
Fort, Eric Gili (2007). Online article on Jay Doblin in *The New IDiom* [Institute of Design (Chicago) student magazine].
Foster, William F. (1919), "A Day with a Sketch-Block at the Front" in *Scribner's* (April).
Foucault, Michel (1973). *The Order of Things*. New York: Vintage Books.
Fox, Frederic E. (1978). Letter to the author dated February 8, 1978.
Fox, Margalit (2005) "Bob Schiffer" in *New York Times* (May 8).
Fox, Milton S. (1942), "Camouflage and the Artist" in *Magazine of Art* (April), pp. 136ff.
___(1943a), "Camouflage" in *School Arts Magazine*, p. 16.
___(1943b), "Camouflage: Theory and Practice: A Summary." Art and Wartime Publications series. No 2. Cleveland OH: Cleveland Museum of Art.
___(1964), Oral History Interview conducted by Harlan Phillips, in the Archives of American Art, Smithsonian Institution.
"Francis B. Hoffman" (1980) in *New York Times* (November 29).
Frank, Dan (2001). *Bohemian Paris: Picasso, Modligliani, Matisse, and the Birth of Modern Art*. New York: Grove Press.
"Frank Eliscu" (n.d.). Online biographical article at <www.frankeliscu.com> as accessed on March 22, 2008.
"Frank H. Schwarz, Mural Painter, 57" (1951) in *New York Times* (September 7), p. 28.
Frayling, Christopher (2007), "Obituary: Richard Guyatt" in *The Guardian* (October 27).
"French Artists Remobilized" (1916), in *Literary Digest* (August 19), p. 411.

dazzle painting. Its purpose is not to prevent a ship from being seen, he explains, but rather to lessen the likelihood of it being struck by a torpedo from a German U-boat. Wilkinson's proposal is accepted, and British artist Cecil King is appointed to assist him.

In advance of America's entry into World War I, a group of US artists (led by Barry Faulkner, Sherry Fry and others) establish the New York Camouflage Society or, as it was also called, the American Association for Camouflage.

•

1918

In March, the US Navy is placed in charge of all ship camouflage designs, while a civilian unit called the Emergency Fleet Corporation (US Shipping Board) is given responsibility for applying those designs to ships. Harold Van Buskirk oversees the Navy camouflage section, which is divided into two subsections. The design subsection, headed by artist Everett L. Warner, is located in Washington DC; while the research subsection, headed by physicist Loyd A. Jones, is located at Eastman Kodak Laboratories in Rochester NY. Other research is conducted by the Submarine Defense Association, under the direction of American civil engineer Lindon W. Bates, assisted by his son Lindell.

Friedmann, Herbert (1942). *The Natural History Background of Camouflage*. Washington DC: Smithsonian Institute. War Background Studies No 5 (December 11).

Frothington, Thomas G. (1971). *The Naval History of the World War: The United States in the War 1917-1918*. Freeport NY: Books for Libraries Press.

Fry, Sherry Edmundsons (1917). "An American Corps for Camouflage," in *American Architect* 112 (July 25), p. 68.

Frye, Northrup (1995). *The Bush Garden: Essays on the Canadian Imagination*. House of Anansi.

___(2004). *Fearful Symmetry: A Study of William Blake*. University of Toronto Press.

Fuller, Grace Hadley, ed. (1940). *Camouflage: A List of References*. Washington DC: Library of Congress, Division of Bibliography.

___(1942). *A List of References on Camouflage*. Washington DC: Library of Congress, Division of Bibliography.

Fusscas, Helen K. (1992). *A World Observed: The Art of Everett Longley Warner 1877-1963*. Exhibition catalog. Old Lyme CT: Florence Griswold Museum.

"**Gabriel** Kohn" (n.d.). Biographical article in Grove Dictionary of Art at GroveArt.com.

Gaede, Marnie (1986). *The Hopi Photographs: Kate Cory 1905-1912* (Albuquerque: University of New Mexico Press.

Gale, Richard (1968). *Call to Arms*. London: Hutchinson.

Gallatin, Albert (1919). *Art and the Great War*. New York: E.P. Dutton.

Gardner, James (1983). *Elephants in the Attic*. London: Orbis.

___(1993). *The ARTful Designer: Ideas off the Drawing Board*. London: Lavis.

Garraty, John, and Mark Carnes, eds. (199). *American National Biography*. New York: Oxford University Press.

Garrels, Gary (2000). *Sol LeWitt: A Retrospective*. New Haven CT: Yale University Press.

Garwood, Darrell (1944). *Artist in Iowa: A Life of Grant Wood*. New York: W.W. Norton.

Gaunt, William (1949). *The March of the Moderns*. London: Jonathan Cape.

Genet, Jean (1966). *Miracle of the Rose*. New York: Grove Press.

"George K. Hartwell" (1949) in *New York Times* (December 13), p. 31.

"George Kenneth Hartwell" (n.d.) online biographical article on Art of the Print website, as accessed on March 11, 2008.

"George L. Stout, at 80; Expert On Restoration of Works

of Art" (1978) in *New York Times* (July 3), p. 22.
"George Sakier" (n.d.) Online biography at Industrial Designers Society of America website, as accessed on April 3, 2008.
"George Sakier" (n.d.) Wikipedia online biography, as accessed on April 3, 2008.
"Gerald H. Thayer: Naturalist, Writer and Artist Was Known for Bird Studies" (1939) in *New York Times* (June 8), p. 32.
"Gerome Brush" (1954) in *New York Times* (September 15), p. 33.
Giardina, Bernadette Passi (n.d.) "Kerr Eby (1889-1946)," online biographical article at <www.oldprintshop.com> as accessed on April 4, 2008.
Gilder, Rosamond (1944), "'You Bet Your Life': Report on a Camouflage Show" in *Theatre Arts* (September), pp.521-527.
Gimpel, René (1966). *Diary of an Art Dealer*. John Rosenberg, trans. New York: Farrar, Straus and Giroux.
Ginzl, David J. (2001). *Barnett: The Story of "Florida's Bank."* Tampa FL: University of Tampa Press.
Giraudy, Daniele (1971), "Correspondance Henri Matisse - Charles Camoin" in *Revue de l'Art* Vol 12, pp. 7-33.
"Girls in Show Raid Freed at Hearing" (1935) in *New York Times* (November 15), p. 21.
Gombrich, E.H. (1969). *Art and Illusion: A Study of the Psychology of Pictorial Representation*. Princeton NJ: Princeton University Press.
___(1979). *The Sense of Order*. Ithaca NY: Cornell University Press.
Gombrich, E.H., Julian Hochberg and Max Black (1972). *Art, Perception and Reality*. Baltimore: Johns Hopkins University Press.
Goodden, Henrietta (2007). *Camouflage and Art: Design for Deception in World War 2*. London: Unicorn Press.
Goodman, Jean (1978). *Edward Seago: The Other Side of the Canvas*. London: Collins.
Goodman, Nelson (1960), "The Way the World Is" in *Review of Metaphysics* (September), pp. 48-56.
Goodman, Paul (1946). *The State of Nature*. New York: Vanguard Press.
Goodman, Percival (1941), "Camouflage Held Important" in *New York Times*. Letter to the editor dated February 4.
Goossen, E.C. (1973). *Ellsworth Kelly*. New York: Museum of Modern Art.
"Gordon Campbell" (n.d.) online biography at <www.findagrave.com/cgi-bin/fg.cgi?page+gr&GRid=10222242>, as accessed on April 7, 2008.
"Gordon Smith, Art Expert, Dies; Built Albright-Knox's Collection" (1979) in *New York Times* (May 22), p. C16.
Gorky, Arshile (1940), "Camouflage" (Grand Central College of Art course description), as reprinted in Harold Rosenberg, *Arshile Gorky: The Man, the Time, the Idea*. New York: Horizon Press (1962), pp. 133-135.
Gould, Stephen Jay (1985), "Red Wings in the Sunset" in *Natural History* (May), pp. 12-24.
___(1992). *Bully for Brontosaurus: Reflections on Natural History*. New York: W.W. Norton.
"Graham Barry Clilverd" (n.d.) online biography at <www.atoftheprint.com> as accessed on April 4, 2008.
Graham, Nan Wood (1993). *My Brother, Grant Wood*. Iowa City: State Historical Society of Iowa.
Grahame, Rachel (2002), "Stelzer, Eunice Minnie (1880-1962)" in *Australian Dictionary of Biography* Vol 16. Melbourne AU: Melbourne University Press, pp 300-301.
Grant, Annette (1969), "Man of 100 Books" in *Newsweek* (May 19).

Beginning in May, camouflage training is required of all soldiers from New York State, under the command of Major John F. O'Ryan.

The first enlistments of the American Camouflage Corps take place on September 4, the first four volunteers for which are US artists Barry Faulkner (cousin of Abbott H. Thayer), Sherry Edmundson Fry, Evert Herter and William Twigg-Smith. The officers heading the unit at Camp American University are Augustus Saint-Gaudens and Dwight Bridge.

As dazzle ship camouflage is increasingly adopted by US naval camoufleurs, US journalists propose other names for it, to distinguish its American variety from that of the British. In news articles, the most widely used alternative is *baffle painting*, but there were various others, among them *jazz painting*, *razzle dazzle*, *eccentric painting*, and *crazy quilt camouflage*.

American painter Thomas Hart Benton, having returned to the US after studying in Europe, enlists in the US Navy and is assigned to make drawings of camouflaged ships that enter the harbor at Norfolk VA (both foreign and domestic). Artist Louis Bouché is given the same assignment.

Gerome de Forest Brush (son of George de Forest Brush), who had collaborated

Graves, Lewis (1920), "Adventures in Prohibition" in *World's Work*. New York: Doubleday, Page and Co.
Graves, Robert Muir (2002). *Classic Golf Hole Design*. Hoboken NJ: John Wiley and Sons.
Gray, Anne (1986a), "Longstaff, William Frederick (Will) (1879 - 1953)" in *Australian Dictionary of Biography* Vol 10. Melbourne AU: Melbourne University Press, pp. 142-143.
___(1986b) "McCubbin, Louis Frederick (1890 - 1952)" in *Australian Dictionary of Biography* Vol 10. Melbourne AU: Melbourne University Press, pp 243-244.
Gray, David R. (2002), "Carrying Candian Troops: The Story of RMS Olympic as a First World War Troopship" in *Canadian Military History* Vol 11 No 1 (Winter), pp. 54-70.
Gregory, Richard L. (1966). *Eye and Brain: The Psychology of Seeing*. New York: McGraw Hill.
___(1970). *The Intelligent Eye*. New York: McGraw Hill.
___(1987). *The Oxford Companion to the Mind*. New York: Oxford University Press.
___ and E.H. Gombrich, eds. (1973). *Illusion in Nature and Art*. New York: Charles Scribner's Sons.
"Grenville Rickard, Architect Here, 66" (1956) in *New York Times* (May 4), p. 25.
Gribaudo, Ezio (1977). *Jean Puy*. Genéve: Petit Palais.
"Griffith Baily Coale (1890-1950)" (n.d.) Naval Historical Center online biography at <www.history.navy.mil/ac/artist/c/coale/coale1.html>, as accessed on June 17, 2008.
Grubbs, David (2005), "Composer's Notes for *Eight Clouds in the Shape of US Presidents*," a performance at the Tate Modern (January 29).
Guthrie, Stuart (1993), *Faces in the Clouds: A New Theory of Religion*. New York: Oxford University Press.

Hall, Donald (1993) *Life Work*. Boston: Beacon Press.
Hall, Peter G. (1989), "Obituary: Patrick Alfred Pierce Moran 1917-1988" in *Biometrics* Vol 45 No 2 (June), pp. 687ff.
Harbord, James G. (1925). *Leaves from a War Diary*. New York: Dodd, Mead.
Hardy, Alister (1953), "Biology and Psychical Research" in *Proceedings of the Society for Psychical Research* Vol 50 No 183 (May).
___(1965). *The Living Stream: A Restatement of Evolution Theory and its Relationship to the Spirit of Man*. New York: Harper and Row.
___(1976a) Letter to the author (August 31).
___(1976b) Letter to the author (September 13).
___(1981a) Letter to the author (May 20).

___(1981b) Letter to the author (July 23).
Harmon, Mamie (n.d.), Mamie Harmon Papers in the Archives of American Art.
"Harold Bowler, 62, A Portrait Painter"(1965) in *New York Times* (July 24), p. 21.
"Harold Sleeper, A City Aide, Dies" (1960) in *New York Times* (November 11), p. 31.
"Harper Goff" (n.d.) Wikipedia online biography, as accessed on October 16, 2008.
Harries, Meirion and Susie (1983). *The War Artists: British Official War Art of the 20th Century.* London: Michael Joseph.
Harris, Stephen L. (2005). *Duty, Honor, Privilege.* Dulles VA: Potomac Books.
Harrison, Charles (1981). *English Art and Modernism 1900-1939.* Bloomington: Indiana University Press.
Harrod, Tanya (1999), "Obituary: Steven Sykes" in *The (London) Independent* (February 24).
Hartcup, Guy (1980). *Camouflage: A History of Concealment and Deception in War.* New York: Charles Scribner's Sons.
Hartgen, Frances Caroline (n.d.). *A Maine Passage: Two Memoirs.*
"Harvey C. Hodgens" (1954) in *New York Times* (August 2), p. 17.
Haskell, Barbara (1985). *Ralston Crawford.* New York: Whitney Museum of American Art.
Havens, George R. (1969). *Frederick J. Waugh: American Marine Painter.* Orono ME: University of Maine Press.
Hays, Robert R. (1999). Letter to the author dated May 1.
Heaton Cooper, William (1984). *Mountain Painter: An Autobiography.* Kendal, Cumbria: Frank Peters Publishing.
Heider, Fritz (1983). *The Life of a Psychologist: An Autobiography.* Lawrence: University Press of Kansas.
Hemingway, Ernest (1922), "A Paris-to-Strasbourg Flight" in *Toronto Daily Star* (September 9) as reprinted in William White, ed., *By-Line: Ernest Hemingway: Selected Articles and Dispatches of Four Decades.* New York: Charles Scribner's Sons (1967).
Henderson, Mary C. (2001). *Mielziner: Master of Modern Stage Design.* New York: Back Stage Books.
Henderson, William Penhallow (n.d.). William Penhallow Henderson Papers. Archives of American Art.
Henle, Mary, et al., eds. (1973). *Historical Conceptions of Psychology.* New York: Springer.
___, ed. (1976) *Vision and Artifact.* New York: Springer.
Henning, Patrick (n.d.) "The Walt Disney Story," online biography at <www.u46.k12.il.us/lhs/henningpatrick/19999.htm> as accessed on March 17, 2008.
"Henry Davenport" (1965) in *Weston Town Crier* (Weston MA) (May 13).
"Henry Poole" (n.d.) biographical data at Tate Archives Online, as accessed on 23 March 2008.
"Herbert Haseltine Dead at 84; Sculptured Statues of Animals" (1962) in *New York Times* (January 9), p. 47.
"Hermann Murphy, Landscape Artist" (1945) in *New York Times* (April 17), p. 23.
"Hiding a Big Gun in the Open with a Coat of Spotted Camouflage Paint" (1917) in *Scientific American* 117 (November 10), p. 349.
Hillier, Bevis (1975). *Austerity Binge.* London: Studio Vista (1975).
___(1983). *The Style of the Century 1900-1980.* London: Herbert Press.
Hobbs, Susan (1982), "Nature into Art The Landscapes of Abbott Handerson Thayer" in *American Art Journal* Vol 14 No 3 (Summer), pp. 5-55.

earlier (1902) with Abbott H. Thayer on a ship camouflage patent, is appointed a dock supervisor for the painting of US merchant ships along the eastern seaboard, where he works at such locations as Norfolk, Boston and New York.

Secretary of the Navy Josephus Daniels invites the preeminent US inventor Thomas A. Edison to devise ways to counteract torpedo attacks by U-boats, including ship camouflage. Edison's unorthodox camouflage of the *SS Ockenfels* (carried out on his behalf by Everett L. Warner) falls apart before it gets out of the harbor.

American artist Charles Burchfield is inducted into the US Army and is assigned to camouflage.

Having served as a dock officer for camouflage ship painting during World War I at the request of Norman Wilkinson, British Vorticist painter Edward Wadsworth produces a series of woodcuts of dazzle camouflage being applied to ships, and, in 1919, a large and magnificent painting of the same subject, titled *Dazzle Ships in Drydock at Liverpool*.

American poet Amy Lowell publishes in *The Dial* a poem about dazzle-painted ships, titled "Camouflaged Troop-Ship: Boston Harbor."

•

Hochberg, Julian E. (1964). *Perception*. Englewood Cliffs NJ: Prentice-Hall.

Hogg, J., ed. (1969). *Psychology and the Visual Arts*. Middlesex UK: Penguin Books.

Holabird, John Auger, Jr. (1993), "Oral History of John Auger Holabird" as interviewed by Susan S. Benjamin. Chicago Architects Oral History Project, Art Institute of Chicago (2003), as accessed on July 21, 2008.

Holder, Jo, and Joan Kerr (n.d.), "Robert Emerson Curtis" in Dictionary of Australian Artists Online, accessed on May 23, 2008.

Holley, Joe (2006), "Louis Dalton Porter; Used Artistic Skills to Trick German Army" in *Washington Post* (July 8), p. B06.

"Homer Saint-Gaudens" (1941) in *Current Biography*, pp. 745-747.

Hoppin, Jason (2006), "Artist William Saltzman dies at 89" in *St Paul (MN) Pioneer Press* (March 20).

Hong, Kay, and Judy Hughes (1994), "The Concept of Concealment: Mimicry in Nature and Hidden Images in Art" in *Wildlife Art News* Vol 13 No 4 (July-August), pp. 58-64.

"Horace C. Pearsons, 85" (2001) in *Cape Cod Times* (September 17).

Horne, Alistair (1964). *The Price of Glory: Verdun 1916*. Middlesex, England: Penguin.

Horner, Harry (1943), "I'm in the Army Now" in *Theatre Arts* (September), pp. 536-537.

Hough, Emerson (1897). *The Story of the Cowboy*. New York: Grosset and Dunlap.

"How an Artist Discovered a Natural Law (1911), in *Scientific American* (January 7), p. 3.

Hreachmack, Patrick (1996). *The Painter's Guide to World War Two Naval Camouflage*. Phoenixville PA: Clash of Arms Publishers.

Hsu, Hsuan L. (2006), "Who Wears the Mask? (on Neil Leach's *Camouflage*)" in *Minnesota Review*, online at <http://www.theminnesotareview.org/journal/ns67/hsu_hsuan_l_1.shtml>, as accessed on May 22, 2008.

Hulton, Ponus, ed. (1987). *The Acimboldo Effect: Transformations of the Human Face from the 16th to the 20th Century*. New York: Abbeville Press.

Hurst, Hugh (1919), "Dazzle-Painting in War Time" in *International Studio* (September), pp. 93-99.

"**Ideas** in Camouflage from Dame Nature" (1918) in *Current Opinion* (October), p. 240.

Innes, Christopher (2005). *Designing Modern America:*

Broadway to Mainstreet. New Haven CT: Yale University Press.
Inverarity, Robert Bruce (n.d.), Robert Bruce Inverarity Papers in the Archives of American Art.
"Irving Gerald Moshinsky" (n.d.) online biographical data at Obitcentral website at www.obitcentral.com/obitsearch/obits/ny/ny-nassau1.html accessed on March 26, 2008.
Irwin, Will (1917), "Camouflage" in *Saturday Evening Post* (September 29).
Ittelson, William H. (1969). *The Ames Demonstrations in Perception*. New York: Hafner Publishing.
Iverson, Erling F. (1940), "Air Raid Protection" in *Architectural Forum* (November), pp. 429-436.

"**James** A. Brennan, A Song Writer, 71" (1956) in *New York Times* (August 25), p. 15.
James, Bessie R. (1920). *For God, For Country, For Home: The National League for Woman's Service*. New York: G.P. Putnam's Sons.
"James Daugherty" (n.d.). Wikipedia online biography, as accessed on August 27, 2008.
"James McIntosh Patrick" (n.d.) , online biographical data at <en.easyart.com/arthistory/artist/James%20McIntosh> as accessed on August 22, 2006.
"James Milton Sessions" (n.d.) online biographical page at <www.jamesmsessions.com/Sessions_Biography.html>, as accessed on March 23, 2008.
James, William (1920). *The Letters of William James*. Henry James, ed. Boston: Atlantic Monthly Press.
"Jan Gordon Dies; Artist, Musician" (1944) in *New York Times* (February 3), p. 19.
"Jan Gordon" (n.d.) online biographical article at <http://janandcoragordon.co.uk/default.aspx> as accessed on March 27, 2008.
Jaques, Florence Page (1973). *Francis Lee Jaques: Artist of the Wilderness World*. New York: Doubleday.
Jensen, Arthur B. (1919). *Jensen System of Modeling: Employing the Superskill Modeling Device*. Chicago: The Jensen Modeling Device Company.
"Jo Mead, 81, interior designer artist" (2000) in *Chicago Sun-Times* (August 14), p. 52.
Johansen, Waldemar (1943), "Military Uses of Art" in *School Arts Magazine*.
"John Campbell Hutton" (n.d.), biographical article at online directory of New Zealand expatriates at <www.teara.govt.nz/1966/E/ExpatriatesBiographies/UnitedKingdom/en> as accessed on March 8, 2008.
"John Graham Kerr, British Zoologist" (1957) in *New York Times* (April 25), p. 31.
"John Gregory" (n.d.) in *National Encyclopedia of American Biography* Vol XX, pp. 523-524.
"John J.A. Murphy, Artist, Dies; Wood Engraver, Book Designer" (1967) in *New York Times* (January 25), p. 42.
"John Lavalle, Portrait Painter And Landscape Artist, Dead" (1971) in *New York Times* (November 15), p. 44.
"John Lewis" (1996) in *The Independent* (December 28).
"John Sanger Dies; Architect Was 63" (1955) in *New York Times* (August 25), p. 23.
"John S. Detlie, 96:; Architect, Hollywood Movie-Set Designer" (2005) in *Washington Post* (December 4), p. C10.
"John Vassos" (n.d.). Online biography at Industrial Designers Society of America website, as accessed on April 3, 2008.
"John Wilde" (n.d.) online biography at Sullivan Goss Gallery website, as accessed on September 22, 2007.

1919
As the war ends, there is an easing of the (largely British) censorship of news photos of dazzle ship camouflage. On January 4, the *London Illustrated News* publishes a large, two-page spread of photographs of dazzle-painted ships.

A "dazzle [costume] ball" is sponsored by the Chelsea Arts Club at the Royal Albert Hall in London on March 12.

•

1920
British poet Alfred Noyes publishes a poem titled "Cubism" in which he alludes to its resemblance to camouflage and to aerial views of the earth.

Solomon J. Solomon, *Strategic Camouflage.*

American artist William Andrew Mackay establishes a camouflage training school.

•

1921
Following a series of debilitating strokes, Abbott H. Thayer dies.

•

1922
Matthew Luckiesh, *Visual Illusions: Their Causes, Characteristics and Applications.*

•

Johnson, Burges (1920), "Camouflage at the Front" in Rebecca Lawrence Lowrie, ed., *Child's Book of the War.* NY: Harper and Brothers.
Johnson, Jory (1999) "Timeline of American Landscape Architecture: 1940 to 1949" in *Landscape Architecture* Vol 89 (May), pp. 84-85.
Johnson, Paul C., ed. (1973). *The Early Sunset Magazine 1898-1928.* San Francisco: California Historical Society.
"Jo Mielziner Dead at 74; Pioneering Set Designer" (1976) in *New York Times* (March 16), p. 38.
Jon Gnagy: An Exhibition of Paintings and Litho Drawings (1964). Exhibition catalog. Idyllwild CA.
Jones, Barbara, and Bill Howell (1972). *Popular Arts of the First World War.* New York: McGraw Hill.
Jones, Helen (1993), "Davies, Natalia (Nat) (1907 - 1951)" in *Australian Dictionary of Biography* Vol 13. Melbourne AU: Melbourne University Press, p. 583.
Jones, Jonathan (2002), "Lessons of War" in *The Guardian* (May 2).
Jones, Loyd A. (1919), "Low Visibility Phase of Protective Coloration" in *Journal of the Franklin Institute* vol 188 (September), pp. 363-387, and (October), pp. 507-533.
Jones, R.V. (1973), "The Theory of Practical Joking—Its Relevance to Physics" in R.L. Weber, ed., *A Random Walk in Science.* New York: Crane, Russak and Company.
___(1978). *Most Secret War.* London: Hamish Hamilton.
"Joseph B. Abrahams" (1969) in *New York Times* (July 2), p. 43.
"Joseph B. Platt, Designer, Dies; Did 'Gone With the Wind' Sets" (1968) in *New York Times* (February).
"Joseph Burgess, 70, A Portrait Painter" (1961) in *New York Times* (January 18), p. 33.
Julesz, Béla (1995). *Dialogues on Perception.* Cambridge MA: MIT Press.
Jungmann, A.M. (1918), "Dame Nature—Instructor in Camouflage" in *Popular Science Monthly* (September), pp. 346-347.

Kahn, Elizabeth Louise (1984). *The Neglected Majority: "Les Camoufleurs," Art History, and World War I.* Lanham MD: University Press of America.
Kamholtz, Jonathan (1988), "Sited and Unsighted Figures: The Sculpture of Tom Czarnopys" in *Dialogue* Vol 11 No 6, pp. 12-16.
"Karl Amend" (1944) in *New York Times* (January 3), p. 21.
Karson, Robin (2007). *A Genius for Place: American Landscapes of the Country Place Era.* Amherst: University of Massachusetts Press.
Keen, Angeline Myra (1932), "Protective Coloration in the Light of Gestalt Theory" in *Journal of General*

Psychology, pp. 200-203.
"Keith Douglas" (n.d.) on Soldier Poets website (BBC Channel 4) at <www.channel4.com/history/microsites/S/soldier_poets/biog_douglas>as accessed on March 29, 2008.
"Keith S. Williams, Painter, Etcher, 45" (1951) in *New York Times* (February 14), p. 29.
Kelland, Clarence B. (1918), "Uncles Sam's Fake Factory" in *National Weekly Collier's* (September 28), pp. 13, 18 and 28.
Kelley, H. Roy (n.d.), Guide to the H. Roy Kelley Papers at Cornell University, online at <http://rmc.library.cornell.edu/EAD/htmldocs/RMM03864.html> as accessed on June 5, 2008.
Kelly, Donald H. (1990), "Forty Years of Image Technology and Vision Research," in *Optics and Photonics News* (July). Briefer version accessed online at <www.ai.sri.com/v~vision/donkelly.html> on April 7, 2008.
Kent, Rockwell (1955). *It's Me, O Lord.* New York: Dodd, Mead and Company.
Kepes, György (1944). *Language of Vision*. Chicago: Paul Theobald.
___(1956). *The New Landscape in Art and Science*. Chicago: Paul Theobald.
___(1978). Letter to the author dated February 20.
Kern, Stephen (1994), "Cubism, Camouflage, Silence, and Democracy: A Phenomenological Approach" in Roger Friedland and Deirdre Boden, eds. *NowHere: Space, Time and Modernity*. Berkeley: University of California Press.
___(2003). *The Culture of Time and Space, 1880-1918*. Cambridge MA: Harvard University Press.
Kerr, John Graham (1919), "Camouflage of Ships in War" in *Nature* (may 15), pp. 204-205.
___(1941a), "Camouflage in Warfare" in *Nature* Vol 147 (June 21), pp. 758-760.
___(1941b), "Use of Paint in Camouflage" in *Nature* (November 1), p. 527.
Kerr, Joan (n.d.) "George Courtney Benson" at the Dictionary of Australian Artists Online as accessed on March 15, 2008.
___(n.d.) "Alec Brierley" at the Dictionary of Australian Artists Online, as accessed on March 15, 2008.
___(n.d.) "Roy Frederick Leslie Dalgarno" at the Dictionary of Australian Artists Online, as accessed on March 12, 2008.
___(n.d.) "G.T.C. Hubble" at the Dictionary of Australian Artists Online, as accessed on March 17, 2008.
___(n.d.) an Artists Online, as accessed on March 27, 2008.
___(n.d.) "Herbert McClintock" at the Dictionary of Australian Artists Online, as accessed on March 19, 2008.
___(n.d.) "Norm Rice" at the Dictionary of Australian Artists Online, as accessed on March 19, 2008.
___(n.d.) "Ray Whiting" at the Dictionary of Australian Artists Online, as accessed on March 27, 2008.
Kienbusch, William (n.d.). William Kienbusch Papers at the Archives of American Art.
King, Paul B. (n.d.). Paul B. King Papers at the Archives of American Art.
Kingsland, Sharon (1978), "Abbott Thayer and the Protective Coloration Debate" in *Journal of the History of Biology* Vol 11 No 2, pp. 223-244.
Klein, Adrian, and J.C. Mottram (1919), "Military Camouflage" in *Nature* (July 10), p. 364.
Knight, Charles R. (2005). *Autobiography of an Artist*. GT Labs.
Knight, Harry (c1965), Oral History Interview conducted by Harlan Phillips, in the Archives of American Art, Smithsonian Institution.

1928

John D. Whiting, *Convoy: A Story of the War at Sea.*

•

1929

American scientist-turned-artist Charles Bittinger, having overseen US Navy camouflage during World War I, applies scientific means in the development of experimental artworks. He makes painted murals, for example, in which one image is visible when viewed in incandescent light, and an alternative image when viewed in ultraviolet light.

•

1932

American psychologist Angeline Myra Keen publishes an article in which she discusses the resemblance between Abbott H. Thayer's "laws of disguise" and the Gestalt psychologists' "laws of perceptual organization."

St. Louis Browns baseball player Goose Goslin steps up to the plate with a "camouflaged bat," consisting of a standard bat with black and white lengthwise stripes. He is not allowed to use it.

•

1935

French philosopher Roger Caillois publishes an essay on mimicry in an intellectual journal. In 1984, it is translated into English and published in *October*

Koestler, Arthur (1964). *The Act of Creation: A Study of the Conscious and Unconscious in Science and Art*. New York: Macmillan.

___(1978). *Janus: A Summing Up*. New York: Random House.

___and J.R. Smythies, eds. (1971). *Beyond Reductionism.* Boston: Beacon Press.

Koffka, Kurt (1935). *Principles of Gestalt Psychology.* New York: Harcourt Brace.

Köhler, Wolfgang (1947). *Gestalt Psychology.* New York: Liveright.

Konody, Paul G. (1919), "Canada at War as Shown on Canvas" in *New York Times* (June 8).

"Konstantin Alexeyevich Korovin" (n.d.), biographical article at Art-Roots.com website, as accessed on May 22, 2008.

Kostelanetz, Richard, ed. (1970). *Moholy-Nagy.* New York: Praeger.

Kozman, Myron (n.d.), Myron Kozman Papers at University of Illinois at Chicago, with online inventory at <http://www.uic.edu/depts/lib/specialcoll/services/rjd/findingaids/MKozmanb.html> as accessed on May 6, 2008.

Krass, Peter (2007). *Portrait of War: The US Army's First Combat Artists and the Doughboys' Experience in WWI.* New York: John Wiley and Sons.

"Kruger-Gray, 62, Painter, Sculptor" (1943) in *New York Times* (May 5), p. 27.

Kuh, Katharine (1969). *Break-Up: The Core of Modern Art.* Greenwich CT: New York Graphic Arts Society.

Kuhn, Thomas (1970). *The Structure of Scientific Revolutions.* Chicago: University of Chicago Press.

Kurth, Peter (2001). *Isadora: A Sensational Life.* Boston MA: Little Brown.

Labatut, Jean (n.d.) Jean Labatut Papers at Princeton University Library.

Laffin, John (1991). *The Western Front Illustrated 1914-1918.* London: Grange Books.

Lahey, Richard (n.d.). Richard Lahey Papers at the Archives of American Art.

Lamb, Trevor (1995). *Color.* UK: Cambridge University Press.

Lauve, Henry de S. (1942). *Camouflage Manual for General Motors Camouflage School.* Detroit MI: General Motors. 22 pp. Includes General Motors Camouflage and War Service Section Camouflage School Organization Chart, with a list of participants. Online under "Ironically, Harley Earl Named His War Service Section: CAMOUFLAGE" at <www.carofthecentury.com> as accessed on March 28, 2008.

Lavery, Brian (2006) *Churchill's Navy: The Ships, Men and Organization, 1939-1945*. Naval Institute Press.

Lawton, Rebecca E. (n.d.), "Heroic America: James Daugherty's Mural Drawings from the 1930s," as accessed online on April 4, 2008.

Leach, Edmund (1976). *Culture and Communication*. UK: Cambridge University Press.

Leach, Neil (2006). *Camouflage*. Cambridge MA: MIT Press.

Lear, Frederick Roy (n.d.), Papers of Frederick Roy Lear at Syracuse University, with an online finding guide at <http://sumweb.syr.edu/archives/papers/facpapr/frlbio.htm> as accessed on March 16, 2002.

"Ledyard Towle" (1973) in *New York Times* (November 11), p. 73.

Lehndorff, Vera, and Holger Trulzsch (1986). *Veruschka: Trans-figurations*. New York: New York Graphic Arts Society, and Little Brown and Company.

Lekakis, Michael (1987). *Michael Lekakis: A Retrospective Exhibition*. Exhibition catalog. New York: Kouros Gallery.

L'Engle, Madeleine (1974). *The Summer of the Great Grand-Mother*. New York: Farrar, Straus and Giroux.

"Leon Underwood" (n.d.). Online biography at Redfern Gallery website at <www.redfern-gallery.com/pages/artistinfo/40.html> as accessed on April 10, 2008.

Lerner, Nathan (1978). Letter to the author (February 11).

Lettice, M.N. (1981), "Clarke, Sir George Sydenham [Baron Sydenham] (1848-1933" in *Australian Dictionary of Biography* Vol 8. Melboure AU: Melbourne University Press, pp. 13-14.

Levitch, Mark (2006). *Pantheon De La Guerre: Reconfiguring a Panorama of the Great War*. University of Missouri Press.

Levy, Julien (1972). Foreword in William C. Seitz, ed., *Arshile Gorky*. New York: Museum of Modern Art / Arno Press.

Lewis, John (1951), "The Wood Engravings of Blair Hughes-Stanton" in *Image* Vol 6 (Spring).

___ (1994). *Such Things Happen: The Life of a Typographer*. Suffolk, England: Unicorn Press.

"Lewis S. Rubenstein" (n.d.), online biographical article on Poughkeepsie Day School website, as accessed on March 2, 2008.

Ley, Ronald (1990). *A Whisper of Espionage: Wolfgang Köhler and the Apes of Tenerife*. Garden City Park NY: Avery.

Leytham, Geoffrey (n.d.), "Alister Clavering Hardy and Spiritual Evolution" at <http://www.scimednet.org/Leadarts/Leytham_Hardy.htm> as accessed on June 1, 2008.

Liberman, Alexander (1968). *The Artist in His Studio*. New York: Viking Books.

"Lieut. D.I. Reynolds Is Killed in Italy" (1918) in *New York Times* (October 5), p. 5.

Lipman, Jean, ed. (1971). *The Collector in America*. New York: Viking.

Lippard, Lucy R., ed. (1992). *Partial Recall*. New York: New Press.

Lisle, B. Orchard (1941), "Discussions, Comments, Notes: Naval Camouflage" in *US Naval Institute Proceedings* (January), p. 97.

"Leslie Thrasher" (2001) in Read, Walt. *The Illustrator in America 1860-2000*. New York: Society of Illustrators.

Long, Christopher (1979), "Fred Beddington: Sharp-shooter and artist, he became leading expert on camouflage" at <www.christopherlong.co.uk/pri/beddington.html> as accessed on April 4, 2008.

"Louis Dalton Porter" (n.d.) online Wikipedia biography accessed on July 13, 2008.

"Louis de Monard" (n.d.), online biographical data at <www.bronze-gallery.com/

magazine as "Mimicry and Legendary Psychasthenia."

•

1937
In connection with his camouflage training school, William Andrew Mackay publishes a *Handbook on Ship Camouflage*.

American artist and camoufleur Charles Bittinger travels to Canton Island in the Pacific with the US Navy and the National Geographic Society to observe and to paint on location a total solar eclipse.

•

1938
Bauhaus painter Oskar Schlemmer begins working with a painting company in Stuttgart, for which, among other things, he designs and carries out building camouflage.

•

1939
A group of Australian civilians, both artists and scientists, led by British-born zoologist William Daken, establish the Sydney Camouflage Group.

•

19139-45
Designer and painter Max Bill, who will later be Director of the Hochschule für Gestaltung in Ulm, serves in the Swiss Army and is assigned to camouflage.

French_Sculptors/Louis_de_monard.html> as accessed on March 16, 2002.
Love, Richard H. (1999). *Carl W. Peters: American Scene Painter: From Rochester to Rockport*. Rochester NY: University of Rochester Press.
Lowell, Amy (1919). *Pictures from the Floating World*. New York: Macmillan.
"Loyd Ancile Jones" (n.d.) in *National Encyclopedia of American Biography* Vol 41, pp. 75-76.
Luce, Donald, and Laura Andrews (1982). *Francis Lee Jaques: Artist-Naturalist*. St Paul: University of Minnesota Press.
Luckiesh, Matthew (1919), "The Principles of Camouflage" in *Scientific American* (January 25, and February 8 and 22).
___(1965). *Visual Illusions: Their Causes, Characteristics and Applications*. New York: Dover Publications.
Lysohir, Marilyn (1988), "The Dark Side of Dazzle" in *Ceramic Monthly* Vol 36 No 6 (June/July/August), pp. 44-47.

MacAulay, Bettina (1993) "Annand, Douglas Shenton (1903 - 1976)," in *Australian Dictionary of Biography*. Vol. 13. Australia: Melbourne University Press, p. 60.
MacCarthy, Fiona (1972). *All Things Bright and Beautiful*. London: Allen and Unwin.
MacGregor, Alasdair Alpin (1961). *Percyval Tudor-Hart 1873-1954: Portrait of an Artist*. London: P.R. Macmillan Limited.
MacIntosh, Heather M. (1998), "Bain, William James Sr. (1896-1985)," biographic essay at <www.historylink.org/essays/output.cfm?file_id+117> as accessed on March 14, 2008.
Mackay, William A. (1919). "Process of Rendering Objects Less Visible Against Backgrounds," US Patent No. 1305296.
___(1937) *Handbook on Ship Camouflage*. Washington DC: US Navy, Bureau of Construction and Repair.
MacKenzie, Alister (1915), "Military Entrenchments" in *Golf Illustrated* Vol 3 No 1, pp. 42-44.
___(1919), "Entrenchments and Camouflage" [Lecture by a British Officer Skilled in Landscape Gardening] in *Professional Memoirs, Corps of Engineers*, US Army and Engineer Department at large Vol 14 No 47, pp. 574-638.
___(date), "Common Sense of Camouflage Defense" in *Military Engineer* Vol XXVI No 145, pp. 42-47.
___(1920). *Golf Architecture: Economy in Course Construction and Green-Keeping*. London: Simpkin, Marshall, Hamilton, Kent and Co. Ltd.
MacMorris, (Leroy) Daniel (n.d.). (Leroy) Daniel MacMorris

Papers at the University of Missouri, described online at <www.umkc.edu/WHMCKC/Collections/IKC0107.HTM>, as accessed on June 4, 2008.
Maddry, Larry (1997). "His Days in Norfolk Forever Colored Artist's Work" in *The Virginian-Pilot*, accessed online on March 13, 2008 at Scholarly Communications Project, University Libraries, Virginia Tech.
"Made Films at Verdun" (1916) in *New York Times* (October 26).
"Major, 'Over 60,' Takes G.I. Course" (1951) in *New York Times* (April 22), p. 27.
Marcham, Frederich George, ed. (1971). *Louis Agassiz Fuertes and the Singular Beauty of Birds*. New York: Harper and Row.
Margrie, Victor (2002), "Robert Goodden" in *The Independent* (April 13).
"Marine Camouflage" (n.d.) Unpublished US Navy memorandum [World War I era], with annotations by Everett Warner.
Marston, John (1942). *Camouflage: Field Training Manual, Second Marine Division*. Washington DC: US Marine Corps.
Martin, Douglas (2002). "Edward Brodney, 92, Who Painted War Scenes" in *New York Times* (August 19).
Masey, Jack, and Conroy Lloyd Morgan (2008). *Cold War Confrontations: US Exhibitions and their Role in the Cultural Cold War*. Birkhauser.
Maskelyne, Jasper (1949). *Magic: Top Secret*. London: Stanley Paul.
Mauch, James William (2000). *Aware: A Retrospective of the Life and Work of Eric Sloane*. Laurys Station PA: Garrigues House.
Mavigliano, George J. (1987) "The Chicago Design Workshop: 1939-1943" in the *Journal of Decorative and Propaganda Arts* Vol 6 (Autumn), pp. 34-47.
"Max(well Spencer) Dupain" (n.d.). Biographical article in Grove Dictionary of Art at GroveArt.com.
"Stanley McCandless, Taught Lighting at Yale Drama School" (1967) in *New York Times* (August 5), p. 21.
McCann, E. Armitage (1931), "Dazzle-Painting a Model" in *Popular Mechanics* (April), p. 89
"M'Clelland Barclay Is Sued for Divorce" (1930) in *New York Times* (January 17), p. 25.
"McClelland Barclay Missing at Sea: Artist Was Doing Pacific Sketches" (1943) in *New York Times* (July 25), p. 1.
"McClelland Barclay Listed as War Victim"(1944) in *New York Times* (July 25), p. 6.
McCloskey, Barbara (2005). *Artists of World War II*. Greenwood Publishing.
McCormick, Ken, and Hamilton Darby Perry, eds. (1990). *Images of War: The Artist's Vision of World War II*. New York: Orion Books.
McElroy, Gil (2008). *Razzle Dazzle: The Uses of Abstraction*. Exhibition catalog. Oshawa, Ontario, Canada: Robert McLaughlin Gallery.
McGill, Douglas G. (1987), "Michael Lekakis, Sculptor of Abstract Works" in *New York Times* (November 19).
"Meldon Mitchell, Commercial Artist" (1967) in *New York Times* (February 14), p. 43.
Menchaca, Juan (1964), as interviewed by Sylvia Loomis on November 13. Oral history transcript in the Archives of American Art.
Méndez Baiges, Maite (2007). *Camuflaje: Engano y Ocultación en El Arte Contemporáneo*. Madrid, Spain: Ediciones Siruela.
Merilaita, S. (1998), "Crypsis Through Disruptive Coloration in an Isopod" in *Proceedings of the Royal Society B* Vol 265, pp. 1059-1064.
Merton, T.R. (1940), "Camouflage in Modern Warfare." Letter to the editor in *Nature* Vol 146 No 3700 (September 28), p. 429-430.
Merriman, Bob (1943), "Camouflage Began With Trojan Horse" in *The Flat Hat* [College

1940
American illustrator McClelland Barclay applies for a patent for a dazzle-like system of camouflage for airplanes. Prototypes are tested but the system is never adopted.

An article in *Time* magazine claims that the first course in industrial camouflage in the US is being conducted at the Kansas City Art Institute by artist Keith Martin.

British zoologist and camouflage instructor Hugh B. Cott publishes a widely-admired book on natural camouflage titled *Adaptive Coloration in Animals*.

Cyril H. Smith, *Camouflage Simply Explained*.

American industrial designer Erling F. Iverson initiates courses in civilian camouflage and air raid precautions at Pratt Institute in New York.

•

1941
British documentary filmmaker Geoffrey Barkus serves as Director of Camouflage for the Middle East Forces of the Royal Engineers. He oversees a now legendary wartime deception, in which he and his camouflage artists create the illusion of a massive build-up of forces at El Alamein in the North African desert.

of William and Mary student newspaper] (April 6), p. 3.
Meryman, Richard (1999), "A Painter of Angels Became the Father of Camouflage" in *Smithsonian Magazine* (April), pp. 116-128.
Metzger, Wolfgang (1975). *Gestze des Sehens*. Frankfurt am Maine: Verlag Waldemar Kramer.
Meyer, Leonard B. (1956). *Emotion and Meaning in Music*. Chicago: University of Chicago Press.
___(1967). *Music, the Arts and Ideas: Patterns and Predictions in Twentieth-Century Culture*. Chicago: University of Chicago Press.
Michael Lekakis: A Retrospective Exhibition (1987). Exhibition catalog. New York: Kouros Gallery.
"Michel Jacobs" (n.d.), biographical article at AskArt website,as accessed on April 4, 2008.
Millikan, Robert (n.d.). Robert Millikan Papers in the Archives of the California Institute of Technology.
"Miss Jackson Weds Chas. L. Thrasher" (1920) in *New York Times* (June 15).
Mitchell, Robert W., and Nicholas S. Thompson, eds. (1986). *Deception: Perspectives on Human and Nonhuman Deceit*. Albany NY: State University of New York Press.
Moholy-Nagy, Laszlo (1947a). *The New Vision and Abstract of an Artist*. New York: George Wittenborn.
___(1947b). *Vision in Motion*. Chicago: Paul Theobald.
Moholy-Nagy, Sibyl (1969). *Moholy-Nagy: Experiment in Totality*. Cambridge MA: MIT Press.
Mooney, Rick (1997), "Dress for Success" in *Farm Journal* (January), p. C-3.
Moore, Helen Watkeys, comp. (1918). *Camouflage Cookery: A Book of Mock Dishes*. New York: Duffield and Company.
Morgan, Ann Lee, ed. (1984). *Contemporary Designers*. Detroit MI: Gale Research Company.
Morison, Elting, ed. (1954). *The Letters of Theodore Roosevelt*. Cambridge MA: Harvard University Press.
Morley, Christopher (1920). *Travels in Philadelphia*. Philadelphia PA: David McKay.
Morton, Edward (n.d.). Conversation with the author (in the late 1970s).
Mount, Charles Merrill (1955). *John Singer Sargent: A Biography*. New York: W.W. Norton.
"Mrs. George Brush" (1949) in *New York Times* (July 30), p. 15.
"Mrs. Gordon Smith, Art Gallery Aide" (1959) in *New York Times* (July 25), p. 17.
Muirhead, Desmond (1995), "Symbols in Golf Course Architecture" in *Executive Golfer* (July).
Muncaster, Martin (1978). *The Wind in the Oak: The Life,*

Work and Philosophy of the Marine and Landscape Artist Claude Muncaster. London: Robin Garton.
Murchie, Guy (1978). *The Seven Mysteries of Life: An Exploration in Science and Philosophy.* Boston: Houghton Mifflin.
Museum of Modern Art (1972). *Britain at War.* New York: Ayer Publishing.
Nash, Eric. P. (2001), "Seymour V. Reit, 83, a Creator of Casper the Friendly Ghost" in *New York Times* (December 17).
"Natural and Artificial Camouflage" (1919) in *Nature* Vol 102 No 2569 (January 23), p. 408.
"Natural Camouflage" (1941) in *Time* (August 4).
Naversen, Ronald (1989). *The Scenographer as Camoufleur.* Dissertation. Carbondale IL: Southern Illinois University.
"Nazis Say Britain's Bombers Are Invisible; Declare 'Secret Varnish' Hid Berlin Raiders" (1940), in *New York Times* (August 30), pp. 1 and 3.
Nemerov, Alexander (1997), "Vanishing Americans: Abbott Thayer, Theodore Roosevelt, and the Attraction of Camouflage" in *American Art* Vol 11 No 2 (Summer), pp. 51-81.
"New York Artist Killed in France" (1918) in *New York Times* (June 28), p. 4.
Newark, Tim (1998). *Brassey's Book of Uniforms.*London: Brassey's.
___, and Jonathan Miller (2007). *Camouflage.* London: Thames and Hudson.
___, Quentin Newark and J.F. Borsarello (1996). *Brassey's Book of Camouflage.* London: Brassey's.
Nibbe, Kevin (n.d.). Artist list for dissertation titled *The Great Opportunity: American Artists and the Great War,* online at <www.klnibbephd.info/htm/dissertation.html> as accessed on March 17, 2008.
Nicolaides, Kimon (1941). *The Natural Way to Draw.* Boston: Houghton Mifflin.
Norris, Margot (1985). *Beasts of the Modern Imagination: Darwin, Nietzche, Kafka, Ernst, and Lawrence.* Baltimore MD: Johns Hopkins University Press.
Novotny, Ann (1976). *Alice's World: The Life and Photography of an American Original, 1866-1952.* New York: Devin-Adair.
"Now It Can Be Told" (1945) in *New York Journal-American* (December 5), p. 19.
Noyes, Alfred (1920). *Collected Poems of Alfred Noyes.* New York: Frederick A. Stokes.

"Object 351" (n.d.), People's Choice Exhibition website at <www.portsmouthmuseums.co.uk/pages/351-1997_639.htm> as accessed on April 4, 2008.
"Observer Hides in Dummy Horse Near Enemy Trench" (1918) in *Popular Mechanics* (January), p. 72.
O'Connell, Kim A. (2000), "Designing for War" in *Landscape Architect* Vol 34 (November) pp. 32-35 & 87-90.
Olson, Stanley (1986). *John Singer Sargent: His Portrait.* New York: St Martin's Press.
Oppenheimer, Melanie (1996), "Keighley, Erna Laura (1891 - 1955)" in *Australian Dictionary of Biography* Vol 14. Melbourne AU: Melbourne University Press, pp. 606-607.
Osborne, Harold (1981). *The Oxford Companion to Twentieth-Century Art.* Oxford, England: Oxford University Press.

Park, Edwards (1985), "A Phantom Division Played a Role in Germany's Defeat" in *Smithsonian Magazine* (April), pp. 138-147.
Parsons, Betty (1981), Oral History Interview conducted by Gerald Silk on June 11, in the Archives of American Art, Smithsonian Institution.

In anticipation of the US entry into World War II, various New York-based theatrical designers form a civilian organization called the Camouflage Society of Professional Stage Designers. They are trained in camouflage (by Baron Nicholas Cherkassoff and Herbert Kniffen) in order to make it more likely that they will serve as wartime camoufleurs.

In Australia, zoologist William Dakin becomes technical director of camouflage for the Ministry of Home Security, and publishes a manual called *Art of Camouflage*.

British Lieutenant Colonel C.H.R. Chesney publishes *The Art of Camouflage*, in which he is openly critical of the use of artists as camouflage experts.

As the first director of the Art and Color Division (later the Styling Division) of General Motors, automotive designer Harley Earl begins a camouflage research and training division at the company. Other camoufleurs at GM include Ledyard Towle, David Jones and McClelland Barclay.

With British artist Stanley William Hayter and several others, American architect Percival Goodman forms a short-lived industrial camouflage firm, the Camouflage Engineering Company.

Parsons, William Barclay (1920). *The American Engineers in France*. New York: D. Appleton.

"Pasteboard Soldier Built to Aid Camouflage" (1917) in *Popular Mechanics* (October), p. 485.

Patoureau, Michel (2001). *The Devil's Cloth: A History of Stripes and Striped Fabric*. New York: Columbia University Press.

"Paul B. King, Noted as Portrait Artist" (1947) in *New York Times* (November 26), p. 23.

Paul, Elliot (1942), "Art and Camouflage" in Frank Crowninshield, ed., *Vogue's First Reader*. New York: Julian Messnér, pp. 256-259.

"Paul F. Brown" (1944) in *New York Times* (December 9), p. 15.

"Paul Gerchik, Army Camouflage Expert" (1998) in *Los Angeles Times* (August 20).

"Paul McCobb, Modern Designer, Dies" (1969) in *New York Times* (March 12), p. 33.

Pearce, Barry (1996), "Jackson, James Ranalph (1882 - 1975)" in *Australian Dictionary of Biography* Vol 14. Melbourne AU: Melbourne University Press, pp. 544-545.

___(2002), "Strachan, David Edgar (1919 - 1970)" in *Australian Dictionary of Biography* Vol 16. Melbourne AU: Melbourne University Press, pp 325-326.

Peixotto, Ernest (1917), "Special Service for Artists in War Time" in *Scribner's Magazine* Vol 62 No 1 (July), pp. 1-10.

___(1919). *The American Front*. New York: Charles Scribner's Sons.

Pennoyer, Albert Sheldon (1954). *Locomotives in Our Lives: Railroad Experiences of Three Brothers for More Than Sixty Years, 1890-1951*. New York: Hastings House.

"Pennsylvania Academy Exhibition" (1918) in *New York Times* (February 10).

Penrose, Roland (1941). *Home Guard Manual of Camouflage*. London: G. Routledge and Sons.

___ (1973). *Picasso: His Life and Work*. New York: Harper and Row.

___(1981). *Scrapbook 1900-1981*. London: Thames and Hudson.

Perry, Lawrence (1919). *Our Navy in the War*. New York: Charles Scribner's Sons.

Peterson, Barbara Bennett (2004), "Mary Pukui" in Susan Ware, et al., *Notable American Women: A Biographical Dictionary*. Cambridge MA: Harvard University Press.

Pfennigwerth, Ian (2007). *The Australian Cruiser Perth 1939-1942*. Australia: Rosenberg Publishing.

"Philip Little, 84, A Marine Artist" (1942) in *New York Times* (April 1), p. 21.

"Plenty of Camouflage on Vessels in New York Harbor"

(1918) in *New York Times* (January 20), p. 63.
Pollock, Lindsay (2006). *The Girl with the Gallery: Edith Gregory Halpert and the Making of the Modern Art Market*. New York : Public Affairs.
Pope, Erica (2007), "Keep Feeling Obliteration: Anne Polashenski dynamically reinterprets the art of camouflage" at <www.lasvegascitylife.com> (January 18).
Porter, Harold E. (1921). *Aerial Observation: The Airplane Observer, the Balloon Observer and the Army Corps Pilot*. New York: Harper.
Portmann, Adolf (1959). *Animal Camouflage*. Ann Arbor MI: University of Michigan Press.
___(1967). *Animal Forms and Patterns*. New York: Schocken Books.
Poulton, E.P. (1902), "The meaning of the white undersides of animals" in *Nature* Vol 65 No 1695, 596-597.
___(1921), "Natural Camouflage" in *Nature* Vol 107 No 2689 (May 12), pp. 338-340.
Pousette-Dart, Nathaniel (1931). *Ernest Haskell: American Etcher, His Life and Work*. New York: T. Spencer Hutson.
Powell, E. Alexander (1917) *Italy at War*. New York: Charles Scribner's Sons.
Power, Rachel (2002), "Duncan, George Bernard (1904 - 1974)" in *Australian Dictionary of Biography* Vol 16. Melbourne AU: Melbourne University Press, pp 68-70.
Preusser, Robert O. (n.d.), Robert O. Preusser Papers in the Archives of American Art.
"Prof K.F. Wittman of Pratt Institute" (1951) in *New York Times* (April 18), p. 31.
"Progress of Camouflage" (1940) in *Nature* 146 (October 12), pp. 482-483.
"Protective Coloration of Fleets" (1916) in *Literary Digest* (August 26), p. 458.
Proudfoot, Helen, "Brown, Alfred John (1893 - 1976)" (1993) in *Australian Dictionary of Biography* Vol 13, pp 271-272. Melbourne AU: Melbourne University Press.
Pumpelly, Raphael (1918). *My Reminiscences*. New York: Henry Holt.
Pycraft, W.P. (1925). *Camouflage in Nature*. London: Hutchinson.

Rabinowitz, Nancy (2005), "Blending into the Background" in [Brattleboro VT] *Weekend Reformer* (September 24-25), p. 21.
"Ralph Samuels, 65, an Ex-Dean Of Photography Institute, Dead" (1972) in *New York Times* (December 21), p. 38.
"Ralph T. Walker Is Dead at 83; Hailed as 'Architect of Century'" (1973) in *New York Times* (January 18), p. 44.
Rasmussen, Cecilia (2002), "Douglas' Dream Took Wing in Santa Monica" in *Los Angeles Times* (August 4).
Rasmussen, Steen Eiler (1964). *Experiencing Architecture*. Cambridge: MIT Press.
"Ray Greenleaf" (1950) in *New York Times* (February 15), p. 27.
Raymond, Jay (n.d.). *The Golf-Father*. Foxrock Books online at www.foxrockbooks.com.
Reed, Walt (2001). *The Illustrator in America 1860-2000*. New York: Society of Illustrators.
"Reginald H. Church" (1942) in *New York Times* (November 8), p. 53.
Regusis, Paul (1990a). Letter to the author dated August 21.
___(1990b). Letter to the author dated September 12.
Reik, Theodor (1949). *Listening with the Third Ear*. New York: Farrar, Straus and Company.
Reinhardt, Ad (1975). *Art-As-Art: The Selected Writings of Ad Reinhardt*. Barbara Rose, ed. New York: Viking Press.
Reiss, Roland (1997/1999). Oral history interview conducted by Paul Karlstrom on August 22 and September 9, 1997, and June 11, 1999, in the Archives of American Art, Smithsonian Institution.

Armenian-born US painter Arshile Gorky launches a camouflage course at Grand Central School of Art in New York.

British artist and camoufleur Edward Seago requisitions hundreds of life-size models of Neville Chamberlain, dressed in battle uniforms and helmets, for use as dummy soldiers.

•

1942

Animators and set designers at the Walt Disney Studios, among others, contribute to the camouflage of US aircraft factories on the West Coast.

Industrial designer Norman Bel Geddes designs camouflage for military and civilian clients, as do other US designers.

Konrad Wittman, *Industrial Camouflage Manual.*

In London, Stanley William Hayter founds an industrial camouflage firm (his second) called the Industrial Camouflage Research Unit, with artists Roland Penrose, John Buckland Wright, Julian Trevelyan and others.

Eric Sloane, *Camouflage Simplified.*

Hungarian-born painter and designer György Kepes begins teaching a camouflage course at the Institute of Design in Chicago. With Laszlo Moholy-Nagy, at the request of the

Reit, Seymour (1978). *Masquerade: The Amazing Deceptions of World War II.* New York: Hawthorn.

"Remarkable Example of Camouflage" (1919) in *Scientific American* (May 10), p. 491.

Reynolds, Peter (2000), "Parkes, Cobden (1892 - 1978)" in *Australian Dictionary of Biography* Vol 15. Melbourne AU: Melbourne University Press, pp. 569-570.

"Richard H. Morton" (n.d.), online biographical data at <www.brown.edu/Facilities/ University_Library/ exhibits/Artists/files/Artists.html> as accessed on March 10, 2008.

"Richard Levin" (n.d.) in Oldclaysmorian Society Newsletter at <www.ocsociety.co.uk/ Newsletter%2001/obitlevin.htm, as accessed on March 22, 2008>.

Richardson, Brenda (1998). *Andy Warhol: Camouflage.* Exhibition catalog. New York: Gagosian Gallery.

Richardson, John (1996). *A Life of Picasso.* Vol II: 1907-1917. New York: Random House.

Richardson, John Adkins (1971). *Modern Art and Scientific Thought.* Urbana IL: University of Illinois Press.

"Richard S. Meryman, Portrait Painter" (1963) in *New York Times* (October 1), p. 39.

"Richmond Knapp Fletcher" (1920) in *Harvard College Class of 1908.* Secretary's Third Report. Decennial Report (April), pp. 170-171.

Rickard, Grenville (1942). "Camouflage: Then and Now," in *The Military Engineer,* 34 (April), pp. 189-197.

"Robert Lockwood" (n.d.), biographical listing on AskArt.com website, as accessed on May 22, 2008.

Roberts, Jan (2000) "Latour, Loma Kyle (1902 - 1964)" in *Australian Dictionary of Biography* Vol 15. Melbourne AU: Melbourne University Press, pp. 59-60.

Robson, Arthur D. (1942), "Camouflage." US Patent No. 2,292,848.

Rock, Irvin (1995). *Perception.* New York: Scientific American Library.

"Rodrigo Moynihan" (n.d.), at Tate Archives Online as accessed at on 23 March 2008.

Rodyenko, Peter (1936) "On Camouflage Training" in *The Reserve Officer* Vol 13 No 9 (September), pp. 5-7.

___(1941) "Camouflage: Modern aerial photography outmodes paint, results in introduction of new techniques" in *Life* magazine (January 13), pp. 39-42.

Roosevelt, Theodore (1919). *African Game Trails.* New York: Charles Scribner's Sons.

Root-Bernstein, Robert Scott (1991). *Discovering.* Cambridge MA: Harvard University Press.

Rose, Kenneth Jon (1983), "Survival by Deception" in *Science Digest* (December), pp. 59ff.

Rosenberg, Harold (1962). *Arshile Gorky: The Man. The Time. The Idea.* New York: Horizon Press.
Rosenberg, Louis Conrad (n.d.), Louis Conrad Rosenberg Collection at the University of Oregon. Accessed online on June 23, 2008.
Rosenblum, Robert (1961). *Cubism and Twentieth-Century Art.* New York: Harry Abrams.
Roshwald, Aviel, and Richard Stites (2002). *European Culture in the Great War: The Arts Entertainment and Propaganda, 1914-1919.* UK: Cambridge University Press.
Roskam, Albert (1987). *Dazzle Painting: Kunst Als Camouflage: Camouflage Als Kunst.* Exhibition catalog. Rotterdam, The Netherlands: Stichting Kunstprojecten en Uitgevergij Van Spijik.
Ross, Alan (1983). *Colors of War: War Art 1939-45.* London: Jonathan Cape.
Rossine, Vladimir Baranoff (n.d.). Wikipedia online biography, as accessed on March 7, 2008.
Rotzoll, Brenda Warner (2000). "Obituaries: Jo Mead 81, Interior Design Artist" in *Chicago Sun-Times* (August 14), p. 52.
Roussanne, F. (1994). "Chronique de l'Institut d'Historie des Conflts Contemporain" in *Guerres Mondiales et Conflits Contemporains* Vol 44 No 175, pp. 147-150.
Rowe, C.H. (1940), "Camouflage in Wartime" in *Nature* (August 3), p. 168.
Rubin, Harry (1943), "The Art of Camouflage" in *School Arts* (September), pp. 9-13.
Rumrill, Alan F., and Carl B. Jacobs, Jr. (2007). *Steps to Great Art: Barry Faulkner and the Art of the Muralist.* Keene NH: Historical Society of Cheshire County. [Includes CD-ROM recording of Barry Faulkner speaking about Abbott H. Thayer and camouflage.]
Rutledge, Martha (1996), "Ellis, Melicent Jane (Jean) (1887 - 1974)" in *Australian Dictionary of Biography* Vol 14. Melbourne AU: Melbourne University Press, pp 97-98.
Rutter, Frank (1999). "The Influence of the War on Art" in H.W. Wilson, ed., *The Great War.* 12, Chapter 263. Trident Press.
Ruxton, Graeme D., Thomas N. Sherratt and Michael P. Speed (2004). *Avoiding Attack: The Evolutionary Ecology of Crypsis, Warning Signals and Mimicry.* UK: Oxford University Press.

Sage, John Hall (1917), "Thirty-Fourth Stated Meeting of the American Ornithologists Union" in *Auk* Vol XXXIV No 1 (January), pp. 76-85.
Saint-Amour, Paul K. (2003), "Modernist Reconnaissance" in *Modernism/Modernity* Vol 10 No 2, pp. 349-380.
Saint-Gaudens, Homer (1919). "Camouflage and Art" in *Art Bulletin* Vol 2 No 1 (September), pp. 23-30.
___(1933), "Camouflage Reminiscences" in *Military Engineer* Vol 25 No 141 (June), pp. 242-248.
___(1941a) "We're the Men of the Fantasy Forces: Military Camouflage" in *Christian Science Monitor* (June 21), p. 6.
___(1941b). *The American Artist and His Times.* New York: Dodd Mead.
___(1941c) "Present Day Camouflage" in *Reserve Officer* (May), pp. 10-13.
Salmon, Evan (2006). *Dazzle Ship: Evan Salmon.* Exhibition brochure. Wollongong City Gallery: Wollongong, New South Wales, Australia.
Schiffman, Nathan (1997). *Abracadabra!: Secret Methods Magicians and Others Use to Deceive Their Audience.* Amherst NY: Prometheus Books.
Schlemmer, Tut, ed. (1972). *The Letters and Diaries of Oskar Schlemmer.* Middletown CT: Wesleyan University Press.
Schoenauer, Norbert (n.d.), "Nobbs at McGill," online biographical essays at <http://

city's mayor, he explores the possibility of disguising the nighttime appearance of the city's landmarks from the air, to protect it from enemy bombing.

Major Robert P. Breckenridge (descended from Confederacy Secretary of War John Caleb Breckenridge) of the US Corps of Engineers publishes *Modern Camouflage: The New Science of Protective Concealment.*

Spanish Surrealist Salvador Dali publishes an article in *Esquire* magazine on "Total Camouflage for Total War," in which he states his intention to influence World War II camouflage, just as Cubism (and his nemesis Picasso) had influenced that of World War I.

•

1943

A tactical deception unit is formed by the US Army for the purpose of deceiving the enemy by, among other things, impersonating other US Army units and staging phony battlefield events. Known as the Ghost Army, they arrive secretly in England in 1944.

Having graduated from Pratt Institute (where he was trained in camouflage), American designer Jay Doblin works as a camouflage designer at Raymond Loewy Associates in New York.

•

cac.mcgill.ca/Nobbs/fontanus-mcgill.htm> as accessed on April 4, 2008.

Schoener, Allon (1989), "Noel Martin—A Designer's Designer" in *Cincinnati Arts* Vol 2 No 1 (Summer), pp.52-61.

Schwabacher, Ethel (1957). *Arshile Gorky.* New York: Macmillan, pp 81-82.

___(1978). Letter to the author dated January 30.

Schwartz, Hillel (1996). *The Culture of the Copy: Striking Likenesses, Unreasonable Facsimiles.* New York: Zone Books.

Scott, Peter (1961). *The Eye of the Wind.* Boston: Houghton Mifflin.

"Sea Camouflage Against U-Boats: The Art of 'Dazzle-Painting" (1919) in *Illustrated London News* (January 4), pp. 20-22.

Seaman, Lloyd (1919), "Masterpieces of Navy Camouflage" in *Popular Mechanics* (1919), pp. 217-219.

"Seeing But Not Seen" (1918), in *Scientific American* Vol 118 (May 18), pp. 451 and 464.

"Seeking AWOL Wartime Heritage" (2005) in *Heritage NSW: Newsletter of the NSW Heritage Office and Heritage Council of NSW* 12 No 1 (Autumn), p. 8.

Seitz, William C. (1972). *Arshile Gorky.* New York: Museum of Modern Art / Arno Press.

Selvig, Forrest (1968), Oral History Interview with William Kienbusch in Archives of American Art.

Serle, Geoffrey (1986), "MacDonald, James Stuart (Jimmy) (1878 - 1952)" in *Australian Dictionary of Biography* Vol 10. Melbourne AU: Melbourne University Press, pp. 251-252.

Shapiro, Theda. (1976). *Painters and Politics.* New York: Elsevier.

"Shared Experience–Art and War: Rodrigo Moynihan CBE RA" (n.d.), online information at <http://www.awm.gov.au/sharedexperience/artist/artist.asp?id=48> as accessed on March 20, 2008.

Shaw, Evelyn Levy (1985). "A Tribute to Meyer Abel" in *Art Academy News* [Cincinnati OH] Vol 3 No 9 (May).

"Sheffield Kagy" (n.d.). Wikipedia online biography, as accessed on March 7, 2008.

Sherratt, Thomas N., Arash Rashed and Christopher D. Beatty (2005), "Hiding in Plain Sight" in *Trends in Ecology and Evolution* Vol 20 No 8 (August), pp. 414-416.

"Shows New Ship Painting" (1917) in *New York Times* (October 20), p. 20.

Siegriest, Louis (1978). Oral history interview on June 21, 1978, in the Archives of American Art, Smithsonian Institution.

Silcox, David P. (1996). *Painting Place: The Life and Work of*

David B. Milne. Toronto ON: University of Toronto Press.
Silvers, Phil (1974). *The Man Who Was Bilko: The Autobiography of Phil Silvers*. New York: W.H.Allen.
Skerrett, Robert G. (1919), "How We Put It Over on the Periscope" in *The Rudder* Vol 35 No 3 (March), pp. 97-102, and Vol 35 No 4 (April), pp. 175-179.
Sloane, Eric (1941), "Elementary Notes on Camouflage" in *Air Progress* (October), p. 15.
___(1942). *Camouflage Simplified*. New York: Devin Adair.
Slosson, Preston W. (1931). *The Great Crusade and After 1914-1928*. New York: Macmillan.
Smith, Corinna Lindon (1962). *Interesting People*. Norman OK: University of Oklahoma Press.
Smith, Cyril H. (c1940). *Camouflaged Simply Explained*. London: Pitman and Sons.
Smith, David Livingston (2004). *Why We Lie: The Evolutionary Roots of Deception and the Unconscious Mind*. New York: Macmillan.
Smith, David R. (1978). Letter to the author dated February 1.
Smith, Dwight La Vern (1979). *The American and Canadian West: A Bibliography*. ABC-Clio.
Smith, Isabel L. (1921), "Camouflage in the United States Navy" in *Daughters of the American Revolution Magazine* Vol LV No. 8 (August), pp. 426-432.
Smith, J. André (1917), "Notes on Camouflage" in *Architectural Record* (November), pp. 468-477.
Smith, Robert (2001), "Esteban Vincente Dies at 97" in *New York Times* (January 12).
Smith, Sydney Ure, ed. (1944). *Douglas Annand: Drawings and Paintings in Australia*. Sydney Ure Smith.
Solomon, Solomon J. (1920a). *Strategic Camouflage*. London.
___(1920b), "The Secret of German Camouflage" in *World's Work* (April), pp. 453-458.
"Solomon Joseph Solomon" (n.d.) Wikipedia online biography, as accessed on November 29, 2007.
"Solomon J. Solomon" (n.d.) on Tate Archives online, accessed on November 29, 2008.
"Some Thoughts on Camouflage" (1939) in *Action Francaise* (July 15).
Sozanski, Edward J. (2002), "Rudolf Staffel, 90, Famous Ceramic Artist" in *Philadelphia Inquirer* (January).
Spalding, Elisabeth (1944). "A Portrait." Transcript of a lecture on Henry McCarter, in the Archives of American Art, Smithsonian Institution.
Spearritt, Katie, and Peter Spearritt (1990), "Trompf, Percival Albert (1902 - 1964)" in *Australian Dictionary of Biography* Vol 12. Melbourne AU: Melbourne University Press, pp. 264-265.
Spearritt, Peter (1993), "Bunning, Walter Ralston (1912 - 1977)" in the *Australian Dictionary of Biography* Vol 13. Melbourne AU: Melbourne University Press, pp. 297-298.
Staerck, ed. (1998) *Allied Photo Reconnaissance of World War II*. Sand Diego CA: Thunder Bay Press.
Stahl, Joan (1995). *American Artists in Photographic Portraits: From the Peter A. Juley and Son Collection, National Museum of American Art, Smithsonian Institution*. New York: Dover Publications.
Stanley, Roy M. II (1998). *To Fool A Glass Eye: Camouflage versus*

1944
Roland Penrose, *Home Guard Manual of Camouflage.*

György Kepes publishes *Language of Vision*, in which he acknowledges his indebtedness to the Gestalt psychologists.

•

1946
As the official artist for Operations Crossroads, camoufleur Charles Bittinger is one of the artists designated to paint the first atomic explosion at the Bikini Atoll.

•

1951
Lancelot Law Whyte, ed., *Aspects of Form.*

•

1952
Geoffrey Barkus, *The Camouflage Story.*

•

1954
German-born Gestalt psychologist and art theorist Rudolf Arnheim, having immigrated to the US in 1940, publishes a major influential book on art in relation to vision, titled *Art and Visual Perception: A Psychology of the Creative Eye.*

•

1957
Julian Trevelyan, *Indigo Days.*

Photoreconnaissance in World War II. Washington DC: Smithsonian Institution Press.
"Stanley Rogers, 73, Author and Artist" (1961) in *New York Times* (January 23), p. 23.
Steegmuller, Francis (1968). *Cocteau: A Biography.* Boston: Nonpareil Books.
Stein, Gertrude (1933). *The Autobiography of Alice B. Toklas.* New York: Random House.
___(1938). *Picasso.* London: P.T. Batsford.
___(1945). *Wars I Have Seen.* New York: Random House.
"Stephen Bone" (n.d.), biographical data at Tate Archives Online, as accessed on March 23, 2008.
Sterling, Keir B., et al. (1997). *Biographical Dictionary of American and Canadian Naturalists and Environmentalists.* Westport CT: Greenwood Press.
"Steven Kyle, 66, Designer, Is Dead" (1979) in *New York Times* (October 18), p. D23.
"Steven Spurrier" (n.d.). Wikipedia online biography, as accessed on October 12, 2008.
Stevens, Martin, I.C. Cuthill et al. (2007), "The Effectiveness of Disruptive Coloration as a Concealment Strategy" in J.M. Alonso, et al., eds., *Progress in Brain Research.* Amsterdam: Elsevier, pp. 49-65.
Stevenson, William (1976). *A Man Called Intrepid: The Secret War.* New York: Harcourt.
Stewart, Ian, and Martin Golubitsky (1992). *Fearful Symmetry: Is God a Geometer?* UK: Penguin Books.
Stokes, Charlotte (1982), "Collage as Jokework" in *Leonardo* Vol 15 No 3, pp. 199-204.
Stockbridge, Frank Parker (1920). *Yankee Ingenuity in the War.* New York: Harper and Brothers.
"Striped Suits and Papier Maché Carcasses as Camouflage" (1918) in *Scientific American* (January 26), p. 31.
Sumrall, Robert F. (1971), "Ship Camouflage (WWI): Deceptive Art" in *US Naval Institute Proceedings* Vol 97 (July), pp. 57-77.
___(1973) "Ship Camouflage (WWII): Deceptive Art" in *US Naval Institute Proceedings* Vol 99 (February), pp. 67-81.
___(1973) Comments and Discussion regarding "Ship Camouflage (WWI) Deceptive Art" in *US Naval Institute Proceedings* Vol 99 (February), pp. 90-91.
Sweetman, Bill (2001). *Lockheed Stealth: The Evolution of an American Arsenal.* St. Paul MN: MBI Publishing.
Sykes, Steven (1990). *Deceivers Ever: The Memoirs of a Camouflage Officer.* Turnbridge Wells, Kent UK: Spellmount Ltd.
Sypher, Wylie (1960). *Rococo to Cubism in Art and Literature.* New York: Vintage Books.

___(1962). *The Loss of Self in Modern Literature and Art*. New York: Random House.

Talbot Kelly, R.B. (1955). *Bird Life and the Painter.* London: Studio Publications, 1955.
___(1980). *A Subaltern's Odyssey: Memoirs of the Great War 1913-1917*. R.G. Loosmore, ed. London: William Kimber.
Tarbel, Harlan (1924). *How to Chalk Talk*. Chicago: T.S. Denison.
Taussig, Michael (2008), "Zoology, Magic and Surrealism in the War on Terror" in *Critical Inquiry* Vol 34 Supplement (Winter), pp. S98-S116.
Taylor, Robyn (2000), "Kohler, Edward Frederick (1890 - 1964)" in *Australian Dictionary of Biography* Vol 15. Melbourne AU: Melbourne University Press, p. 38.
"T. Fred Slack" (2003), Online obituary at <http://goliath.ecnext.com/coms2/gi_0199-4234156/OBITUARIES-Vitals-Obituary.html> as accessed on March 22, 2008.
"Thatched Barn" (n.d.). Wikipedia online article, as accessed on March 23, 2008.
Thayer, Abbott H. (1896a), "The Law Which Underlies Protective Coloration" in *Auk* Vol 13 (April), pp. 124-129.
___(1896b), "Further Remarks on the Law Which Underlies Protective Coloration" in *Auk* Vol 13 (October), pp. 318-320.
___(1902), "The Meaning of the White Undersides of Animals" in *Nature* (April 24), p. 597.
___(1911), "Concealing-Coloration: A Demand for Investigation of My Tests of the Effacive Power of Patterns" in *Auk* Vol 28 (October), pp. 460-464.
___(1918), "Camouflage" in *The Scientific Monthly* Vol VII, pp. 481-494.
Thayer, Gerald H. (1908), "The Concealing Coloration of Animal: New Light on an Old Subject" in *Century Magazine* (May), pp. 249-261.
___(1909). *Concealing Coloration in the Animal Kingdom: An Exposition of the Laws of Disguise Through Color and Pattern: Being a Summary of Abbott H. Thayer's Discoveries.* New York: Macmillan. [Second edition was published in 1918.]
___(1920). *The Seven Parsons and the Small Iguanodon: An Inverted Legend.* New York: G. Putnam's Sons.
___(1923), "Camouflage in Nature and War" in *Brooklyn Museum Quarterly* Vol 10, pp. 147-169.
"Theatre for Studying Camouflaged Ship Models" (1919) in *Scientific American Supplement* (December 13), pp. 348-349.
"Thomas Casilear Cole" (n.d.) Online information at UnivNorthCarolinaManuscripts accessed on May 23, 2008.
Thone, Frank (1941), "For Better Camouflage" in *Science News Letter* (March 1), p. 142.
Tierney, John (2008), "Tapping Into What a Deer Sees, and Doesn't" in *New York Times* (September 23).
Tinbergen, Niko (1968) *Curious Naturalists.* Garden City NY: Anchor Books.
Titchener, E.B. (1910), "Psychological Literature" in *American Journal of Psychology* Vol 21 No 3 (July), pp. 500-504.
Toch, Maximilian (1918), "The Fine Art of Military Camouflage" in *Munsey's Magazine* Vol 64 No 1 (June), pp. 5-8.
___(1919). "Discussion" in *Transactions of the Illuminating Engineering Society* Vol 14 (July 21), pp. 230-232.
___(1931), "Adventures in Camouflage" in *The Military Engineer* Vol 23 (July-August), pp. 307-309.

1958

American artist (and WWII camouflage instructor) Daniel MacMorris extracts portions (by cutting them out) of a massive World War I-era French painted panorama, titled *Panthéon de la Guerre*, and rebuilds it as a much smaller, edited mural, now on public view at the National World War I Museum in Kansas City.

Peter Scott, *The Eye of the Wind*.

•

1967

British artist Robin Darwin (grandson of Charles Darwin) becomes the first rector and vice-provost of the Royal College of Art, where he works with a number of artists who had designed camouflage during World War II, at which time he himself had served as secretary to the British Camouflage Directorate.

•

1978

Seymour Reit, *Masquerade: The Amazing Camouflage Deceptions of World War II*.

•

1980

Guy Hartcup, *Camouflage: A History of Concealment and Deception in War*.

•

Toliver, Oleta Stewart, ed. (1994). *An Artist at War: The Journal of John Gaitha Browning*. Denton TX: University of North Texas Press.

"Tom Gourdie" (n.d.). Online biography at <www.artistsfootsteps.co.uk> accessed on March 22, 2008.

Toombs, Jane (2002). *Nightingale Man*. Amherst Junction WI: Hard Shell Word Factory.

Towle, H. Ledyard (1917), "What the American 'Camouflage' Signifies" in *New York Times* (June 3), p. 14.

Tracy, Brandon (1919), "A New Camouflage Art" in *Popular Mechanics* (1919), pp. 366-367.

Tracy, Evarts (1919), "Memoranda on the Camouflage Service of the US Army" in *Professional Memoirs, Corps of Engineers, US Army* Vol 11. (March-April), pp. 175-184.

"Training in Camouflage" (1942) in *New York Times* (August 17), p. 14.

Trevelyan, Julian (1957). *Indigo Days*. London: MacGibbon and Kee, 1957.

"Tricks of the War-Trade" (1915) in *Literary Digest* (July 3), pp. 33-34.

"Troth Announced of Virginia Moore" (1937) in *New York Times* (February 12), p. 20.

Tulk, Alfred James (n.d.). Alfred James Tulk Papers at Archives of American Art.

Turbayne, Colin M. (1971). *The Myth of Metaphor.* Columbia: University of South Carolina Press.

Turner, Jane, ed. (1996). *The Dictionary of Art*. London and New York: Grove Dictionaries.

"Two Paintings in One" (1921) in *Literary Digest* (March 12), p. 25.

Underhill, Nancy D.H. (1988), "Smith, Sydney George Ure (1887-1949)" in *Australian Dictionary of Biography* Vol 11. Melbourne AU: Mlebourne University Press, pp. 662-663.

University of Arizona Art Gallery (1965). *His Golden Year: A Retrospective Exhibition of Watercolors, Oils and Graphics by Charles E. Burchfield*. Exhibition catalog. Tucson: University of Arizona Press.

Upright, Diane (1987). *Ellsworth Kelly: Works on Paper.* New York: Harry Abrams.

US Committee on Public Information (1918). *War Cyclopedia*. Washington DC: Government Printing Office.

US War Department (1942). *Civilian Defense: Protective Concealment*. Washington DC: US Office of Civilian Defense.

"Use of Paint in Camouflage" (1941) in *Nature* (November 1), p. 527.

Valle, Pietro (n.d.) "The Cadyville Sauna by Dan Hisel" at <architettura.supereva.com/artland/20021229/index_en.htm>, as accessed on March 20, 2008.
"Value of Dazzle Painting" (1919) in *Literary Digest* (July 19), p. 31.
Van Buskirk, Harold (1919), "Camouflage" in *Transactions of the Illuminating Engineering Society* Vol 14 (July 21), pp. 225-229.
"Van Buskirk Youth Dies" (1936) in *New York Times* (August 14), p. 17.
"Van Buskirk Victor in Veterans' Fencing; Last-Minute Entrant Leads Four Rivals" (1937) in *New York Times* (May 7), p. 36.
Vandenburg, Margaret (2001). *An American in Paris: A Novel.* San Francisco CA: Cleis Press.
Verstegen, Clare (1990), "Camouflage: A Master's Thesis" in *Surface Design Journa*l (Fall), pp. 15-17.
"Vincent A. Hartgen" (2003). Obituary in the *Penn State Gazette*, online at <www.upenn.edu/gazette/0503/0503obits.html> as accessed on March 27, 2008.
Vogel, Donald S. (2000). *Memories and Images: The World of Donald Vogel and Valley House Gallery.* University of North Texas Press.
Vonnegut, Kurt (1988). *Bluebeard.* New York: Dell.

Wadsworth, Barbara (1989). *Edward Wadsworth: A Painter's Life*. Salisbury, England: Michael Russell.
Waldron, Raymond S., Jr. (n.d.), Raymond S. Waldron, Jr., Papers at the Kellen Archives Center Collections at Parsons College at <http://library.newschool.edu/speccoll/kellen/parcoll.php#waldron>, as accessed on June 6, 2008.
Walker, John A. (1983), "Dream-Work and Art-Work" in *Leonardo* Vol 16 No 2, pp. 109-114.
Wallace, Emily Duane (1942), "Camouflage Exhibit at the Franklin Institute" in *Scientific Monthly* Vol 54 (May), pp. 484-487.
Wallach Art Gallery (2001). *Percival Goodman: Architect, Planner, Teacher, Painter.* New York: Columbia University.
"Walt Disney: A Short Biography" (n.d.) on JustDisney.com website at < http://www.justdisney.com/walt_disney/biography/w_bio_short.html> as accessed on March 17, 2008.
"Walter Arnett" (n.d.) Biographical article at <www.brown.edu/Facilities/University_Library/exhibits/Artists/files/Artists.html> as accessed on March 10, 2008.
"War and the Art Schools" (1940) in *Art Digest* Vol 15 No 1, p. 28.
Warburton, Nigel (2003). *Erno Goldfinger: The Life of an Architect*. London: Routledge.
Ward, John Lawrence (2003). *Edwin Dickinson: A Critical History of His Paintings*. University of Delaware.
Warner, Everett L. (1919a), "The Science of Marine Camouflage Design" in *Transactions of the Illuminating Engineering Society*. Vol XIV No 5 (July 21), pp. 215-219.
___(1919b), "Fooling the Iron Fish: The Inside Story of Marine Camouflage" in *Everybody's Magazine* (November), pp. 102-109.
___(1919c), "Marine Camouflage." Unpublished typescript.
___(n.d.), "Summary of Points to be Made in a General Lecture on Marine Camouflage." Unpublished typescript in two parts.
___(1944), "Ship Camouflage Manual for Pattern Design Application." Unpublished typescript, prepared for the US Bureau of Ships.

1981
Roy R. Behrens, *Art and Camouflage: Concealment and Deception in Nature, Art and War.*

•

1984
Elizabeth Louise Kahn, *The Neglected Majority: "Les Camoufleurs," Art History and World War I.*

•

1990
Stephen Sykes, *Deceivers Ever: The Memoirs of a Camouflage Officer.*

•

2002
Roy R. Behrens, *False Colors: Art, Design and Modern Camouflage.*

•

2004
Hardy Blechman, *DPM: Disruptive Pattern Material: An Encyclopedia of Camouflage*

•

2007
Henrietta Goodden, *Camouflage and Art: Design for Deception in World War II.*

Tim Newark, *Camouflage.*

Maite Méndez Baiges, *Camuflaje: Engano y Ocultaciónen El Arte Contemporáneo.*

Warner, Thomas E. (c1998-2008). Correspondence with the author.

Warren, Lynn (1988). *Tom Czarnopys.* Exhibiton brochure. Chicago: Museum of Contemporary Art.

Warren, Virginia Lee (1965), "How to Serve Dinner for Eight in a Guest Room" in *New York Times* (September 6), p. 8.

"Wartime Activities of College Art Staffs" (1946) in *College Art Journal* Vol 5 No 4 (May), p. 364.

Waterhouse, Jill (1993), "Cumpston, Gladys Maeva (1887-1975)" in *Australian Dictionary of Biography* Vol 13. Melbourne AU: Melbourne University Press, p. 543.

Watson, Robert, ed. (2004). *Life in the White House.* Buffalo NY: SUNY Press.

Watson, Steven and Catherine J. Morris, eds. (2000). *An Eye on the Modern Century: Selected Letters of Henry McBride.* New Haven CT: Yale University Press.

Weinberg, Louis (1927). *Color in Everyday Life.* New York: Dodd, Mead and Company.

Wells, H.G. (1917). *Italy, France and Britain at War.* New York: Macmillan.

Wescher, Herta (1971). *Collage.* Robert E. Wolf, trans. New York: Abrams, pp. 110-12.

Weschler, Judith (1978). *György Kepes: The MIT Years 1945-1977.* Cambridge MA: MIT Press.

Whaley, Bart, and Jeffrey Busby (2000), "Detecting Deception: Practice, Practitioners, and Theory" in *Trends in Organized Crime* Vol 6 No 1 (Fall), pp. 73-105.

Wharton, Don (1942), "Fooling Enemy Airmen" in *Reader's Digest* (May), pp. 33-35.

White, Charles Wilbert (1965), Oral History Interview conducted by Betty Lochrie Hoag on March 9, in the Archives of American Art, Smithsonian Institution.

White, Nelson C. (1951). *Abbott H. Thayer: Painter and Naturalist.* Hartford CT: Connecticut Printers.

Whiting, John D. (1928). *Convoy: A Story of the War at Sea* (New York: Books Inc.).

Whitney-Ganson, Joseph (1917), "Artillery Camouflage on the West Front" in *New York Times* (November 11), p. 68.

Whyte, Lancelot Law, ed. (1966). *Aspects of Form.* Bloomington IN: Indiana University Press.

Wickler, Wolfgang (1968). *Mimicry in Plants and Animals.* New York: McGraw-Hill.

Wigginton, Eliot, ed. (1979). *Foxfire 5.* Garden City NY: Anchor Books.

"Wilford Conrow, A Portraitist, 77" (1957) in *New York Times* (November 26), p. 30.

Wilkinson, David M., and Thomas N. Sherratt (2008), "The Art of Concealment" in *Biologist* Vol 55 No 1 (February), pp. 10-15.

Wilkinson, Norman (1922), "Naval Camouflage" article in *Encyclopedia Britannica* 12th edition Vol 1, pp. 546-547.
___(1969). *A Brush with Life*. London: Seeley Service.
"William A. Mackay, Mural Painter, 63" (1939) in *New York Times* (July 28), p. 11.
"William D. Timmons, 62" (1960) in *New York Times* (May 31), p. 31.
"William H. Magoffin" (1942) in *New York Times* (February 3), p. 19.
"William Heaton Cooper" (n.d.), biographical article at <http://www.heatoncooper.co.uk/willbio.htm> as accessed on December 23, 2001.
Williams, David (1989). *Liners in Battledress*. St Catharines, ON: Vanwell Publishing.
___(2002). *Naval Camouflage 1914-1945: A Complete Visual Reference*. US Naval Institute Press.
Williams, Hiram (1984). *Notes for a Young Painter*. Englewood Cliffs NJ: Prentice-Hall.
Winsor, John (1996), "General Patterns Army" in *The Independent* (August 4).
"Winter, Muralist, Takes His Own Life" (1949) in *New York Times* (April 8), p. 22.
"With the Sappers" in the *Journal of the South African Military History Society* at <www.rapidttp.com/milhist/121as.html> as accessed on March 17, 2008.
Witt, David L. (2002). *Modernists in Taos: From Dasburg to Martin*. Santa Fe NM: Red Crane Books.
Wolanin, Barbara A. (1983). *Arthur B. Carles: Painting with Color*. Philadelphia: Pennsylvania Academy of the Fine Arts.
Wolfram, Eddie (1975). *History of Collage: An Anthology of Collage, Assemblage and Event Structures*. New York: Macmillan, pp. 62-63.
Wolinski, Natacha (1998), "Art et Camouflage: Les Grandes Manoeuvres" in *Beaux Arts* magazine No 169 (June 1), pp. 58-62.
"Women Engaged in the Art of Camouflage" (1942), in *New York Times* (April 20), p. 18.
Woodham, Jonathan (1995) "Reviews: James Gardner, the ARTful Designer: Ideas off the Drawing Board" in *Journal of Design History* Vol 8 No 4 (1995), pp. 313-315.
"World War II Veteran Bob Hayes Served Stateside" (2006) in *St. Mary's Today Online Edition* (May) at <www.stmarystoday.com/News/VeteranBobHayes.html> as accessed on April 7, 2008.
Yardley, Edward (1996), "Passion for the Sea" in *Antique Dealer and Collectors Guide* Vol 49 No 12 (July 1996), pp. 36-39.
Yarker, Gwen (n.d.), "Herbert Barnard John Everett, 1876-1949," online biographical article at Maritime Art Greenwich website at <http://www.nmm.ac.uk/mag/pages/mnuInDepth/Essay.cfm?Essay=documents/essay/HERbert_barnard_john_everett5.htm&Title=Herbert%20Barnard%20John%20Everett,%201876-1949> as accessed on March 1, 2008.
"Yesteryear: And the rest is...art history" (2004) in *UTA Magazine* Online (Fall), as accessed on April 4, 2008.
Zakia, Richard D. (1979). *Perception and Photography*. Rochester NY: Light Impressions.
___(1997). *Perception and Imaging*. Newton MA: Focal Press.
Zarin, Cynthia (2004), "The Storyteller: Fact, fiction, and the books of Madeleine L'Engle" in the *New Yorker* (April 12).
Zerubavel, Eviatar (1993). *The Fine Line: Making Distinctions in Everyday Life*. Chicago: University of Chicago Press.
Zusne, Leonard (1975). *Names in the History of Psychology*. New York: John Wiley, pp. 441-442.

index

about the author

Roy R. Behrens is a writer and designer who has taught graphic design, illustration and design history for more than 35 years, at the University of Wisconsin-Milwaukee, the Art Academy of Cincinnati, and, since 1990, at the University of Northern Iowa.

For many years he was associated with *Ballast Quarterly Review* (a "periodical commonplace book" which he founded in 1985); *North American Review*; *Leonardo: International Journal of the Arts, Sciences and Technology*; *Gestalt Theory*; *Print Magazine*, and other publications. He has written hundreds of essays and articles for these and other periodicals, including *Lotus Internationale*, *Revista de Occidente*, *Philosophical Transactions of the Royal Society B*, *Tate Etc.*, and so on.

As a graphic designer, he has illustrated books, short stories and poetry by Jerzy Kosinski, Margaret Atwood, Joyce Carol Oates, Joseph Langland, and other American writers. In recent years, he designed Dean and Geraldine Schwarz's massive book on *Marguerite Wildenhain and the Bauhaus: An Eyewitness Anthology*, chosen as a Recommended Book for 2008 by the blog Design Observer. Described by *Communication Arts* magazine as "one of the most original thinkers in design," he was a nominee for the National Design Awards in 2003.

Among his other recent books are FALSE COLORS: *Art, Design and Modern Camouflage* (2002), and COOK BOOK: *Gertrude Stein, William Cook, and Le Corbusier* (2005). He is married to artist Mary Snyder Behrens, with whom he is co-proprietor of Bobolink Books.

Related titles

FALSE COLORS: *Art, Design and Modern Camouflage* (2002). Roy R. Behrens. Published by Bobolink Books. 9 x 6 ins. 224 pp. Illus (b&w). ISBN 0-97132-440-9. Paperbound.

An increasingly popular volume about the little known use of artists, designers and architects as army, navy and civilian camouflage experts (called camoufleurs) during World Wars I and II. Described and illustrated are a wide range of attempts—some ingenious, others absurd—at fooling the eye by artists from France, United Kingdom, Germany, the US and other countries.

"I have long admired Professor Behrens' original research."

—*Rudolf Arnheim*

"An exhilarating kaleidoscope of artists fooling around, making visual puns, tricking society, and conjuring up novel solutions."

—*Kevin Kelly*

"...highly original, well-researched, and fascinating..."

—*Modernism / Modernity*

"[The author's] romp through history and ideas... makes this book both refreshing and provocative."

—*Graphic Impressions*

"[The goodness of this book] lies in its freshness, its range of information, and its sound technical theories about how we see."

—*Harper's Magazine*

Titles from **Bobolink Books** can be purchased through Amazon.com (new and used), or go to www.bobolinkbooks.com.

COOK BOOK: *Gertrude Stein, William Cook, and Le Corbusier (2005).* Roy R. Behrens. Published by Bobolink Books. 7 x 5 ins. 96 pp. Illus (b&w). ISBN 0-9713244-1-7. Paperbound.

This is *not* a cookbook—it's a Cook book, a book about a man named Cook. It is a biographical sketch of one of Gertrude Stein's dearest and most loyal friends, a largely unknown artist named William Edwards Cook (1881-1959).

Born and raised in Iowa, he left the US in 1903 to study in France, and continued living overseas (in France, Italy, USSR and Spain) for the rest of his life. A close friend of Stein and Alice B. Toklas, it was Cook who taught Stein how to drive. They also vacationed together on the Spanish island of Majorca.

In addition, when his father died in 1928, he used part of his inheritance to commission the then unknown young architect Le Corbusier to design a starkly Modernist home (called Villa Cook), the "first truly cubic house," on the outskirts of Paris. He also painted a portrait of Pope Pius X, served as an undercover agent for the US Secret Service, and, as a Red Cross worker in the USSR, witnessed the aftermath of the Russian Revolution.

Based largely on hundreds of pages of unpublished correspondence between Stein, Toklas and Cook, this is a candid and often amusing account of their warm and enduring friendship. But it is also about the mystery that Paris held for the "lost generation" of expatriate artists, writers and musicians, about Stein's and Cook's shared interest in automobiles and bullfighting, about Stein's outspoken fondness for Iowa (although she had never visited there), and about everybody's painful memories of growing up, leaving home, and never being able to return.

Titles from **Bobolink Books** can be purchased through Amazon.com (new and used), or go to www.bobolinkbooks.com.

This [book] is as good as topnotch Behrens gets!
—*Guy Davenport*

[Professor] Behrens' *Cook Book* jumps out at my eyes, my ears. It comes from everywhere, never drags those even blocks of print that dull the mind. Look at it, read it, let it tease you: it's researched with all the care that keeps its sense of humor and its visual and voice delights. Travel with it, leave home, go and explore the many ways for a book to be a house for living.
—*Ulla Dydo*

"[A] delightful little book ...a high point in recent scholarship on Gertrude Stein."
—*Paul Padgette*

...

"Before food was invented, books about cooking were unknown."
—*Fred Allen*

Marguerite Wildenhain and the Bauhaus: An Eyewitness Anthology (2008). **Edited by Dean and Geraldine Schwarz. Designed by Roy R. Behrens. Published by South Bear Press. 11 x 8.5 ins. 770 pp. 837 illus (402 b&w, 436 color). ISBN 978-0-9761381-2-9. Clothbound.**

"[This book] is a copious archive of resources, many of which are published here for the first time."

—Ceramics Monthly

A phenomenal storehouse of documents on the life and work of master potter Marguerite Wildenhain, as well as the origins of the European concept of *Handwerk*, the role of pottery at the Weimar Bauhaus (the most famous art school in history), and the continuing, prospering legacy of studio pottery in Europe and the US. Includes scores of Bauhaus-related images, interwoven with essays, memoirs, diaries, letters, interviews and other candid, written texts, many of which are published here for the first time.

Titles from **South Bear Press** can be purchased through Amazon.com (new and used), or go to www.southbearpress.org.

...

"2008 Recommended Book"

—DesignObserver.com

"Those wishing to reconstruct both the European milieu that shaped Wildenhain's work and its fruition on American soil, will find [in this book] an embarrassment of riches..."

—American Craft Magazine